Dharmendra K. Gupta
S. Murugan • Aswetha Iyer
Editors

Nano-Ecosystems

Engineering Nature at the Nanoscale

Springer

Editors
Dharmendra K. Gupta
Ministry of Environment, Forest and Climate Change
Indira Paryavaran Bhavan
Jorbagh, New Delhi, India

S. Murugan
Division of Biotechnology
Karunya Institute of Technology and Sciences
Coimbatore, Tamil Nadu, India

Aswetha Iyer
Division of Biotechnology
Karunya Institute of Technology and Sciences
Coimbatore, Tamil Nadu, India

ISBN 978-981-95-5032-6 ISBN 978-981-95-5033-3 (eBook)
https://doi.org/10.1007/978-981-95-5033-3

This Springer imprint is published by the registered company Springer Nature Singapore Pte Ltd.
The registered company address is: 152 Beach Road, #21-01/04 Gateway East, Singapore 189721, Singapore

Preface

The term "nano-ecosystem" is not merely a metaphor but embodies the integration of nanoscale materials, devices, and processes into ecological systems, aiming to improve environmental functionality, energy efficiency, biodiversity, and sustainability. From mimicking biological structures to developing nanodevices for pollution monitoring and control, the contributions in this volume reflect the profound potential of nanotechnology in addressing some of the most pressing environmental challenges of our time. The ever-evolving landscape of nanotechnology has led to the birth of novel interdisciplinary fields that intersect with ecology, environmental sustainability, and systems thinking. One such emerging domain is that of nano-ecosystems, which represents the dynamic interface between engineered nanomaterials and natural ecosystems. This book, *Nano-Ecosystems: Engineering Nature at the Nanoscale*, brings together a diverse set of ideas, technologies, and approaches that reimagine how nanoscale innovations can coexist with and enhance environmental processes, rather than disrupt them.

The chapters presented here encompass a wide spectrum of themes: the foundational principles of nano-ecosystem engineering, bio-inspired nanoarchitectures, energy harvesting systems, environmental sensors, and nanotechnology-based approaches for carbon capture, soil revitalization, waste management, and habitat restoration. Special emphasis is also placed on forward-looking technologies such as nanorobotics for microplastic removal, the integration of artificial intelligence in ecological modeling, and ethical considerations in deploying nano-ecosystems at scale. This book not only explores the transformative applications of nanotechnology but also provides critical perspectives on its limitations, risks, and societal implications. It is a product of collaboration between contributors from across India and other parts of the world, reflecting a truly global perspective on sustainable nano-innovation.

The editors of this book extend their sincere gratitude to all the contributing authors for their thoughtful research, valuable time, and dedication to advancing this interdisciplinary field. Their insights have shaped this book into a comprehensive and timely resource for students, researchers, industry professionals, and policymakers working at the intersection of nanotechnology and environmental science.

As editors, *Dr. Dharmendra K. Gupta, Dr. Murugan Sevanan, and Aswetha Iyer* honored to present this volume, with the hope that it serves as both a reference and an inspiration for future research and responsible innovation in the field of nano-ecosystem engineering.

Jorbagh, New Delhi, India — Dharmendra K. Gupta
Coimbatore, Tamil Nadu, India — S. Murugan
Coimbatore, Tamil Nadu, India — Aswetha Iyer

Contents

Editors and Contributors

About the Editors

Dharmendra K. Gupta is Director(S)/Scientist-F at Ministry of Environment, Forest and Climate Change, New Delhi, India. His field of research includes abiotic stress by radionuclide's/heavy metals and xenobiotics in plants; antioxidative system in plants, environmental pollution (radionuclide's/heavy metals) remediation through plants and microbes (Phytoremediation/Bioremediation). He has been awarded several international awards by entities such as the Japan Society for the Promotion of Science (JSPS); the Belgian Science Policy Office (BELSPO); JAE-Doc., Spain; TWAS-CNPq, Italy; Royal Society, UK; MASAV, Israel; Department of Science & Technology, India; and Jawaharlal Nehru Foundation, India. He has published more than 130 internationally peer-reviewed original research/review articles/book chapters and 29 books and served as Book Series Editor for 2 book series from Springer, Switzerland, and 1 from World Scientific, Singapore. He has successfully completed 10 multidisciplinary research projects from international and national bodies.

S. Murugan is an Associate Professor in the Department of Biotechnology at Karunya Institute of Technology and Sciences, a distinguished Deemed University in Coimbatore, India. He boasts a rich academic background, holding an MSc in Microbiology from Manonmaniam Sundaranar University, India, and a PhD in Microbiology from Bharathiar University, India. His illustrious career spans over 25 years, during which he has exhibited profound dedication to education. His scholarly contributions are remarkable, encompassing 80 research articles, 5 articles about career development, and 7 general articles in esteemed international journals. Additionally, he has showcased his erudition by crafting more than 15 insightful book chapters.

Aswetha Iyer is an accomplished researcher with a strong focus on environmental biotechnology. By becoming an Assistant Professor at the age of 23, she has come a long way with a keen interest in contributing to academic research. With a background in Forensic Biotechnology and Botany, she remains at the forefront of developing scalable nano-biological technologies to address the most pressing contaminants in modern industry. Her academic contributions are extensively documented through publications that bridge the gap between microbiological discovery and circular economy applications. Notable works include her research on the metalloproteomics of metal-resistant bacteria and the taxonomic profiling of textile industry effluent from Tiruppur for water recycling. She has also won the Best Research Award 2024 for her outstanding research on developing a cost-effective, eco-friendly framework for achieving environmental compliance through cutting-edge nano-microbial science at a prestigious national conference sponsored by ICAR and MoES in Tamil Nadu, India.

Contributors

M. Abiyoga Department of Microbiology, Rathnavel Subramaniam (RVS) College of Arts and Science, Coimbatore, Tamil Nadu, India

P. Amudha Department of Biochemistry, Vels Institute of Science, Technology & Advanced Studies (VISTAS), Chennai, Tamil Nadu, India

A. M. Anandkumar Department of Botany, Nallamuthu Gounder Mahalingam College, Pollachi, Tamil Nadu, India

Kiruthika Balasubramanian New Jersey Academy of Science, Kean University, Union, NJ, USA

Janani Chandran Department of Biotechnology, Srimad Andavan Arts and Science College (Autonomous), Trichy, Tamil Nadu, India

Al-Nakib Chowdhury Department of Chemistry, Bangladesh University of Engineering and Technology, Dhaka, Bangladesh

P. Divyadharshini Department of Industrial Biotechnology, Bharath Institute of Higher Education and Research, Chennai, Tamil Nadu, India

Sukanya Ghosh GBPUAT, Pantnagar, Uttarakhand, India

Abu Bin Imran Department of Chemistry, Bangladesh University of Engineering and Technology, Dhaka, Bangladesh

M. Jaison BCKV, Mohanpur, West Bengal, India

D. Jansirani Department of Biotechnology, P.S.R Engineering College, Sivakasi, Tamil Nadu, India

S. Jeyabharathi Department of Microbiology, Cauvery College for Women (Automonous), (Affiliated to Bharathidasan University), Trichy, Tamil Nadu, India

D. Jhoshitha Department of Industrial Biotechnology, Bharath Institute of Higher Education and Research, Chennai, Tamil Nadu, India

Angeline Julius Center for Material Engineering and Regenerative Medicine, Bharath Institute of Higher Education and Research, Chennai, Tamil Nadu, India

R. Karthika Department of Biotechnology, K. S. Rangasamy College of Arts and Science (Autonomous), Tiruchengode, Namakkal, Tamil Nadu, India

J. Krishnaraj Department of Automobile Engineering, Hindusthan College of Engineering and Technology, Coimbatore, Tamil Nadu, India

Gobalan Krishnasamy Department of Biotechnology, Jamal Mohamed College, Trichy, Tamil Nadu, India

S. T. Kumaravel Department of Mechanical Engineering, Paavai Engineering College, Namakkal, Tamil Nadu, India

Anas Mahmud Department of Chemistry, Bangladesh University of Engineering and Technology, Dhaka, Bangladesh

Suresh Malakondaiah Center for Material Engineering and Regenerative Medicine, Bharath Institute of Higher Education and Research, Chennai, Tamil Nadu, India

P. A. Mathina Department of Electronics and Communication Engineering, P.S.R Engineering College, Sivakasi, Tamil Nadu, India

Rad Mosharrof Mim Department of Chemistry, Bangladesh University of Engineering and Technology, Dhaka, Bangladesh

A. Mohamed Rizwana Department of Biotechnology, P.S.R Engineering College, Sivakasi, Tamil Nadu, India

A. Noorjahan Rehana Department of Electronics and Communication Engineering, P.S.R Engineering College, Sivakasi, Tamil Nadu, India

Sudharsan Parthasarathy Department of Forestry, Nagaland University (Hqrs.), Lumami, Nagaland, India

Pooja Pillai Department of Industrial Biotechnology, Bharath Institute of Higher Education and Research, Chennai, Tamil Nadu, India

P. Saravanakumari Department of Microbiology, Rathnavel Subramaniam (RVS) College of Arts and Science, Coimbatore, Tamil Nadu, India

K. Sheela PG & Research Department of Biochemistry, Tirupattur, Tamil Nadu, India

M. Sivadharani Department of Microbiology, Cauvery College for Women (Automonous), (Affiliated to Bharathidasan University), Trichy, Tamil Nadu, India

Prathiksha Srikanthan Department of Industrial Biotechnology, Bharath Institute of Higher Education and Research, Chennai, Tamil Nadu, India

Archana Sundaram Department of Biotechnology, Srimad Andavan Arts and Science College (Autonomous), Trichy, Tamil Nadu, India

M. Tamilmozhi Department of Bioscience, Sri Krishna Arts and Science College, Coimbatore, Tamil Nadu, India

Selvendran Thamaraiselvan Department of Biotechnology, Srimad Andavan Arts and Science College (Autonomous), Trichy, Tamil Nadu, India

Siva Vijayakumar Tharumasivam Department of Biotechnology, Center for Research, Dhanalakshmi Srinivasan University, Trichy, Tamil Nadu, India

M. Thenmozhi Department of Biotechnology, Selvam College of Technology, Namakkal, Tamil Nadu, India

Purushothaman Thirumurthy Department of Biotechnology, SNMV College, Coimbatore, Tamil Nadu, India

R. Vidya Department of Biochemistry, Vels Institute of Science, Technology & Advanced Studies (VISTAS), Chennai, Tamil Nadu, India

R. Yuvarajan PG & Research Department, Department of Biotechnology, Mahendra Arts and Science College (Autonomous), Kallipatti, Namakkal, Tamil Nadu, India

The Concept of Nano-Ecosystem and Its Potential

Sudharsan Parthasarathy, Siva Vijayakumar Tharumasivam, Purushothaman Thirumurthy, and Gobalan Krishnasamy

Abstract

The notion of the nano-ecosystem is a novel approach at the nexus of ecological studies and nanotechnology, offering a synergistic framework in which interactions at the nanoscale resemble those of ecological systems. This novel paradigm seeks to improve the functioning, sustainability, and integration of nanomaterials and devices into wider biological and environmental settings by utilizing the concepts of ecosystems in the design and management process. Researchers can create more durable and adaptable technology by equating artificial nanosystems with natural ecosystems. Nano-ecosystems have the potential to be used in many different disciplines, such as energy generation, environmental remediation, and medicine. These technologies provide intriguing answers for efficient energy conversion and storage, tailored medicine administration, and pollution management. A balanced approach to innovation is required in light of the significant ethical and environmental concerns that the advent of nano-ecosystems brings to light. In an effort to give a thorough review of this new topic, this chapter examines the fundamental ideas behind nano-ecosystems, as well as any possible advantages and implementation issues.

S. Parthasarathy (✉)
Department of Forestry, Nagaland University (Hqrs.), Lumami, Nagaland, India

S. V. Tharumasivam (✉)
Department of Biotechnology, Center for Research, Dhanalakshmi Srinivasan University, Samayapuram, Trichy, Tamil Nadu, India
e-mail: Sivavijayakumart.set@dsuniversity.ac.in

P. Thirumurthy
Department of Biotechnology, SNMV College, Coimbatore, Tamil Nadu, India

G. Krishnasamy (✉)
Department of Biotechnology, Jamal Mohamed College, Trichy, Tamil Nadu, India

D. K. Gupta et al. (eds.), *Nano-Ecosystems*,
https://doi.org/10.1007/978-981-95-5033-3_1

Keywords

Nano-ecosystem · Environmental remediation · Nanotechnology · Ecological systems · Sustainability · Nanomaterials · Adaptive technologies

1 Introduction to Nano-Ecosystem

A nano-ecosystem refers to a highly integrated and complex system at the nanoscale, where various nanostructures and nanoparticles interact with each other and with their environment. At the core of a nano-ecosystem are the interactions between nanoparticles, nanodevices, and nanoscale materials (Bhattacharya & Mukherjee, 2023). These interactions are governed by quantum mechanical principles and exhibit unique properties not seen at larger scales. A hallmark of nano-ecosystems is the ability of nanoscale components to self-assemble into structured and functional systems. This self-organization can be driven by various forces, including van der Waals forces, electrostatic interactions, and covalent bonding (Bhushan, 2017). Nano-ecosystems are dynamic, meaning they can change and adapt in response to external stimuli. This adaptability is crucial for applications in areas such as drug delivery, where nanocarriers must respond to the body's environment to release their payloads effectively (Buzea et al., 2007). The study and application of nano-ecosystems require a multidisciplinary approach, combining knowledge from fields such as chemistry, physics, biology, materials science, and engineering. This interdisciplinary nature enables the design of novel nanostructures and the exploration of their potential applications. Nano-ecosystems often exhibit a hierarchical structure, where smaller components come together to form larger, more complex systems.

This hierarchical organization can enhance the functionality and efficiency of the overall system. Efficient energy transfer and conversion processes are critical within nano-ecosystems. These processes can be harnessed for various applications, including energy storage, catalysis, and sensing (De Volder et al., 2013). Nano-ecosystems can process information at the nanoscale, enabling the development of advanced computing systems and sensors. The integration of information processing capabilities at the nanoscale is a key area of research in nanotechnology (Drori & Lavie, 2023). The concept of nano-ecosystems opens up a wide range of potential applications, from medical therapies and environmental remediation to advanced materials and nanorobotics. Understanding and harnessing the principles of nano-ecosystems can lead to significant technological advancements (Gao, 2017). By exploring and defining the nano-ecosystem, researchers can develop innovative solutions that leverage the unique properties of nanoscale materials and interactions. The potential of nano-ecosystems lies in their ability to mimic the complexity and functionality of natural systems, leading to breakthroughs in various scientific and technological fields (Godwin, 2015).

1.1 Historical Background

Concept of a nano-ecosystem is a relatively recent development in the field of nanotechnology, but its roots can be traced back to several key scientific advancements and discoveries. The concept of manipulating matter at the atomic level was famously proposed by physicist Richard Feynman in his 1959 lecture, "There's Plenty of Room at the Bottom" (Gurwitz, 2014). Feynman's visionary ideas laid the groundwork for the development of nanotechnology. In the 1980s, K. Eric Drexler introduced the idea of molecular nanotechnology in his book *Engines of Creation*, where he envisioned the possibility of molecular assemblers creating complex structures atom by atom. Scanning Tunnelling Microscope (STM) was invented in 1981 by Gerd Binnig and Heinrich Rohrer. The STM allowed scientists to visualize and manipulate individual atoms on surfaces, marking a significant milestone in nanotechnology (Ju-Nam & Lead, 2008). Atomic Force Microscope (AFM) was developed in 1986 by Binnig, Calvin Quate, and Christoph Gerber. The AFM further enhanced the ability to study and manipulate nanoscale materials. The discovery of fullerenes (C60) in 1985 by Harold Kroto, Richard Smalley, and Robert Curl, followed by the identification of carbon nanotubes by Sumio Iijima in 1991, introduced novel nanomaterials with unique properties (Hood, 2004).

In the early 1980s, researchers began to explore quantum dots, semiconductor nanoparticles with tunable electronic and optical properties, paving the way for applications in imaging, electronics, and more. Self-Assembling Monolayers (SAMs) was introduced in the 1980s. The SAMs demonstrated how molecules could spontaneously organize into ordered structures on surfaces, a key concept in nano-ecosystems (Iannacci, 2021). The work of Jean-Marie Lehn, who won the Nobel Prize in Chemistry in 1987, advanced the understanding of how molecules can form complex structures through non-covalent interactions. The development of nano-ecosystems has been driven by the convergence of multiple scientific disciplines, including chemistry, physics, biology, and materials science. This interdisciplinary approach has facilitated the creation of complex nano-systems with diverse functionalities (Stanger, 2011). The early twenty-first century has seen significant advancements in nanomedicine, where nano-ecosystems are employed for targeted drug delivery, imaging, and diagnostics. Examples include liposomes, dendrimers, and other nanocarriers. Nano-ecosystems are being explored for environmental remediation, such as water purification and pollutant removal, as well as energy applications like nanostructured solar cells and batteries (Tripathi et al., 2024). Many such recent research is focused on developing more sophisticated nano-ecosystems with enhanced functionalities, such as responsive nanomaterials that can adapt to environmental changes or perform complex tasks (Nasir, 2010).

1.2 Fundamental Principles of Nano-Ecosystem

The concept of a nano-ecosystem revolves around several fundamental principles that define its structure, functionality, and behavior. At the nanoscale, quantum mechanical effects become significant, influencing the behavior of particles and materials. These effects include quantum tunneling, quantum entanglement, and discrete energy levels. The high surface-to-volume ratio of nanomaterials leads to enhanced reactivity and unique properties, such as increased strength, catalytic activity, and altered optical characteristics. Nano-ecosystems often rely on self-assembly processes, where molecules and nanoparticles spontaneously organize into ordered structures without external guidance (Senterre & Wagner, 2014). Self-assembly represents a bottom-up approach to nanofabrication, contrasting with traditional top-down methods. This allows for the creation of complex structures with high precision and minimal defects (Lane, 2022). Nano-ecosystems exhibit hierarchical structures, where smaller components (atoms, molecules, nanoparticles) assemble into larger, more complex systems. This multiscale organization enhances functionality and enables intricate designs. Hierarchical structures can include nested systems, where smaller subsystems operate within larger ones, each contributing to the overall functionality of the nano-ecosystem (Lin, 2010). Nano-ecosystems are designed to be dynamic and responsive to external stimuli, such as changes in temperature, pH, light, or magnetic fields. This adaptability is crucial for applications like drug delivery and environmental sensing (Mueller-Dombois et al., 1994). These systems often incorporate feedback mechanisms that enable them to adjust their behavior in response to environmental changes, similar to how biological systems maintain homeostasis. Efficient energy transfer and conversion are essential in nano-ecosystems, whether for harvesting solar energy, catalyzing chemical reactions, or powering nanodevices (Pisciotta, 2022). This involves optimizing processes at the nanoscale to minimize energy loss and maximize output. For example, nano-ecosystems including components designed for energy harvesting, such as nanostructured solar cells or thermoelectric materials, found to convert environmental energy into usable forms (Sutherland et al., 2024). Nano-ecosystems can incorporate information processing capabilities at the nanoscale, enabling advanced computing, data storage, and sensing functionalities. This includes the development of nanoscale logic circuits, memory devices, and quantum computing elements (Lua et al., 2022). An effective communication between nanoparticles and nanodevices is essential for coordinated actions within the nano-ecosystem. This can be achieved through various signaling mechanisms, such as chemical, electrical, or optical signals (Pisciotta, 2022). Many nano-ecosystems are inspired by biological systems, mimicking natural processes and structures to achieve high efficiency and functionality. Examples include artificial photosynthesis, nanorobots inspired by cellular machinery, and self-healing materials. Bioinspired nano-ecosystems often emphasize sustainability, using environmentally friendly materials and processes to minimize ecological impact and enhance compatibility with natural systems (Mohanta et al., 2024).

Hence, collaborative research efforts enable the design of innovative and multifunctional nano-ecosystems (Fig. 1). Techniques from various fields are combined to fabricate and characterize nano-ecosystems, such as AFM, Transmission Electron Microscopy (TEM) and Scanning Electron Microscopy (SEM), many spectroscopic techniques and molecular modellings (Donaldson & Poland, 2013).

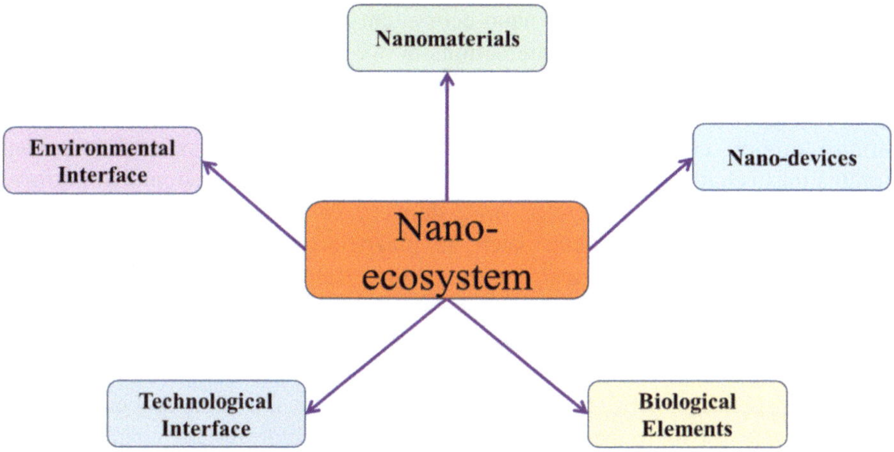

Fig. 1 Nano-ecosystem and their multifunctional roles

Table 1 Key components of a nano-ecosystem

Component	Description
Nanomaterials	Engineered materials with at least one dimension in the nanoscale, such as nanoparticles, nanotubes, and nanowires
Nano-devices	Devices built using nanomaterials, including sensors, actuators, and other functional components
Biological Elements	Incorporation of biological molecules, cells, or organisms for bio-inspired or bio-integrated functions
Environmental Interface	Interaction with environmental factors like water, soil, and air
Technological Interface	Integration with existing technologies and systems for enhanced functionality and performance

1.3 Key Components of a Nano-Ecosystem

A nano-ecosystem comprises various interconnected and interacting components that collectively enable its functionality and adaptability (Table 1). These components are carefully designed and integrated to perform specific tasks at the nanoscale. Nanostructured particles such as gold (Au) nanoparticles, quantum dots, and magnetic nanoparticles exhibit unique optical, electronic, and magnetic properties (Ebbesen, 2008). Another important class of nanocomponents includes cylindrical structures like carbon nanotubes and semiconductor nanowires, valued for their electrical, thermal, and mechanical characteristics. Nanocomposites, consisting of nanoparticles embedded within a matrix, combine the properties of both constituents to achieve enhanced performance (Kamaraj et al., 2023). Surface-modified nanoparticles, functionalized with specific groups or coatings, improve stability, biocompatibility, and targeting capabilities. Self-assembled monolayers, which are thin layers of molecules that spontaneously organize on surfaces, create functional interfaces for diverse applications. Biomolecules such as proteins, DNA, and RNA are also integrated into nano-ecosystems for biosensing, bioimaging, and therapeutic uses (Gao, 2017). Interactions between nanomaterials and biological cells are engineered for targeted drug delivery and tissue engineering. Nanoscale devices capable of detecting and responding to physical, chemical, or biological stimuli support environmental monitoring, medical diagnostics, and industrial applications (Godwin, 2015; Gurwitz, 2014). Nanorobots are designed to perform specific tasks, including targeted drug delivery or environmental remediation. Energy-conversion components such as piezoelectric nanogenerators convert mechanical energy to electrical energy (Hood, 2004), while quantum dots and organic photovoltaics efficiently harvest solar energy. Drug-delivery vehicles, including liposomes, dendrimers, and polymeric nanoparticles, enable the transport of therapeutic agents and advanced release mechanisms such as magnetic targeting and pH-responsive delivery enhance precision (Ju-Nam & Lead, 2008). Nanocatalysts, owing to their high surface area and unique surface characteristics, improve chemical reactions for industrial catalysis and environmental cleanup. Nanozymes, or nanomaterials designed to mimic natural enzymes, facilitate efficient catalytic processes (Gobalan et al., 2023). Additional functional components include nanoscale sensors that detect environmental changes and nanoactuators that produce mechanical responses to external stimuli. Communication mechanisms within nano-ecosystems coordinate interactions between nanoparticles, including optical communication through plasmonic and photonic systems (Klaine et al., 2008). Information-processing elements such as quantum dots for quantum computing and nanowires for neuromorphic computing support advanced computational capabilities. High-density data storage technologies, such as resistive RAM (ReRAM) and Phase-Change Memory (PCM), further extend nano-ecosystem functionality. Structural frameworks provide mechanical support, especially in tissue engineering and regenerative medicine. (Francis et al., 2024). Nanoscale filtration devices purify water, air, and other substances by removing contaminants at the molecular level. Smart materials that change properties in response to environmental conditions improve adaptability and resilience. The safe design and selection of nanomaterials are critical to minimizing biological and environmental risks. Ethical and safety guidelines govern the development and deployment of nano-ecosystems. Lab-on-a-chip systems integrate multiple laboratory functions onto a single miniaturized platform for rapid diagnostics. Nanoscale integrated circuits support advanced electronic applications. Computational modelling tools further assist in simulating and predicting the performance of nano-ecosystems under various conditions.

2 Foundations of Nano-Ecosystems Technology

According to Laurent et al. (2008), nano-ecosystems are built upon a broad range of technological foundations that enable the design, fabrication, and implementation of nanoscale systems with unprecedented precision and functionality. These core technologies span multiple scientific and engineering domains, each contributing uniquely to the development of nano-ecosystems. Thin-film fabrication techniques, such as those used to engineer controlled-property films of carbon nanotubes and graphene, play a critical role (Nel et al., 2006). Sol–gel processing allows the hydrolysis and condensation of metal alkoxides or salts to create the corresponding nanoparticles and nanostructures, respectively. Electrospinning generates nanofibers from polymer solutions for applications in filtration and tissue engineering. Template-assisted synthesis, using nanoporous membranes or patterned substrates, directs the formation of nanomaterials (Oberdörster et al., 2005). Advanced characterization tools support the structural and functional assessment of nano-ecosystems. Microscopic techniques like TEM provide atomic-level images and structural data, while SEM reveals nanoscale surface topography. Atomic Force Microscopy (AFM) enables both imaging and manipulation at the nanoscale by measuring forces between a probe and the sample surface. X-Ray Diffraction (XRD) determines crystalline structure and phase identity. Spectroscopic techniques, such as Raman, Fourier Transform-Infrared (FTIR), and UV-Vis spectroscopy, analyze optical characteristics and chemical composition (Roco, 2003). Self-assembly mechanisms, where molecules spontaneously organize into ordered nanoscale structures, form an essential foundation. Molecular Beam Epitaxy (MBE) deposits atoms or molecules onto substrates to create crystalline layers with precise control of thickness and composition. Lithographic techniques, including electron-beam lithography and nanoimprint lithography, produce nanoscale patterns for device fabrication (Kalyanaraman et al., 2024). Focused Ion Beam (FIB) milling enables direct creation of nanoscale structures, while photolithography uses light to transfer patterns from photomasks onto photosensitive materials and remains vital for nanoscale circuit manufacturing (Savage & Diallo, 2005). Surface engineering and functionalization are also key pillars. The addition of biomolecules or functional groups enhances stability, biocompatibility, and target specificity of nanomaterials. Conjugation of nanoparticles with antibodies or biomolecules supports sensing, imaging, and targeted drug delivery applications. Nanoparticles engineered for therapeutic purposes can target specific tissues or cellular environments.

Nanocomposites represent another essential domain. Incorporating nanoparticles into polymer matrices improves mechanical strength, electrical conductivity, and barrier properties (Schulte & Salamanca-Buentello, 2007). Nanoparticle-reinforced metal matrices offer enhanced conductivity, wear resistance, and thermal stability, while nanoparticle-embedded ceramics provide improved chemical stability, mechanical strength, and heat resistance. Nanoscale sensing technologies enable detection of physical, chemical, and biological changes, supporting industrial process control, environmental monitoring, and diagnostic applications (Wiesner et al., 2006). Nanoelectronics leverage nanoscale materials to create smaller, faster, and more energy-efficient circuits, while nano-photonics utilizes nanoscale light manipulation for high-speed imaging, sensing, and communication. Biomedical nanosystems, including drug-delivery platforms, implantable sensors, and molecular-level diagnostic devices, integrate nanotechnology with biological environments (Ju-Nam & Lead, 2008). Computational technologies serve as the theoretical backbone of nano-ecosystem development. Molecular Dynamics (MD) simulations predict nanoscale behavior by modelling atomic interactions. Density Functional Theory (DFT) provides insight into electronic structure and guides the discovery of new nanomaterials. Finite Element Analysis (FEA) models mechanical behavior and stress distribution in nanoscale structures (Godwin, 2015; Gurwitz, 2014). Advanced fabrication techniques such as 3D nano-printing, with methods including direct laser writing and two-photon polymerization, enable the production of complex 3D nanostructures. Precise manipulation approaches, including magnetic manipulation and optical tweezers, allow accurate positioning and assembly of nanoscale components (Gao, 2017). Ethical, Legal, and Social Implications (ELSI) frameworks address concerns related to privacy, environmental impact, and societal acceptance of nanotechnology. Ensuring safety standards and regulatory compliance remains essential for the responsible development and deployment of nano-ecosystems across various sectors (Kamaraj et al., 2023).

2.1 Nano-Ecosystems Inspired by Biology

Biological systems serve as a powerful source of inspiration for nano-ecosystems, which draw upon natural structures and principles to create innovative nanotechnologies. According to Giridharan et al. (2024), biological inspirations offer insights into efficient and adaptable mechanisms refined over millions of years. Self-assembly, driven by molecular forces such as hydrogen bonding and hydrophobic interactions, enables biomolecules like proteins to fold into specific three-dimensional structures. Similarly, the programmable base-pairing of DNA allows it to fold into intricate nanostructures that function as scaffolds or templates for nanomaterial construction (Wiesner et al., 2006). Biological energy systems, such as the electron transport chain and ATP synthesis, inspire nanoscale energy-production devices and nanogenerators. Ribosomes act as conceptual models for nanoscale assembly lines, guiding the efficient and precise stepwise synthesis of molecules (Xia et al., 2008). Cell membranes contribute to the design of drug-delivery systems and biomimetic nanoscale membranes by replicating selective permeability and molecular transport processes. Synthetic vesicles, inspired by natural lipid bilayers, mimic biological membranes to encapsulate and transport therapeutic agents (Roco, 2003). Nature also inspires the creation of hierarchical nanomaterials and nanocomposites that combine lightweight structure, strength, and flexibility. Diatom and silica shells influence the development of silica-based nanomaterials and nanoscale patterning techniques, replicating complex natural architectures. Photosynthetic machinery such as chloroplasts inspires nanoscale energy-harvesting systems and artificial photosynthesis devices, emulating light-harvesting complexes and electron transport chains. Biological photoreceptors serve as models for nanophotonic technologies and light-responsive materials.

Biological motion offers additional guidance. Molecular motors such as myosin and kinesin demonstrate controlled nanoscale movement, inspiring the development of nanoscale actuators and motors. The motion of cilia and flagella informs nanorobotic propulsion systems for targeted delivery and mobility (Savage & Diallo,

2005). Neural signaling mechanisms inspire nanoscale communication devices and sensors capable of highly sensitive detection and information processing. Olfactory and gustatory receptors guide the design of receptor-based nano-biosensors that detect specific chemicals and environmental signals (Zhang & Webster, 2009). The immune system informs the creation of responsive nano-ecosystems capable of self-healing, disease detection, and targeted drug release. Homeostasis mechanisms provide models for feedback loops and regulatory systems that maintain stability in dynamic nano-ecosystems (Zhao & Castranova, 2011). The Extracellular Matrix (ECM) replicates biochemical and structural cues that guide cell growth and differentiation, inspiring nanoscale scaffolds for tissue engineering and regenerative medicine. Natural materials such as spider silk influence the development of nanofibers and adhesive materials that mimic biological strength, flexibility, and biocompatibility. Viral structures offer templates for gene-delivery vectors and targeted drug-delivery nanocarriers (Xia et al., 2008). Cell membrane–coated nanoparticles mimic natural membranes to enhance biocompatibility and evade immune detection in therapeutic applications.

By incorporating biological inspirations into nano-ecosystems, researchers aim to leverage the efficiency, adaptability, and sustainability of living systems, advancing nanotechnology applications in healthcare, energy, environmental remediation, and beyond (Laurent et al., 2008).

2.2 Biological Nano-Ecosystems in Medicine

Nanotechnology within nano-ecosystems has the potential to revolutionize healthcare by addressing long-standing challenges in disease detection, treatment, and monitoring. By leveraging the unique properties of nanomaterials and nanoscale devices, nano-ecosystems enable personalized therapies, enhanced imaging, and targeted drug delivery (Bhattacharya & Gupta, 2020). Nano-ecosystems enable precise drug delivery by using nanoparticles and nanocarriers such as liposomes, dendrimers, and polymeric nanoparticles. These systems deliver drugs directly to diseased tissues or specific cells, increasing therapeutic efficiency, reducing side effects on healthy tissues, and enabling controlled release profiles. Functionalized nanoparticles, including quantum dots and magnetic nanoparticles, enhance contrast in imaging modalities such as Magnetic resonance Imaging (MRI), Computed Tomography (CT), and fluorescence imaging (Cassee & Peters, 2018). This improves sensitivity, specificity, and resolution, supporting early disease detection and accurate localization. Nano-ecosystems also integrate therapeutic agents into nanoscale formulations that range from anticancer drugs and gene therapies to peptides. By overcoming biological barriers and enabling targeted delivery, they enhance treatment outcomes (Colvin, 2003). Nanostructured scaffolds provide platforms for tissue engineering and regenerative medicine by mimicking the extracellular matrix and promoting cell adhesion, proliferation, and differentiation.

Nano-sensors can be integrated into diagnostic devices for rapid and sensitive detection of disease biomarkers such as proteins and nucleic acids. This enables

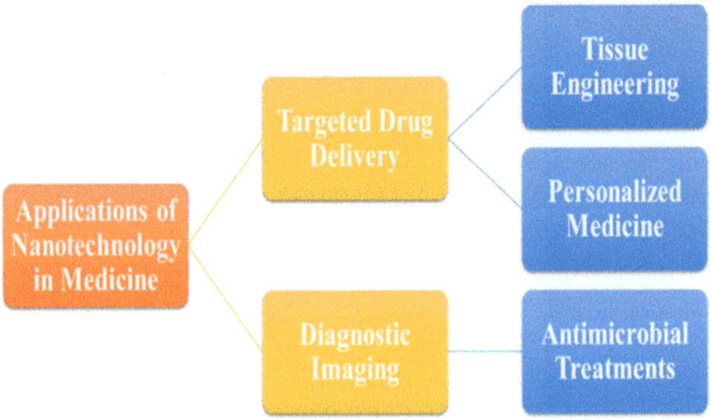

Fig. 2 Applications of different nano-ecosystems in medicine

early diagnosis, personalized treatment planning, and real-time monitoring. Theranostic platforms combine both therapeutic and diagnostic functions within a single nanoscale system, enabling simultaneous imaging and treatment to improve patient outcomes (Bhattacharya & Gupta, 2020). Nano-sensors also support the development of advanced biomarkers that help identify disease-specific molecular signatures and treatment responses. This facilitates personalized medicine by allowing therapies to be tailored according to patient-specific profiles. Intra-arterial nanosensors and nano-electronic devices enable real-time monitoring of vital indicators such as blood glucose, drug concentration, and physiological parameters. These systems support closed-loop therapeutic control by regulating drug release based on patient needs. Emerging nano-delivery technologies overcome biological barriers to deliver drugs to previously inaccessible regions, offering new opportunities for treating complex diseases, including neurological disorders and cancers (Colvin, 2003). Nanoparticle-based vaccine delivery systems enhance immune responses by controlling antigen presentation and adjuvant release. These systems improve vaccine stability, effectiveness, and storage efficiency, offering solutions to global health challenges.

Collectively, these applications (Fig. 2) demonstrate the transformative potential of nano-ecosystems in healthcare, providing faster, more accurate, and more effective solutions for disease diagnosis, treatment, and management (Demokritou, 2013).

2.3 Nano-Ecosystems and Environment

Although nano-ecosystems offer significant technological benefits, they also pose potential environmental risks (Table 2) that require careful assessment and management. Understanding these implications is essential for the responsible development and deployment of nanotechnologies for the environment. Throughout their lifecycle, nano-ecosystems may release nanoparticles into the environment. Due to their extremely small size and resistance to natural degradation, these particles can

Table 2 Environmental and health impacts of nanomaterials

Impact category	Description	Examples of effects
Ecotoxicity	Potential toxicity of nanomaterials to aquatic life, soil organisms, and plants	Reduced growth in algae, toxicity to fish
Bioaccumulation	Accumulation of nanomaterials in organisms and food chains	Presence of nanoparticles in aquatic organisms and groundwater seeping
Persistence	Long-term environmental presence and potential chronic effects	Persistent silver (Ag) and gold (Au) nanoparticles in soil
Human Health Exposure	Inhalation, ingestion, or dermal contact with nanomaterials	Respiratory issues, skin irritation, cancer of various types
Air and Water Quality	Reduced air and water quality due to nanomaterial release and contamination	Airborne nanoparticle pollution, water contamination

persist for long periods. Their interactions with soil, water, and air organisms may influence ecological functions, potentially disrupting biodiversity. Nanoparticles may also bioaccumulate and bio-magnify through food chains, posing risks to higher organisms (Donaldson & Stone, 2004). Nanoparticles entering aquatic systems through effluents or waste streams may affect water quality and aquatic species, while those deposited in soil can alter microbial communities and nutrient-cycling processes. Humans may be exposed to nanoparticles through inhalation, ingestion, or dermal contact, leading to respiratory discomfort, skin irritation, or long-term health effects depending on exposure level and particle characteristics. Risk assessment is challenging due to limited knowledge of nanoparticle behavior in natural environments.

This highlights the need for strict regulations and standards governing the production, use, disposal, and monitoring of nano-ecosystems. Improved detection and monitoring technologies, along with remediation strategies using nanomaterials are essential (Handy, 2008). Incorporating sustainability principles into nano-ecosystem design, such as using biodegradable or environmentally benign nanoparticles, can help reduce ecological impacts. Public engagement and transparent communication are critical to addressing environmental concerns and building trust. Continued research on nanoparticle transport, accumulation, and effects in diverse ecosystems is needed to inform safe and responsible applications (Hristozov & Malsch, 2009). Ultimately, minimizing the environmental risks of nano-ecosystems will require coordinated efforts from researchers, policymakers, industry stakeholders, and the public to ensure that the benefits of nanotechnology are realized without compromising ecosystem or human health.

2.4 Nano-Ecosystems: A Key to Long-Term Sustainability

Nano-ecosystems play a crucial role in advancing global sustainability goals by offering innovative solutions across diverse sectors. These technologies utilize nanoparticles and nanoscale devices to enhance societal well-being, reduce environmental impacts, and improve resource efficiency. In renewable energy, nanomaterials enhance light absorption and charge separation in photovoltaic cells, significantly improving solar panel efficiency (Kah et al., 2013). They also enable high-capacity energy storage systems, including advanced batteries and supercapacitors. In environmental protection, nanoparticles can efficiently remove pollutants from air, water, and soil, while nanofiltration membranes and adsorbents support wastewater treatment and potable water generation. Manufacturing nanomaterials in accordance to green chemistry reduces resource consumption and waste, while improving mechanical, thermal, and barrier properties. They also enhance catalytic selectivity and efficiency, supporting eco-friendly industrial processes (Kirschvink et al., 1997).

Agriculture benefits from nano-sensors are that of monitoring soil health, nutrient levels, and pest conditions in real time, enabling optimized crop yields with minimal inputs. Nanoscale detection tools further enhance food safety by rapidly identifying contaminants and pathogens (Lee et al., 2007). In healthcare, nano-diagnostics and targeted drug delivery systems personalize treatments based on biomarkers and genetic profiles. Nanostructured biomaterials and scaffolds also contribute to regenerative medicine and more efficient healthcare resource use. Nano-ecosystems support climate action through carbon-capture nanomaterials and nano-sensors that monitor greenhouse gas emissions and other environmental parameters (Lee et al., 2007). In infrastructure and smart cities, nanomaterials reinforce structures and enable real-time monitoring of integrity, energy use, transportation systems, and resource distribution. Circular economy goals are supported through recyclable nanomaterials and improved recycling processes that reduce landfill loads and conserve resources. Nanoparticles also enhance environmental and human health protection by selectively neutralizing hazardous substances. Promoting awareness among the public, researchers, and policymakers is essential to highlight nanotechnology's role in sustainability (Liu & Lal, 2010). Collaboration and knowledge sharing are crucial to embedding sustainability principles into the research, design, and deployment of nano-ecosystems. Strong regulatory frameworks and ethical guidelines ensure responsible and safe adoption, while dialogue with stakeholders helps evaluate advantages, risks, and social impacts (Maynard, 2006). By integrating nano-ecosystems into sustainable development strategies, it becomes possible to support economic growth, environmental stewardship, and social equity while addressing global challenges effectively (McNeil, 2005).

3 Nano-Ecosystems and Their Conventional Counterparts

Nano-ecosystems and conventional ecosystems differ significantly in scale, complexity, operational environments, and defining characteristics. Nano-ecosystems operate at the microscopic scale, typically between 1 and 100 nanometers, and consist of nanoscale components that interact within highly controlled environments (Nandhakumar et al., 2024). Their structures are engineered with precise dimensions and predefined functions. In contrast, conventional ecosystems function at macroscopic scales, ranging from local habitats to global biomes, and include diverse living organisms interacting with abiotic factors such as climate and soil. These ecosystems encompass communities of producers, consumers, decomposers, and the networks of interactions between them (Ahamed et al., 2010).

Their properties and behaviors are governed by controlled synthesis, assembly methods, molecular recognition, surface interactions, and quantum effects. Such systems can be programmed to respond to specific environmental or biological cues, enabling adaptive behaviors such as triggered drug release or environmental sensing. Conventional ecosystems, however, are shaped by biological processes such as mutualism, competition, and predation, which influence biodiversity, resilience, and community structure. While nano-ecosystems are intended for specialized technological applications, such as nanoelectronics, environmental monitoring, and nanomedicine, conventional ecosystems perform essential ecological functions including air and water purification, nutrient cycling, and climate regulation. Through provisioning, regulating, and cultural services, natural ecosystems support biodiversity, environmental stability, and human well-being. These ecosystems require conservation and sustainable management due to their sensitivity to disturbances and climate change impacts. The complexity of nano-ecosystems arises from technological advancements, interdisciplinary design, and the incorporation of responsive materials, digital integration, and dynamic sensing capabilities. In contrast, natural ecosystems exhibit complexity shaped over geological time through evolutionary dynamics, genetic diversity, and ecological interactions (Xia et al., 2008). Both systems, however, share the ability to adapt based on interactions among their components. While conventional ecosystems rely on natural processes to maintain life and ecological balance, nano-ecosystems use engineered nanomaterials and technologies to perform targeted functions (Giridharan et al., 2024).

4 Nano-Ecosystems: Difficulties and Boundaries

Despite their promising applications, nano-ecosystems must overcome several challenges and limitations before their full potential can be realized. These issues span technological, regulatory, ethical, and sociological domains. Nanomaterials exhibit unique characteristics based on their composition, size, shape, and surface chemistry; however, the lack of standardized characterization methods makes it difficult to compare and reproduce findings across studies. Comprehensive risk assessment and mitigation strategies are essential, given the potential toxic effects of nanoparticles on organisms and ecosystems (Nel et al., 2017). Human exposure to nanoparticles through inhalation, ingestion, or dermal contact also raises concerns, as long-term health effects remain insufficiently understood. Robust regulatory frameworks are needed to ensure safe management, disposal, and monitoring of nano-ecosystems. Limited understanding of nanoparticle behavior in different environmental matrices complicates risk evaluation and regulatory decision-making (Oberdörster, 2010).

Large-scale commercialization poses additional barriers: producing nanomaterials at industrial levels while maintaining consistent quality and reasonable costs is challenging, as nanoscale properties may shift during mass manufacturing, affecting system reliability. Ethical, privacy, and security concerns further complicate the deployment of nanotechnologies. Ensuring equitable access to nano-ecosystem benefits while minimizing potential risks is crucial. Progress also depends on interdisciplinary collaboration, uniting chemistry, biology, engineering, ethics, and policy, to create well-rounded solutions (Panyala et al., 2008). Clear communication among stakeholders supports transparency, understanding, and informed decision-making. Sustainability challenges include developing long-term strategies for recycling and end-of-life management of nano-ecosystems to reduce environmental burdens. Lifecycle assessments are necessary to evaluate environmental and societal impacts comprehensively. Addressing technical difficulties such as designing nanoscale systems with consistent performance and meeting targeted specifications is essential for advancing fields like nanomedicine, nanoelectronics, and environmental nanotechnology. Global cooperation is also critical. Harmonizing international standards, regulations, and guidelines will enable safer and more efficient development and deployment of nano-ecosystems (Powers et al., 2007). Public engagement and awareness initiatives can encourage critical understanding of both the benefits and risks of nanotechnology. Strengthening education and training in nanotechnology ethics, safety, and sustainability is equally important. Ultimately, researchers, policymakers, industry leaders, and communities must work collectively to address these challenges. Such cooperation will ensure that nano-ecosystems contribute positively to sustainable development while minimizing environmental, societal, and health-related risks (Rauscher et al., 2019).

5 Emerging Areas of Study in Nano-Ecosystems

Research and development in nanotechnology are rapidly advancing due to the growing interest in nano-ecosystems across multiple scientific disciplines. Progress in nanoparticle-based drug delivery systems has enabled highly precise therapeutic targeting and improved treatment efficacy while minimizing adverse effects. The integration of diagnostic and therapeutic functions within single nano-systems also supports the development of personalized medical interventions (Raj et al., 2012). Wearable and flexible electronics incorporating nanomaterials are being created for real-time health and environmental monitoring. Additionally, nanoscale sensors offer sensitive and rapid detection of biomarkers, contaminants, and pathogens. Research continues to explore interactions between nanomaterials and biological systems to better understand their benefits, risks, and mechanisms in medical and environmental contexts. This includes investigations into immune responses to nanomaterials, which are crucial for designing immunomodulatory therapies and advanced vaccine delivery systems. Standardized testing methods and risk-management frameworks are being refined to accurately assess the health, safety, and environmental impacts of nanomaterials, emphasizing the need for responsible production, use, and disposal of nano-ecosystems. In materials science, efforts are directed toward developing self-repairing nanostructured materials for enhanced durability and creating lightweight, high-performance composites for construction, automotive, and aerospace applications. Computational modelling and simulations are increasingly used to optimize nano-system design and predict behavior, while machine learning tools accelerate the discovery of new materials through large-scale data analysis. Global collaboration among academic institutions, industry, and regulatory bodies is essential for advancing innovation, establishing standards, and aligning nanotechnology development with international sustainability goals.

6 The Potential of Nano-Ecosystems in the Future

Nano-ecosystems hold immense promise for addressing global challenges and driving technological innovation. Looking ahead, numerous opportunities highlight the transformative potential of nanotechnology. Breakthroughs in nanomedicine are enabling highly personalized regenerative therapies, advanced diagnostics, and targeted drug delivery tailored to individual genetic and pathological profiles. The development of nanorobots further opens pathways for cellular-level treatments, minimally invasive surgical procedures, and precision drug administration (Panyala et al., 2008). In the energy sector, integrating nanomaterials into photovoltaic systems significantly enhances efficiency while reducing costs, supporting the widespread adoption of solar power. Nanotechnology-enabled batteries and supercapacitors offer greater energy density, faster charging, and longer operational lifetimes, strengthening renewable energy storage solutions (Stone et al., 2010). Environmental remediation is another area of major progress, with innovative nanomaterials enabling effective cleanup of contaminated sites, water purification, and improved air quality. Additionally, nanotechnology facilitates carbon capture and conversion into valuable products, contributing directly to climate change mitigation (Smita et al., 2012). Agriculture stands to benefit from nano-sensors and smart nanomaterials that monitor soil health, crop growth, and pest activity in real time, thereby improving productivity and sustainability. Advanced nanoscale detection systems also enhance food safety by rapidly identifying contaminants and pathogens across supply chains (Oberdörster, 2010).

In materials science, nanotechnology supports the creation of lightweight, high-strength materials for aerospace, automotive, construction, and consumer applications. It also drives major improvements in additive manufacturing by enabling the production of complex, high-precision structures (Shvedova et al., 2008). In electronics and computing, nanoscale components and quantum devices are unlocking faster processing, higher energy efficiency, and entirely new sensing capabilities. Harnessing quantum phenomena at the nanoscale may revolutionize computational capacity and enable solutions beyond classical computing (Savolainen et al., 2010). In public health, nanoscale platforms are being designed to enhance vaccine delivery, strengthen disease prevention, and provide rapid diagnostic tools, particularly in resource-limited settings. The expansion of nano-ecosystems also raises ethical considerations, including privacy, equitable access, and societal impacts. Establishing robust regulatory frameworks is essential to ensure safe development, deployment, and end-of-life management of nanotechnologies. Interdisciplinary education will play a critical role in fostering responsible innovation and equipping future professionals for nanotechnology research, development, and commercialization (Sahu & Hayes, 2017). Strengthening global partnerships will accelerate technological progress, facilitate knowledge exchange, and support collective responses to global challenges.

Ultimately, aligning nanotechnology with international sustainability goals can promote equitable development, environmental stewardship, and improved societal well-being. The future of nano-ecosystems will be defined by their ability to solve pressing global problems, enhance human life, and transform entire sectors.

7 Conclusion

Nano-ecosystems represent the forefront of nanotechnology and have the potential to revolutionize numerous sectors, including healthcare, energy, environmental remediation, and advanced manufacturing. This emerging paradigm employs nanoscale systems, nanodevices, and engineered nanomaterials to address complex challenges with unprecedented efficiency and precision. By enabling control and manipulation at the molecular level, nano-ecosystems integrate artificial nanomaterials and nanoscale technologies designed to mimic or enhance natural processes. Advances in nanoscience, coupled with interdisciplinary contributions from engineering, biology, chemistry, and physics, provide the foundation for these systems. Key applications include personalized healthcare; renewable-energy advancements; and sustainable environmental solutions such as pollution reduction, water purification, and long-term ecological protection. In industrial settings, nano-ecosystems support the development of smart materials, nanocomposites, and cutting-edge additive manufacturing, enabling lightweight, adaptive, and high-performance components. Despite their transformative potential, nano-ecosystems face challenges related to environmental and health risks, including ecotoxicity, as well as regulatory and ethical considerations. Issues such as production scalability, consistency, and public perception further complicate large-scale deployment. Addressing these concerns will require strong interdisciplinary collaboration, ethical governance, and the establishment of robust global standards. Looking forward, nano-ecosystems are poised to drive innovations in sustainable energy, smart agriculture, precision medicine, nanoelectronics, quantum technologies, and digital transformation. International cooperation, enlightened regulation, and responsible implementation will be essential to maximizing societal benefits while minimizing risks.

In conclusion, nano-ecosystems offer a powerful path toward global sustainability, environmental stewardship, and enhanced human well-being. Realizing their full potential will depend on continued investment in research, innovation, and responsible technological adoption. Through multidisciplinary collaboration and informed decision-making, society can harness the transformative capabilities of nano-ecosystems for current and future generations.

References

Ahamed, M., AlSalhi, M. S., & Siddiqui, M. K. J. (2010). Silver nanoparticle applications and human health. *Clinica Chimica Acta, 411*(23–24), 1841–1848.

Bhattacharya, S., & Gupta, P. (2020). Nanotechnology and environmental health and safety: Issues for consideration. *Environmental Research, 186*, 109515.

Bhattacharya, K., & Mukherjee, S. (2023). Toxicology of nanomaterials: An interdisciplinary challenge. *Toxicology Research, 8*(5), 589–593.

Bhushan, B. (2017). *Nanotechnology for environmental remediation*. Springer International Publishing.

Buzea, C., Pacheco, I. I., & Robbie, K. (2007). Nanomaterials and nanoparticles: Sources and toxicity. *Biointerphases, 2*(4), MR17-MR71.

Cassee, R., & Peters, A. (2018). Nanoparticle exposure and public health. *Science, 361*(6398), 181–182.

Colvin, V. L. (2003). The potential environmental impact of engineered nanomaterials. *Nature Biotechnology, 21*(10), 1166–1170.

De Volder, M. F., Tawfick, S. H., Baughman, R. H., & Hart, A. J. (2013). Carbon nanotubes: Present and future commercial applications. *Science, 339*(6119), 535–539.

Demokritou, P. (2013). An in vivo and in vitro toxicity assessment of realistic nanoscale CeO2 inhalation exposures. *Nanotoxicology, 7*(8), 1338–1350.

Donaldson, K., & Poland, C. A. (2013). Nanotoxicity: Challenging the myth of nano-specific toxicity. *Current Opinion in Biotechnology, 24*(4), 724–734.

Donaldson, K., & Stone, V. (2004). Current hypotheses on the mechanisms of toxicity of ultrafine particles. *Annals of the Institute of Occupational Health of Finland, 38*(3), 127–136.

Drori, I., & Lavie, D. (2023). How do innovation ecosystems emerge? The case of nanotechnology in Israel. *Journal of Management Studies., 61*, 3754.

Ebbesen, M. (2008). Ethics in nanotechnology: Starting from scratch? *Bulletin of Science, Technology & Society, 28*(6), 451–462.

Francis, R., Kalyanaraman, R., Boominathan, V., Parthasarathy, S., Chavaan, A., Ansari, I. A., et al. (2024). Piperine's potential in treating polycystic ovarian syndrome explored through in-silico docking. *Scientific Reports, 14*(1), 21834.

Gao, F. (2017). Overview of nanomaterials for water remediation. *Materials, 10*(6), 757.

Giridharan, B., Chinnaiah, A., Saravanan, K. M., Parthasarathy, S., Meenakshi Sundaram, K. K., Tharumasivam, S. V., et al. (2024). Characterization of novel antimicrobial peptides from the epidermis of Clarias batrachus catfish. *International Journal of Peptide Research and Therapeutics, 30*(2), 11.

Gobalan, K., Sudharsan, P., Reshma Devi, R., Koperuncholan, M., Sumithra, D., Prabhu, K., et al. (2023). Anti-inflammatory activity of Ocimum tenuiflorum L. leaf extract loaded Chitosan nanoparticle as a drug carrier–an in vitro study. *European Chemical Bulletin, 12*(1), 1062–1084.

Godwin, H. A. (2015). The future of nanotechnology risk assessment. *ACS Nano, 9*(1), 340–346.

Gurwitz, D. (2014). Ethical and social aspects of personalized medicine. *Expert Review of Personalized Medicine, 1*(1), 111–118.

Handy, R. D. (2008). The ecotoxicology of nanoparticles and nanomaterials: Current status, knowledge gaps, challenges, and future needs. *Ecotoxicology, 17*(5), 315–325.

Hood, E. (2004). Nanotechnology: Looking as we leap. *Environmental Health Perspectives, 112*(13), A740–A749.

Hristozov, D., & Malsch, I. (2009). Hazards and risks of engineered nanoparticles for the environment and human health. *Sustainability, 1*(4), 1161–1194.

Iannacci, J. (2021). The WEAF Mnecosystem (water, earth, air, fire micro/nano ecosystem): A perspective of micro/nanotechnologies as pillars of future 6G and tactile internet (with focus on MEMS). *Microsystem Technologies, 27*(10), 3943–3951.

Ju-Nam, Y., & Lead, J. R. (2008). Manufactured nanoparticles: An overview of their chemistry, interactions and potential environmental implications. *Science of the Total Environment, 400*(1–3), 396–414.

Kah, M., Beulke, S., Tiede, K., & Hofmann, T. (2013). Nanopesticides: State of knowledge, environmental fate, and exposure modeling. *Critical Reviews in Environmental Science and Technology, 43*(16), 1823–1867.

Kalyanaraman, R., Boominathan, V., Francis, R., Chandran, J., Parthasarathy, S., Giridharan, B., & Tharumasivam, S. V. (2024). Exploring potential antiaging compounds from medicinal plants and binding affinity of resveratrol from vitis vinifera with htert gene. *Biochemical and Cellular Archives, 24*(1), 1047.

Kamaraj, P., Sridhar, V. V., Tharumasivam, S. V., Parthasarathy, S., Bupesh, G., Raju, N. K., et al. (2023). Carbon nanoparticles fabricated microfilm: A potent filter for microplastics debased water. *Environmental Pollution, 336*, 122502.

Kirschvink, J. L., et al. (1997). Environmental magnetism: A powerful tool for the geosciences. *Earth and Planetary Science Letters, 152*(1–4), 1–6.

Klaine, S. J., Alvarez, P. J. J., Batley, G. E., Fernandes, T. F., Handy, R. D., Lyon, D. Y., et al. (2008). Nanomaterials in the environment: Behavior, fate, bioavailability, and effects. *Environmental Toxicology and Chemistry, 27*(9), 1825–1851.

Lane, R. M. (2022). *Ecology for learning and teaching: Integrating education science into teacher's Pedagogies using an adaptation of Bronfenbrenner's Ecological Framework* (Doctoral dissertation, ResearchSpace@ Auckland).

Laurent, S., Forge, D., Port, M., Roch, A., Robic, C., Vander Elst, L., & Muller, R. N. (2008). Magnetic iron oxide nanoparticles: Synthesis, stabilization, vectorization, physicochemical characterizations, and biological applications. *Chemical Reviews, 108*(6), 2064–2110.

Lee, K. J., et al. (2007). In vivo imaging of transport and biocompatibility of single silver nanoparticles in early development of zebrafish embryos. *ACS Nano, 1*(2), 133–143.

Lin, S. (2010). *Understanding the biological and environmental implications of nanomaterials*. Clemson University.

Liu, R., & Lal, R. (2010). Nanoparticles and their biological and environmental applications. *Journal of Environmental Quality, 39*(6), 1896–1908.

Lua, W. K. H., Yau, P. C., Seow, C. K., & Wong, D. (2022, August). Lightweight CNN-based deep neural networks application in safety measurement. In *2022 5th international conference on pattern recognition and artificial intelligence (PRAI)* (pp. 455–459). IEEE.

Maynard, A. D. (2006). Nanotechnology: The next big thing, or much ado about nothing? *Annals of Occupational Hygiene, 50*(1), 1–12.

McNeil, S. E. (2005). Nanotechnology for the biologist. *Journal of Leukocyte Biology, 78*(3), 585–594.

Mohanta, Y. K., Biswas, Y., Mohanta, K., & Mahanta, S. (2024). Graphene-based nanomaterials. *IEEE Transactions on Terahertz Science and Technology, 14*(5).

Mueller-Dombois, D., et al. (1994). *Vegetation types: A consideration of available methods and their suitability for various purposes*. Technical Report No. 49, Island Ecosystem IRP, U.S. International Biological Program.

Nandhakumar, R., Ramesh, R. D., Parthasarathy, S., Kennedy, R. K., Karthikeyan, M., & Tharumasivam, S. V. (2024). Assessment of the role of graphene's impact on food biotechnology. In *Graphene-based nanomaterials* (pp. 61–70). CRC Press.

Nasir, A. (2010). Nanotechnology and dermatology: Part II—Risks of nanotechnology. *Clinics in Dermatology, 28*(5), 581–588.

Nel, A., Xia, T., Madler, L., & Li, N. (2006). Toxic potential of materials at the nanolevel. *Science, 311*(5761), 622–627.

Nel, A. E., et al. (2017). Nanomaterial risk assessment: Emerging issues and challenges. *Philosophical Transactions of the Royal Society A: Mathematical, Physical and Engineering Sciences, 375*(2095), 20160332.

Oberdörster, G. (2010). Safety assessment for nanotechnology and nanomedicine: Concepts of nanotoxicology. *Journal of Internal Medicine, 267*(1), 89–105.

Oberdörster, G., Oberdörster, E., & Oberdörster, J. (2005). Nanotoxicology: An emerging discipline evolving from studies of ultrafine particles. *Environmental Health Perspectives, 113*(7), 823–839.

Panyala, N. R., et al. (2008). Silver or silver nanoparticles: A hazardous threat to the environment and human health? *Journal of Applied Biomedicine, 6*(3), 117–129.

Pisciotta, S. (2022). *Development of a fault injection environment for the evaluation of hardening techniques on GPGPU via the NVBitFI framework* (Doctoral dissertation, Politecnico di Torino).

Powers, K. W., et al. (2007). Characterization of the size, shape, and state of dispersion of nanoparticles for toxicological studies. *Nanotoxicology, 1*(1), 42–51.

Raj, S., et al. (2012). Nanotechnology in cosmetics: Opportunities and challenges. *Journal of Pharmacy and Bioallied Sciences, 4*(3), 186–193.

Rauscher, H., et al. (2019). An overview of concepts and terms used in the European Commission's definition of nonmaterial. *Particuology, 42*, 104–111.

Roco, M. C. (2003). Broader societal issues of nanotechnology. *Journal of Nanoparticle Research, 5*(3), 181–189.

Sahu, S. C., & Hayes, A. W. (2017). *Toxicology and risk assessment of nanomaterials: Emerging issues and challenges*. Wiley.

Savage, N., & Diallo, M. S. (2005). Nanomaterials and water purification: Opportunities and challenges. *Journal of Nanoparticle Research, 7*(4–5), 331–342.

Savolainen, K., et al. (2010). Risk assessment of engineered nanomaterials and nanotechnologies—A review. *Toxicology, 269*(2–3), 92–104.

Schulte, P. A., & Salamanca-Buentello, F. (2007). Ethical and scientific issues of nanotechnology in the workplace. *Environmental Health Perspectives, 115*(1), 5–12.

Senterre, B., & Wagner, M. (2014). *Mapping Seychelles habitat-types on Mahé, Praslin, Silhouette, La Digue and Curieuse*. Consultancy Report, Government of Seychelles, United Nations Development Programme, Victoria, Seychelles, 119.

Shvedova, A. A., et al. (2008). Inhalation vs. aspiration of single-walled carbon nanotubes in C57BL/6 mice: Inflammation, fibrosis, oxidative stress, and mutagenesis. *American Journal of Physiology-Lung Cellular and Molecular Physiology, 295*(4), L552–L565.

Smita, S., et al. (2012). Nanoparticles in the environment: Assessment using the causal diagram approaches. *Environmental Health, 11*(1), 1–13.

Stanger, N. R. (2011). Moving "eco" back into socio-ecological models: A proposal to reorient ecological literacy into human developmental models and school systems. *Human Ecology Review, 18*, 167–173.

Stone, V., et al. (2010). Nanomaterials for environmental studies: Classification, reference material issues, and strategies for physicochemical characterization. *Science of the Total Environment, 408*(8), 1745–1754.

Sutherland, C. R., Henderson, A., Giosio, D., Trotter, A., & Smith, G. (2024). Nvgstcapture-1.0 Modified for Autofocus and Peripheral Hardware Interfaces. Available at SSRN 4834786.

Tripathi, A., Mishra, G., & Kant, K. (2024). Plant pathogens and detection approaches with gold nanoparticles. In *Nanotechnology in plant health* (pp. 155–163). CRC Press.

Wiesner, M. R., Lowry, G. V., Alvarez, P., Dionysiou, D., & Biswas, P. (2006). Assessing the risks of manufactured nanomaterials. *Environmental Science & Technology, 40*, 4336.

Xia, T., Kovochich, M., Liong, M., Madler, L., Gilbert, B., Shi, H., et al. (2008). Comparison of the mechanism of toxicity of zinc oxide and cerium oxide nanoparticles based on dissolution and oxidative stress properties. *ACS Nano, 2*(10), 2121–2134.

Zhang, L., & Webster, T. J. (2009). Nanotechnology and nanomaterials: Promises for improved tissue regeneration. *Nano Today, 4*(1), 66–80.

Zhao, X., & Castranova, V. (2011). Toxicology of nanomaterials used in nanomedicine. *Journal of Toxicology and Environmental Health, Part B, 14*(8), 593–632.

Principles of Nano-Ecosystem Engineering

R. Vidya, A. M. Anandkumar, and P. Amudha

Abstract

The integration of nanotechnology with ecological principles has emerged as a promising frontier in addressing pressing environmental challenges. This chapter elucidates the foundational principles underlying nano-ecosystem engineering, a multifaceted approach aimed at harnessing the potential of nanomaterials to promote sustainability and mitigate ecological degradation. Central to nano-ecosystem engineering is the concept of synergy between nanotechnology and ecology, where nanomaterials are strategically designed and applied to augment ecological processes while minimizing adverse environmental impacts. The chapter explores key strategies for achieving this synergy, including the design of nanostructured materials tailored to interact with specific environmental components, such as soil, water, and organisms. Moreover, the chapter delves into the importance of understanding the mechanisms governing the interactions between nanomaterials and the environment, including transport, fate, and toxicity. By elucidating these mechanisms, researchers can develop informed strategies to maximize the benefits of nanotechnology while minimizing potential risks. Furthermore, the chapter highlights the potential applications of nano-ecosystem engineering across various environmental contexts, including pollution remediation, resource management, and sustainable agriculture. Overall, this chapter provides a comprehensive overview of the principles of nano-ecosystem engineering, offering insights into the design, application, and implications of nanotechnology in ecological contexts. By adopting these principles, researchers and

R. Vidya (✉) · P. Amudha
Department of Biochemistry, Vels Institute of Science, Technology & Advanced Studies (VISTAS), Chennai, Tamil Nadu, India

A. M. Anandkumar
Department of Botany, Nallamuthu Gounder Mahalingam College, Pollachi, Tamil Nadu, India

D. K. Gupta et al. (eds.), *Nano-Ecosystems*,
https://doi.org/10.1007/978-981-95-5033-3_2

practitioners can lay the groundwork for creative solutions to environmental challenges, nurturing a future that is both more sustainable and resilient.

Keywords

Nanostructured materials · Environmental challenges · Pollution remediation · Sustainable agriculture

1 Introduction

In recent decades, the convergence of nanotechnology and ecology has opened up new avenues of exploration and innovation in environmental science. This interdisciplinary field, often termed nano-ecosystem engineering, encompasses the study and application of nanotechnology principles to understand, manipulate, and enhance ecological systems. At its core, nano-ecosystem engineering seeks to leverage the unique properties of nanomaterials to address pressing environmental challenges and promote sustainability (Bayda et al., 2020). Nanotechnology, defined as the manipulation of matter at the nanoscale, has revolutionized various industries, from electronics to medicine. However, its potential in ecological contexts has only begun to be realized. The scale at which nanotechnology operates (ranging from 1 to 100 nm) intersects with fundamental processes in nature, offering unprecedented opportunities to interact with and influence ecosystems at their most basic levels. The principles underlying nano-ecosystem engineering draw from both nanoscience and ecology. Nanoscience provides the tools and techniques to synthesize, characterize, and manipulate nanomaterials, while ecology offers insights into the complex interactions among organisms and their environment. By integrating these disciplines, researchers can design innovative solutions that mitigate environmental degradation, enhance ecosystem resilience, and promote sustainable development (Salem, 2023). One of the defining features of nano-ecosystem engineering is its focus on understanding and harnessing the natural processes that govern ecosystems. Nature has evolved sophisticated mechanisms at the nanoscale to facilitate essential functions such as nutrient cycling, energy transfer, and waste remediation. By studying these processes, scientists can develop nanomaterials and nanostructures that mimic or augment natural systems, leading to more efficient and environmentally friendly technologies. The applications of nano-ecosystem engineering are diverse and far-reaching. Environmental sensing and monitoring, for instance, benefit from the high sensitivity and selectivity of nanomaterial-based sensors, enabling real-time detection of pollutants and contaminants. Similarly, nanotechnology plays a crucial role in remediation strategies by enhancing the efficiency of pollutant removal and facilitating the degradation of harmful substances (Mobeen et al., 2022).

In agriculture, nano-ecosystem engineering holds promise for improving crop yields while minimizing environmental impact. Nanoparticle-based fertilizers and pesticides can deliver nutrients and pest control agents more effectively, reducing the need for excessive chemical inputs and mitigating soil and water contamination.

Furthermore, nanomaterials can enhance soil quality and promote plant growth through mechanisms such as nutrient uptake facilitation and water retention. Biomimicry, the practice of emulating nature's designs and processes, is another key aspect of nano-ecosystem engineering. By drawing inspiration from biological systems, researchers can develop innovative nanotechnologies that are not only effective but also sustainable and environmentally benign. For example, the hierarchical structures found in natural materials like butterfly wings and lotus leaves have inspired the design of self-cleaning surfaces and antifouling coatings. Despite its immense potential, nano-ecosystem engineering also raises important ethical and environmental considerations. The use of engineered nanomaterials in ecological contexts must be accompanied by rigorous risk assessment and regulatory oversight to ensure safety for both humans and the environment. Furthermore, concerns regarding the long-term impacts of nanotechnology on ecosystems, such as unintended consequences and ecosystem disruption, must be addressed through comprehensive research and stakeholder engagement (Bhandari et al., 2023).

2 Classification of Nanoparticles

Nanoparticles are classified based on their composition such as organic, inorganic, and carbon-based nanoparticles.

2.1 Carbon-Based Nanoparticles

This classification includes nanoparticles that are solely made of carbon atoms. Nanoparticles within this classification are fullerenes, carbon black nanoparticles, and carbon quantum dots (Ealia & Saravanakumar, 2017). The symmetrical closed-cage structure of fullerenes is a distinguishing characteristic of these carbon molecules. As an illustration, C60 fullerenes are composed of 60 carbon atoms organized in a manner like a football. Additionally, other forms of fullerenes, namely C70 and C540, have also been recognized (Meirzadeh et al., 2023). The carbon black nanoparticles exhibit a clustered structure that bears a resemblance to grapes, consisting of spherical particles that are densely fused (Long et al., 2013). Carbon quantum dots are carbon nanoparticles that exhibit a distinct, quasi-spherical shape and possess dimensions less than 10 nm. Carbon nanoparticles possess the distinctive physicochemical features observed at the nanoscale, while also incorporating the unique traits of sp2-hybridized carbon bonds (Oh et al., 2010). The nanoparticles have remarkable electrical conductivity, elevated strength, an affinity for electrons, and possess optical, thermal, and sorption characteristics (Khan et al., 2019). Consequently, carbon-based nanoparticles are utilized in several domains such as medicine delivery, energy storage, bioimaging (Chandra et al., 2011), photovoltaic devices, and environmental sensing. These applications facilitate the monitoring of microbial ecology and the identification of microbial pathogens (Liu et al., 2018). Nanodiamonds and carbon nano-onions are examples of carbon-based nanoparticles that are more complex in nature. These materials are widely recognized for their minimal toxicity and compatibility with living organisms, making them suitable for a range of applications, including drug delivery and tissue engineering (Mochalin et al., 2012; Ahlawat et al., 2021).

2.2 Organic Nanoparticles

This category encompasses nanoparticles composed of proteins, carbohydrates, lipids, polymers, or other organic compounds (Pan & Zhong, 2016). Prominent examples within this category include dendrimers, liposomes, micelles, and protein complexes. These nanoparticles are typically nontoxic, biodegradable, and in some instances, such as liposomes, may feature a hollow core. Organic nanoparticles are sensitive to thermal and electromagnetic radiation, including heat and light (Ealia & Saravanakumar, 2017). Moreover, they often form through noncovalent intermolecular interactions, rendering them inherently more labile and facilitating clearance from the body (Ng & Zheng, 2015). Various parameters influence the potential applications of organic nanoparticles, such as composition, surface morphology, stability, and carrying capacity. Presently, organic nanoparticles find extensive use in the biomedical field, particularly in targeted drug delivery and cancer therapy.

2.3 Inorganic Nanoparticles

This category encompasses nanoparticles that are not composed of carbon or organic materials. Typical examples within this category include metal, ceramic, and semiconductor nanoparticles NPs. Metal nanoparticles consist solely of metal precursors and can be monometallic, bimetallic, or polymetallic (Toshima & Yonezawa, 1998; Nascimento et al., 2018). Bimetallic nanoparticles can be formed from alloys or arranged in different layers, such as core–shell structures. These exhibit unique optical and electrical properties due to localized surface plasmon resonance characteristics. Some metal nanoparticles also possess distinctive thermal, magnetic, and biological properties, making them increasingly vital materials for a wide range of applications, including physical, chemical, biological, biomedical, and pharmaceutical uses. Achieving precise control over the size, shape, and facets of metal nanoparticles through synthesis methods is crucial for creating advanced materials (Dreaden et al., 2012). Semiconductor nanoparticles are composed of semiconductor materials, which exhibit properties that lie between those of metals and nonmetals. These nanoparticles possess wide bandgaps and display significant alterations in their properties through bandgap tuning compared to bulk semiconductor materials. Consequently, semiconductor nanoparticles play crucial roles in applications such as photocatalysis, optics, and electronic devices (Gupta & Tripathi, 2012). Ceramic nanoparticles are inorganic solids composed of carbonates, carbides, phosphates, and oxides of metals and metalloids, such as titanium and calcium. They are typically synthesized through heat treatment and subsequent cooling processes and can exist in various forms, including amorphous, polycrystalline, dense, porous, or hollow structures. Ceramic nanoparticles are predominantly utilized in biomedical applications due to their high stability and load capacity. However, they also find applications in catalysis, dye degradation, photonics, and optoelectronics (Moreno-Vega et al., 2012).

3 Characterization of Nanoparticles

Various characterization techniques have been employed to analyze the range of physicochemical properties exhibited by nanoparticles. The techniques encompassed in this category consist of X-ray diffraction (XRD), X-ray photoelectron spectroscopy (XPS), Infrared (IR), Scanning Electron Microscopy (SEM), Transmission Electron Microscopy (TEM), Brunauer–Emmett–Teller (BET), and particle size analysis (Joudeh & Linke, 2022). The structural characteristics of nanoparticles are commonly studied using these techniques (Mourdikoudis et al., 2018). Scanning Electron Microscopy (SEM) is a powerful imaging technique widely used for the characterization of materials, including nanoparticles. Unlike conventional optical microscopy, SEM provides detailed surface morphology information with significantly higher resolution. In SEM, a focused electron beam scans the surface of a sample, and interactions between the beam and the sample produce various signals, such as secondary electrons, backscattered electrons, and characteristic X-rays. These signals are detected and used to generate high-resolution images of the sample surface.

SEM offers several advantages for nanoparticle characterization. It allows researchers to visualize the size, shape, and distribution of nanoparticles, providing crucial insights into their morphology and aggregation behavior. Additionally, SEM can distinguish between individual nanoparticles and aggregates, enabling precise analysis of nanoparticle dispersion and stability. Furthermore, SEM can be coupled with Energy-Dispersive X-ray Spectroscopy (EDS) for elemental analysis, allowing researchers to identify the chemical composition of nanoparticles and their surrounding environment. Overall, SEM plays a vital role in understanding the physical characteristics of nanoparticles, aiding in the development and optimization of nanoparticle-based materials for a wide range of applications, including nanotechnology, materials science, biomedicine, and environmental science (Vladár & Hodoroaba, 2020). Transmission Electron Microscopy (TEM) stands as a cornerstone in nanoscale imaging and characterization, offering unparalleled resolution and insight into the structure and properties of materials. In TEM, a focused beam of electrons penetrates through a thin specimen, interacting with its atoms to produce a detailed image. Unlike SEM, which primarily captures surface features, TEM provides cross-sectional views, enabling examination of internal structures with atomic-scale precision. TEM's exceptional resolution, often reaching subnanometer scales, makes it indispensable for studying nanoparticles and nanomaterials. By visualizing individual atoms and their arrangement within a material, TEM allows for precise analysis of nanoparticle size, shape, crystallinity, and defects.

Additionally, TEM facilitates the investigation of nanoscale phenomena such as quantum confinement effects, lattice distortions, and surface reconstructions. Furthermore, TEM can elucidate the composition and chemical bonding of nanoparticles through techniques like EDS and Electron Energy-Loss Spectroscopy (EELS). These methods provide elemental and chemical information, aiding in the identification of nanoparticle composition, surface functionalization, and chemical reactions. TEM's versatility extends beyond materials science, finding applications in fields such as biology, medicine, and environmental science. It enables researchers to explore biological structures, study disease pathology, and analyze environmental samples at the nanoscale. In summary, TEM serves as an indispensable tool for advancing our understanding of nanomaterials and their diverse applications across various disciplines, driving innovation and discovery in the nanotechnology era (Smith, 2015).

X-ray Diffraction (XRD) is a pivotal technique for analyzing the crystallographic structure of materials. By directing X-rays onto a sample, XRD measures the diffraction pattern produced as the X-rays interact with the crystal lattice. From this pattern, valuable information about the material's crystal structure, phase composition, lattice parameters, and grain size can be deduced. It is particularly indispensable for characterizing crystalline nanoparticles, offering insights into their phase purity, degree of crystallinity, and crystallographic orientation. EDX complements XRD by providing elemental analysis capabilities. Integrated with SEM or TEM, EDX detects characteristic X-rays emitted from a sample upon electron beam excitation. This allows for quantitative analysis of elemental composition, mapping of elemental distributions, and identification of impurities within nanoparticles. The EDX approach is invaluable for elucidating the chemical composition and stoichiometry of nanomaterials, essential for understanding their properties and potential applications (Holder & Schaak, 2019).

X-Ray Photoelectron Spectroscopy (XPS), also known as Electron Spectroscopy for Chemical Analysis (ESCA), offers detailed information about the elemental composition, chemical state, and electronic structure of materials. By irradiating a sample with X-rays, XPS measures the kinetic energy of emitted photoelectrons, enabling identification of elements and determination of their chemical environments. XPS is widely used to study surface composition, surface contamination, and chemical bonding in nanomaterials, providing crucial insights into surface chemistry and reactivity (Korin et al., 2017).

Infrared (IR) and Raman spectroscopy are vibrational spectroscopic techniques employed for analyzing molecular composition and structural properties of nanomaterials. IR spectroscopy detects absorption of infrared radiation by molecular vibrations, revealing information about functional groups, molecular conformations, and chemical bonds in nanoparticles. Raman spectroscopy, on the other hand, measures inelastic scattering of monochromatic light by molecular vibrations, offering complementary insights into molecular structure, crystallinity, and defects in nanomaterials (López-Lorente & Mizaikoff, 2016).

Brunauer–Emmett–Teller (BET) analysis is utilized to determine the specific surface area, pore size distribution, and porosity of nanomaterials. Based on gas adsorption–desorption isotherms, BET analysis calculates parameters such as surface area, pore volume, and pore diameter, crucial for understanding the physical properties and performance of nanomaterials in applications such as catalysis, adsorption, and drug delivery (Ambroz et al., 2018).

Zeta potential analysis, performed using a zeta size analyzer, assesses the surface charge and colloidal stability of nanoparticles in suspension. By measuring the electrophoretic mobility of nanoparticles in an electric field, zeta potential analysis provides insights into particle interactions, aggregation tendencies, and dispersion stability, essential for optimizing nanoparticle formulations and preventing aggregation in various applications (Clogston & Patri, 2011).

4 Applications of Nano-Ecosystem

Nano-based structures play a vital role in various fields such as biomedicine, agricultural, cosmetics, environmental, electronics, food, and automobile industry (Table 1).

Table 1 Various types of nanomaterials and its applications

Nanoparticle type	Application	References
Gold (Au) nanoparticles	Biomedical imaging, drug delivery, catalysis, and biosensors	Hu et al. (2020)
Silver (Ag) nanoparticles	Antibacterial, wound healing, textile industry, and water purification	Zhang et al. (2016)
Iron oxide (Fe_2O_3) nanoparticles	Magnetic resonance imaging contrast agent, drug delivery system, environmental remediation, and hyperthermia cancer treatment	Attia et al. (2022)
Quantum dots	Display technologies, biological imaging, sensing applications, and photovoltaics	Cotta (2020)
Silica (Si) nanoparticles	Drug delivery, biosensors, food additives, and catalysis	Nayl et al. (2022)
Polymer nanoparticles	Coating and paints, environmental applications, tissue engineering, and drug delivery system	Adhikari (2021)
Zinc oxide (ZnO) nanoparticles	Antibacterial coatings, photocatalyst, and sunscreens	Raha and Ahmaruzzaman (2022)
Carbon nanotubes	Drug delivery system, energy storage, electronics, and reinforcement in composites	Patel et al. (2020)
Lipid nanoparticles	Vaccine delivery, cosmetic foundations, gene therapy, and drug delivery	Musielak et al. (2022)
Titanium dioxide (TiO_2) nanoparticles	Water purification, sunscreens, antimicrobial coatings, and photovoltaics	Ali et al. (2018)
Cerium oxide (CeO_2) nanoparticles	Biomedical applications, UV blockers, fuel additives, and antioxidant agents	Singh et al. (2020)

4.1 Environmental Applications

The application of nano-ecosystem engineering in the environment holds immense potential to address pressing environmental challenges and promote sustainable management of natural resources. Nano-ecosystem engineering involves the deliberate design and manipulation of nanomaterials and their interactions within ecosystems to achieve specific environmental goals. Some key applications of nano-ecosystem engineering in the environment include (Beni & Jabbari, 2022; Guerra et al., 2018):

(a) Environmental remediation: Nano-ecosystem engineering offers innovative approaches for remediation of contaminated environments. Nanomaterials such as nanoparticles, nanofibers, and nanocomposites can be designed to efficiently adsorb, degrade, or immobilize pollutants such as heavy metals, organic pollutants, and emerging contaminants. For example, engineered nanoparticles with high surface area and reactivity can be used for the removal of heavy metals from water bodies through adsorption or catalytic degradation processes. Similarly, nanocomposite materials can be deployed for in situ remediation of contaminated soil and groundwater, enhancing the efficiency and effectiveness of traditional remediation techniques.

(b) Water treatment and purification: Nano-ecosystem engineering plays a crucial role in advancing water treatment and purification technologies. Nanomaterial-based filtration membranes, adsorbents, and photocatalysts offer superior performance in removing contaminants from water, including bacteria, viruses, organic pollutants, and microplastics. Nanoparticles such as titanium dioxide (TiO2) and graphene oxide (GO) exhibit photocatalytic properties that can degrade organic pollutants and disinfect water under ultraviolet (UV) irradiation, providing a sustainable solution for water purification.

(c) Sustainable agriculture: Nano-ecosystem engineering contributes to sustainable agriculture practices by improving crop productivity, nutrient utilization efficiency, and soil health. Nanomaterials such as nanofertilizers, nanoencapsulated agrochemicals, and nanosensors enable precise and targeted delivery of nutrients, pesticides, and herbicides to plants, reducing chemical input requirements and minimizing environmental impacts. Furthermore, nanomaterial-based soil amendments and nanoscale delivery systems enhance soil fertility, water retention, and microbial activity, promoting soil health and sustainable crop production.

(d) Environmental monitoring and sensing: Nano-ecosystem engineering facilitates the development of advanced sensors and monitoring devices for real-time detection and analysis of environmental parameters. Nanomaterial-based sensors offer high sensitivity, selectivity, and rapid response times, enabling continuous monitoring of air, water, soil, and biological systems. Nanoscale sensors can detect a wide range of environmental pollutants, pathogens, and contaminants, providing early warning systems for environmental threats and facilitating timely intervention and remediation efforts.

(e) Climate change mitigation: Nano-ecosystem engineering contributes to climate change mitigation efforts through the development of nanomaterial-based technologies for carbon capture, storage, and utilization. Nanoparticles such as Metal–Organic Frameworks (MOFs) and porous carbon materials offer high surface area and porosity for efficient adsorption and sequestration of greenhouse gases such as carbon dioxide (CO_2). Furthermore, nanomaterial-based catalysts enable the conversion of CO_2 into value-added products such as fuels and chemicals through renewable energy-driven processes, contributing to the reduction of greenhouse gas emissions and the transition to a low-carbon economy.

4.2 Cosmetic Industry Applications

Nano-ecosystem in cosmetics refers to the integration of nanotechnology into cosmetic formulations, revolutionizing the effectiveness and safety of beauty products. This application encompasses the use of nanoparticles, nanocarriers, and nanostructured materials to enhance various aspects of cosmetics, from ingredient delivery to product performance. The following are key areas where nano-ecosystems are applied in cosmetics (Cardoza et al., 2022; Gupta et al., 2022):

(a) Enhanced delivery systems: Nano-ecosystems enable the encapsulation of active ingredients in nanoparticles or nanocarriers, improving their stability and targeted delivery to the skin. This ensures better penetration of beneficial compounds, such as vitamins, antioxidants, and peptides, leading to increased efficacy of cosmetic products.
(b) Improved texture and feel: Nanostructured materials can modify the texture and consistency of cosmetic formulations, resulting in smoother textures, better spreadability, and enhanced sensory experience for consumers. Nanoparticles can also impart desirable properties like matte finish or long-lasting wear to makeup products.
(c) Sunscreen technology: Nanoparticles such as titanium dioxide and zinc oxide are widely used in sunscreens for their ability to provide broad-spectrum UV protection without leaving a white residue on the skin. Nano-sized particles allow for even distribution and improved coverage, enhancing the effectiveness of sun protection in cosmetics.
(d) Antiaging formulations: Nano-ecosystems play a crucial role in developing antiaging cosmetics by facilitating the delivery of ingredients that promote collagen production, skin regeneration, and wrinkle reduction. Nanocarriers ensure the targeted delivery of these actives to the deeper layers of the skin, where they can exert their beneficial effects.
(e) Natural ingredients enhancement: Nanoencapsulation can improve the stability and bioavailability of natural ingredients in cosmetics, preserving their efficacy and prolonging shelf life. This allows for the incorporation of botanical extracts, essential oils, and other natural compounds into cosmetic formulations without compromising their integrity.

(f) Reduced irritation and sensitivity: Nano-ecosystems help minimize skin irritation and allergic reactions by controlling the release of active ingredients and reducing their contact with the skin surface. This is particularly beneficial for individuals with sensitive skin or those prone to allergic reactions to traditional cosmetic ingredients.

4.3 Agricultural Applications

The application of nano-ecosystem engineering in the agricultural industry represents a promising frontier for addressing key challenges related to food security, sustainability, and environmental stewardship (Fig. 1). Nano-ecosystem engineering involves the deliberate design and manipulation of nanomaterials and their interactions within agricultural ecosystems to enhance crop productivity, resource efficiency, and environmental resilience. Some key applications of nano-ecosystem engineering in agriculture include (Neme et al., 2021; Liu et al., 2021)

(a) Nanofertilizers and nutrient management: Nano-ecosystem engineering enables the development of nanofertilizers and nutrient delivery systems tailored to meet the specific needs of crops and soil conditions. Nanomaterial-based fertilizers enhance nutrient uptake efficiency, reduce nutrient losses through leaching and volatilization, and promote controlled release of nutrients to plants over time. Nanoparticle-based carriers and delivery systems facilitate targeted and sustained release of essential nutrients, micronutrients, and growth-promoting compounds, optimizing nutrient utilization and minimizing environmental impacts.

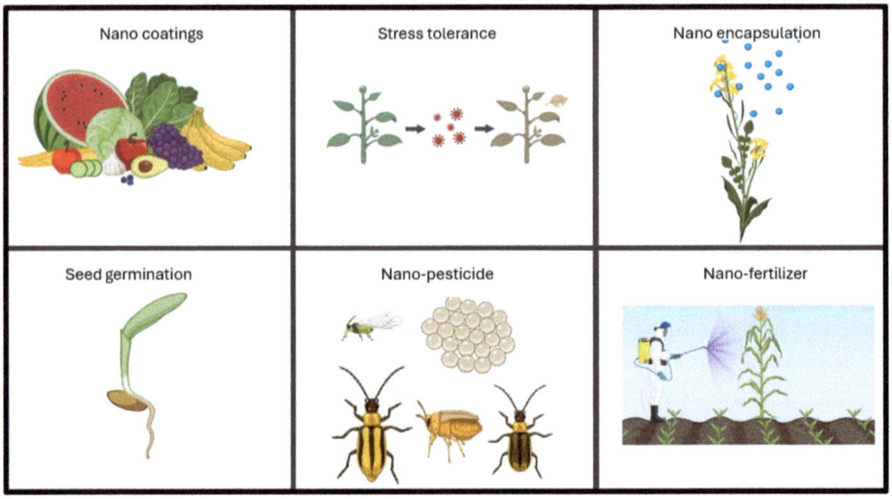

Fig. 1 Application of nanosystem in agriculture

(b) Precision agriculture and smart farming: Nano-ecosystem engineering contributes to the advancement of precision agriculture and smart farming technologies, enabling real-time monitoring, analysis, and management of agricultural systems. Nanoscale sensors, biosensors, and imaging technologies offer high sensitivity, selectivity, and spatial resolution for monitoring soil moisture, nutrient levels, pest infestations, and crop health parameters. Nanomaterial-based nanosensors and smart delivery systems enable precise application of agrochemicals, water, and nutrients based on plant requirements, reducing input costs, minimizing environmental pollution, and improving overall farm productivity.
(c) Crop protection and pest management: Nano-ecosystem engineering provides innovative solutions for crop protection and pest management, offering alternatives to conventional pesticides and agrochemicals. Nanoparticle-based formulations, nanoencapsulation, and nanocarriers enhance the efficacy, stability, and targeted delivery of biopesticides, plant growth regulators, and natural plant defense inducers. Furthermore, nanomaterial-based insecticides, fungicides, and herbicides offer novel modes of action, reducing the development of pesticide resistance in pest populations and minimizing adverse effects on nontarget organisms and ecosystems.
(d) Soil health and environmental sustainability: Nano-ecosystem engineering promotes soil health and environmental sustainability through the development of nanomaterial-based soil amendments, biostimulants, and bioremediation technologies. Nanoparticles such as nanoclays, nanohumus, and nanoscale organic matter enhance soil structure, water retention, nutrient cycling, and microbial activity, improving soil fertility and resilience to environmental stressors. Additionally, nanomaterial-based bioremediation strategies facilitate the degradation and detoxification of soil pollutants, contaminants, and agrochemical residues, restoring soil health and ecosystem function.
(e) Nanotechnology in seed technology and crop improvement: Nano-ecosystem engineering offers novel approaches for seed treatment, genetic modification, and crop improvement through nanotechnology-based innovations. Nanoparticle coatings, seed priming, and seed enhancement technologies enhance seed germination, seedling vigor, and stress tolerance, improving crop establishment and yield potential. Furthermore, nanomaterial-based gene delivery systems enable targeted gene editing, RNA interference, and genetic transformation, accelerating the development of genetically modified crops with enhanced traits such as disease resistance, drought tolerance, and nutrient efficiency.

4.4 Food Industry Applications

The application of nano-ecosystem engineering in the food industry presents a range of innovative solutions to enhance food quality, safety, and sustainability. Nano-ecosystem engineering involves the deliberate design and manipulation of nanomaterials and their interactions within food systems to achieve specific objectives. Some key applications of nano-ecosystem engineering in the food industry include (Mohammad et al., 2022; Biswas et al., 2022).

(a) Food packaging and preservation: Nano-ecosystem engineering enables the development of advanced food packaging materials with enhanced barrier properties, antimicrobial activity, and freshness preservation capabilities. Nanomaterial-based coatings, films, and nanoparticles can be incorporated into packaging materials to improve shelf life, reduce spoilage, and inhibit microbial growth in food products. Additionally, nanoscale sensors and indicators embedded in packaging materials enable real-time monitoring of food quality, safety, and freshness, providing consumers with valuable information about product integrity and shelf life.
(b) Food safety and quality control: Nano-ecosystem engineering contributes to food safety and quality control through the development of rapid, sensitive, and reliable detection methods for foodborne pathogens, contaminants, and adulterants. Nanomaterial-based biosensors, immunoassays, and DNA probes offer high sensitivity, specificity, and speed for detecting microbial contaminants, chemical residues, and allergens in food products. Furthermore, nanoscale imaging techniques, such as atomic force microscopy (AFM) and scanning electron microscopy (SEM), enable detailed characterization of food microstructure, texture, and composition, facilitating quality assurance and regulatory compliance.
(c) Nutrient delivery and fortification: Nano-ecosystem engineering facilitates the delivery and bioavailability of nutrients, vitamins, and bioactive compounds in food products through the development of nanoscale delivery systems and encapsulation technologies. Nanoparticle-based carriers, liposomes, and emulsions protect sensitive bioactive ingredients from degradation during processing and storage, enhancing their stability and efficacy. Additionally, nanoscale encapsulation enables targeted delivery of nutrients and functional ingredients to specific tissues or organs in the body, improving nutrient absorption and bioavailability.
(d) Functional foods and nutraceuticals: Nano-ecosystem engineering drives innovation in the development of functional foods and nutraceuticals with enhanced health benefits and therapeutic properties. Nanomaterial-based formulations, nanoparticles, and nanoemulsions enable the incorporation of bioactive compounds, antioxidants, and phytochemicals into food products, promoting health and wellness. Furthermore, nanoscale delivery systems enhance the stability, solubility, and bioavailability of bioactive ingredients, maximizing their therapeutic effects and physiological benefits.
(e) Food waste reduction and sustainable packaging: Nano-ecosystem engineering offers solutions for reducing food waste and enhancing sustainability in the food industry through the development of biodegradable packaging materials, edible coatings, and active packaging systems. Nanomaterial-based packaging

materials and coatings extend the shelf life of perishable food products, reducing spoilage and waste. Additionally, nanoscale sensors and indicators embedded in packaging materials enable real-time monitoring of food quality and safety, facilitating inventory management and reducing food losses along the supply chain.

4.5 Electronics Industry (Hossain et al., 2023, 2022)

(a) Nanoelectronics and semiconductor devices: Nano-ecosystem engineering enables the development of nanoscale electronic components, such as transistors, diodes, and sensors, with enhanced performance and functionality. Nanomaterials such as carbon nanotubes, graphene, and quantum dots offer unique electrical, optical, and mechanical properties that can be exploited to fabricate high-performance semiconductor devices with reduced size, power consumption, and heat dissipation.
(b) Nanocomposites and functional materials: Nano-ecosystem engineering facilitates the integration of nanomaterials into electronic packaging materials, printed circuit boards, and flexible electronics to enhance mechanical strength, thermal conductivity, and electromagnetic shielding properties. Nanocomposites, nanoparticle-based inks, and conductive nanofibers enable the development of lightweight, flexible, and durable electronic devices with improved reliability and performance.
(c) Nanophotonics and optoelectronics: Nano-ecosystem engineering drives innovation in nanophotonic devices, such as light-emitting diodes (LEDs), photovoltaic cells, and optical sensors, by exploiting the unique optical properties of nanomaterials. Nanoparticles, nanowires, and quantum dots enable the fabrication of highly efficient and tunable optoelectronic devices with enhanced light absorption, emission, and detection capabilities.
(d) Nanoelectromechanical systems (NEMS): Nano-ecosystem engineering enables the development of nanoscale electromechanical systems, such as nanoactuators, nanosensors, and nanomechanical resonators, with unprecedented sensitivity, accuracy, and responsiveness. Nanomaterial-based NEMS devices offer new opportunities for sensing, actuation, and control in electronic systems, enabling applications such as nanoscale manipulation, biomedical sensing, and environmental monitoring.

4.6 Automotive Industry (Shafique & Luo, 2019)

(a) Lightweight materials and structural components: Nano-ecosystem engineering contributes to the development of lightweight materials and structural components for automotive applications, such as carbon nanotube-reinforced composites, graphene-based coatings, and nanoceramic coatings. These nanomaterials offer exceptional strength-to-weight ratios, corrosion resistance, and thermal stability, enabling the fabrication of lightweight vehicles with improved fuel efficiency and performance.

(b) Energy storage and conversion: Nano-ecosystem engineering enables the development of advanced energy storage and conversion technologies for automotive applications, such as lithium-ion batteries, supercapacitors, and fuel cells. Nanomaterials such as nanostructured electrodes, nanocomposite electrolytes, and nanoparticle catalysts enhance energy storage density, charge/discharge rates, and cycling stability, enabling the widespread adoption of electric and hybrid vehicles.

(c) Smart and connected vehicles: Nano-ecosystem engineering drives innovation in smart and connected vehicle technologies, such as nanosensors, nanoelectronics, and nano-communications systems, enabling real-time monitoring, control, and communication within automotive systems. Nanomaterial-based sensors and actuators enable autonomous driving, vehicle-to-vehicle communication, and predictive maintenance, enhancing safety, efficiency, and convenience for drivers and passengers.

(d) Environmental sustainability: Nano-ecosystem engineering contributes to environmental sustainability in the automotive industry by enabling the development of nanomaterial-based catalysts, filters, and coatings for emissions control, air purification, and fuel efficiency enhancement. Nanomaterials such as catalytic nanoparticles, nanoporous membranes, and nanostructured coatings reduce pollutant emissions, improve air quality, and mitigate environmental impact, promoting sustainable transportation solutions for future generations.

4.7 Biomedical Applications

The application of nano-ecosystem engineering in biomedicine has revolutionized diagnosis, treatment, and monitoring of diseases, offering novel solutions to complex medical challenges (Fig. 2). Nano-ecosystem engineering involves the deliberate design and manipulation of nanomaterials and their interactions within biological systems to achieve specific objectives. Some key applications of nano-ecosystem engineering in biomedicine include (Ramos et al., 2017; Silva et al., 2023)

(a) Drug delivery systems: Nano-ecosystem engineering enables the development of nanoscale drug delivery systems, such as liposomes, nanoparticles, and polymer–drug conjugates, for targeted and controlled drug delivery. These nanomaterial-based carriers protect drugs from degradation, enhance their solubility, and facilitate their accumulation at specific sites within the body, reducing side effects and improving therapeutic outcomes. Furthermore, nanoscale delivery systems enable the encapsulation of multiple drugs, imaging agents, and therapeutic payloads in a single carrier, enabling combination therapy and personalized medicine approaches.

(b) Theranostic nanomedicine: Nano-ecosystem engineering drives innovation in theranostic nanomedicine, which combines diagnostic and therapeutic func-

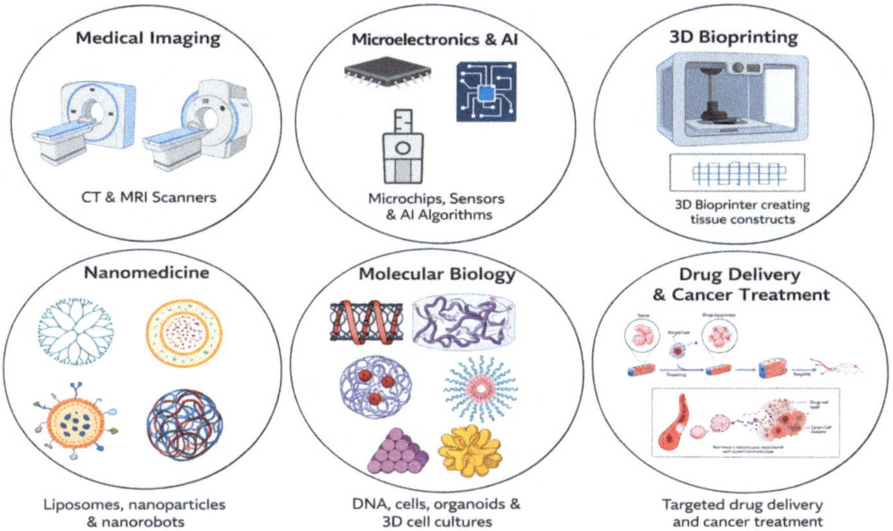

Fig. 2 Applications of nano-ecosystem engineering in biomedicine

tionalities within a single nanomaterial platform. Theranostic nanoparticles, such as quantum dots, magnetic nanoparticles, and gold nanoparticles, enable simultaneous imaging and treatment of diseases, providing real-time feedback on treatment efficacy and disease progression. These multifunctional nanomaterials offer new opportunities for personalized medicine, targeted therapy, and precision medicine approaches in cancer treatment, cardiovascular disease, and neurological disorders.

(c) Regenerative medicine and tissue engineering: Nano-ecosystem engineering facilitates the development of nanomaterial-based scaffolds, matrices, and implants for regenerative medicine and tissue engineering applications. Nanoscale biomaterials, such as nanofibers, nanocomposites, and hydrogels, mimic the structure and mechanical properties of native tissues, promoting cell adhesion, proliferation, and differentiation. Furthermore, nanomaterial-based growth factors, cytokines, and bioactive molecules enhance tissue regeneration, wound healing, and organ repair, offering new solutions for tissue engineering, organ transplantation, and regenerative therapies.

(d) Nanoparticle-based imaging and diagnostics: Nano-ecosystem engineering drives innovation in nanoparticle-based imaging and diagnostic technologies, enabling early detection, accurate diagnosis, and monitoring of diseases. Nanomaterials such as quantum dots, gold nanoparticles, and superparamagnetic iron oxide nanoparticles enable high-resolution imaging modalities, such as fluorescence imaging, magnetic resonance imaging (MRI), and computed tomography (CT), for visualizing anatomical structures and molecular processes in vivo. Furthermore, nanoscale contrast agents and molecular probes enable targeted imaging of specific biomarkers, enabling early detection of cancer, cardiovascular disease, and neurological disorders.

(e) Nano-based biosensors and point-of-care devices: Nano-ecosystem engineering contributes to the development of nanomaterial-based biosensors and point-of-care devices for rapid and sensitive detection of biomolecules, pathogens, and disease markers. Nanoparticle-based biosensors, such as plasmonic nanoparticles, carbon nanotubes, and quantum dots, offer high sensitivity, specificity, and multiplexing capabilities for detecting nucleic acids, proteins, and small molecules in biological samples. Furthermore, nanomaterial-based point-of-care devices enable decentralized testing, remote monitoring, and real-time diagnostics, facilitating early intervention and personalized treatment strategies in healthcare settings.

5 Challenges in Nano-Ecosystem Engineering

5.1 Waste Handling in Nano-Ecosystem Engineering

Nano-ecosystem engineering, while offering numerous benefits, also presents challenges in waste handling and disposal. As nanomaterials are increasingly utilized in various applications, the management of waste containing nanoparticles becomes crucial to prevent environmental contamination and potential health hazards. Waste generated during the synthesis, manufacturing, and use of nanomaterials may contain nanoparticles in various forms, including suspended particles, agglomerates, or surface-coated materials (Younis et al., 2018).

One approach to waste handling involves implementing strict containment measures and engineering controls during nanomaterial synthesis and processing to minimize the release of nanoparticles into the environment. Additionally, effective filtration and capture systems can be employed to remove nanoparticles from waste streams before disposal. However, challenges remain in developing cost-effective and scalable methods for nanoparticle removal and waste treatment. Furthermore, the disposal of waste containing nanomaterials raises concerns about long-term environmental impacts and potential risks to ecosystems and human health. Nanoparticles released into the environment may accumulate in soil, water bodies, and sediments, posing risks to aquatic organisms, soil microorganisms, and terrestrial ecosystems. Therefore, appropriate risk assessment and management strategies are essential to mitigate the environmental impact of nanoparticle-containing waste (Zahra et al., 2022).

5.2 Environmental Toxicities

Nanoparticles have unique physicochemical properties that can lead to interactions with living organisms and ecosystems, potentially causing adverse effects on environmental health. Environmental toxicities associated with nanoparticles vary depending on factors such as particle size, shape, surface chemistry, and composition (Kumah et al., 2023). Common environmental toxicities include

(a) Ecotoxicity: Nanoparticles may exert toxic effects on aquatic organisms, such as fish, algae, and invertebrates, through mechanisms such as oxidative stress, membrane damage, and interference with cellular functions. Additionally, nanoparticles released into soil can affect soil microbial communities and disrupt ecosystem processes.
(b) Bioaccumulation and biomagnification: Nanoparticles may bioaccumulate in organisms and biomagnify through food chains, leading to increased concentrations at higher trophic levels. This phenomenon can result in elevated exposure and potential toxicity to organisms at the top of the food chain, including humans.
(c) Genotoxicity and mutagenicity: Some nanoparticles have been shown to induce DNA damage, mutations, and chromosomal aberrations in living organisms, raising concerns about potential long-term health effects and genetic consequences in exposed populations.
(d) Endocrine disruption: Certain nanoparticles have been implicated in disrupting endocrine systems in animals, affecting hormone regulation, reproductive health, and development.

6 Future Challenges in Nano-Ecosystem Engineering

Despite significant progress in nano-ecosystem engineering, several challenges remain to be addressed to ensure the sustainable and responsible use of nanomaterials in environmental applications (Pokrajac et al., 2021; Köhler, 2021). Key challenges and future directions include

(i) Risk assessment and regulation: Continued research is needed to improve our understanding of the environmental fate, behavior, and toxicity of nanoparticles. Robust risk assessment methodologies and regulatory frameworks must be developed to evaluate the environmental impact of nanomaterials and ensure their safe use.
(ii) Sustainable synthesis and manufacturing: Efforts should focus on developing green synthesis methods and sustainable manufacturing processes to minimize waste generation, energy consumption, and environmental footprint associated with nanoparticle production.
(iii) Nanomaterial design for environmental compatibility: Nanomaterials should be designed with environmental compatibility in mind, considering factors such as biodegradability, bioavailability, and eco-friendly synthesis routes.
(iv) Monitoring and remediation technologies: Advanced monitoring and remediation technologies are needed to detect and mitigate nanoparticle contamination in environmental matrices, such as water, soil, and air. Innovative approaches, such as nanoremediation, hold promise for the efficient removal and degradation of nanoparticle pollutants.
(v) Public awareness and engagement: Increasing public awareness and engagement on the potential risks and benefits of nanotechnology is essential to foster responsible stewardship and informed decision-making regarding the use of nanomaterials in environmental applications.

In summary, waste handling, environmental toxicities, and future challenges in nano-ecosystem engineering underscore the importance of adopting a proactive and interdisciplinary approach to ensure the safe and sustainable integration of nanotechnology into environmental management practices. By addressing these challenges collaboratively, researchers, policymakers, industry stakeholders, and the public can work together to harness the potential of nanotechnology while minimizing environmental risks and maximizing environmental benefits.

7 Conclusion and Future Perspectives

In conclusion, the principles of nano-ecosystem engineering provide a holistic framework for understanding and manipulating interactions between nanomaterials and ecosystems. By integrating principles such as scale-dependent behavior, emergent properties, and ecological dynamics, researchers can develop sustainable approaches for harnessing the potential of nanotechnology while minimizing environmental risks. From the design and synthesis of nanomaterials to their application in environmental remediation, agriculture, and healthcare, nano-ecosystem engineering offers promising solutions to address pressing global challenges. Looking ahead, the field of nano-ecosystem engineering holds immense potential for innovation and discovery. Emerging trends such as green synthesis methods, biomimetic design principles, and advanced characterization techniques are expected to revolutionize the development of environmentally friendly nanomaterials. Furthermore, interdisciplinary collaboration and knowledge exchange will drive progress in understanding the complex interactions between nanomaterials and ecosystems, paving the way for sustainable solutions to complex environmental problems.

In the future, regulatory frameworks and policies will play a crucial role in guiding the safe and responsible use of nanomaterials in ecosystems. Continued efforts to assess the environmental impact and risks associated with nanotechnology will inform regulatory decision-making and ensure the protection of ecosystems and human health. Additionally, public engagement and education will be essential to foster trust and transparency in the development and application of nanotechnology. Overall, the principles of nano-ecosystem engineering offer a promising pathway toward a sustainable future, where nanotechnology can be harnessed to address environmental challenges while preserving the integrity and resilience of ecosystems. By embracing these principles and embracing a multidisciplinary approach, researchers can unlock the full potential of nanotechnology to create a healthier and more sustainable planet for future generations.

References

Adhikari, C. (2021). Polymer nanoparticles-preparations, applications and future insights: A concise review. *Polymer-Plastics Technology and Materials, 60*(18), 1996–2024. https://doi.org/10.1080/25740881.2021.1939715

Ahlawat, J., Asil, S. M., Barroso, G. G., Nurunnabi, M., & Narayan, M. (2021). Application of carbon nano onions in the biomedical field: Recent advances and challenges. *Biomaterials Science, 9*(3), 626–644. https://doi.org/10.1039/D0BM01476A

Ali, I., Suhail, M., Alothman, Z. A., & Alwarthan, A. (2018). Recent advances in syntheses, properties and applications of TiO_2 nanostructures. *RSC Advances, 8*(53), 30125–30147. https://doi.org/10.1039/C8RA06517A

Ambroz, F., Macdonald, T. J., Martis, V., & Parkin, I. P. (2018). Evaluation of the BET theory for the characterization of meso and microporous MOFs. *Small Methods, 2*(11), 1800173. https://doi.org/10.1002/smtd.201800173

Attia, N. F., Abd El-Monaem, E. M., El-Aqapa, H. G., Elashery, S. E. A., Eltaweil, A. S., El Kady, M., Khalifa, S. A. M., Hawash, H. B., & El-Seedi, H. R. (2022). Iron oxide nanoparticles and their pharmaceutical applications. *Applied Surface Science Advances, 11*, 100284. https://doi.org/10.1016/j.apsadv.2022.100284

Bayda, S., Adeel, M., Tuccinardi, T., Cordani, M., & Rizzolio, F. (2020). The history of nanoscience and nanotechnology: From chemical–physical applications to nanomedicine. *Molecules, 25*(1), 112. https://doi.org/10.3390/molecules25010112

Beni, A. A., & Jabbari, H. (2022). Nanomaterials for environmental applications. *Results in Engineering, 15*, 100467. https://doi.org/10.1016/j.rineng.2022.100467

Bhandari, G., Dhasmana, A., Chaudhary, P., Gupta, S., Gangola, S., Gupta, A., Rustagi, S., et al. (2023). A perspective review on green nanotechnology in agro-ecosystems: Opportunities for sustainable agricultural practices & environmental remediation. *Agriculture, 13*(3), 668. https://doi.org/10.3390/agriculture13030668

Biswas, R., Mahabub Alam, Animesh Sarkar, Md Ismail Haque, Md. Moinul Hasan, & Mominul Hoque. (2022). Application of nanotechnology in food: Processing, preservation, packaging and safety assessment. *Heliyon, 8*(11), e11795. https://doi.org/10.1016/j.heliyon.2022.e11795

Cardoza, C., Nagtode, V., Pratap, A., & Mali, S. N. (2022). Emerging applications of nanotechnology in cosmeceutical health science: Latest updates. *Health Sciences Review, 4*, 100051. https://doi.org/10.1016/j.hsr.2022.100051

Chandra, S., Das, P., Bag, S., Laha, D., & Pramanik, P. (2011). Synthesis, functionalization and bioimaging applications of highly fluorescent carbon nanoparticles. *Nanoscale, 3*(4), 1533–1540. https://doi.org/10.1039/C0NR00735H

Clogston, J. D., & Patri, A. K. (2011). Zeta potential measurement. In S. E. McNeil (Ed.), *Characterization of nanoparticles intended for drug delivery* (pp. 63–70). Humana Press. https://doi.org/10.1007/978-1-60327-198-1_6

Cotta, M. A. (2020). Quantum dots and their applications: What lies ahead? *ACS Applied Nano Materials, 3*(6), 4920–4924. https://doi.org/10.1021/acsanm.0c01386

Dreaden, E. C., Alkilany, A. M., Huang, X., Murphy, C. J., & El-Sayed, M. A. (2012). The golden age: Gold nanoparticles for biomedicine. *Chemical Society Reviews, 41*(7), 2740–2779. https://doi.org/10.1039/C1CS15237H

Ealia, S. A. M., & Saravanakumar, M. P. (2017). A review on the classification, characterisation, synthesis of nanoparticles and their application. *IOP Conference Series: Materials Science and Engineering, 263*(3), 032019. https://doi.org/10.1088/1757-899X/263/3/032019

Guerra, F. D., Attia, M. F., Whitehead, D. C., & Alexis, F. (2018). Nanotechnology for environmental remediation: Materials and applications. *Molecules, 23*(7), 1760. https://doi.org/10.3390/molecules23071760

Gupta, S. M., & Tripathi, M. (2012). An overview of commonly used semiconductor nanoparticles in photocatalysis. *High Energy Chemistry, 46*(1), 1–9. https://doi.org/10.1134/S0018143912010134

Gupta, V., Mohapatra, S., Mishra, H., Farooq, U., Kumar, K., Ansari, M. J., Aldawsari, M. F., Alalaiwe, A. S., Mirza, M. A., & Iqbal, Z. (2022). Nanotechnology in cosmetics and cosmeceuticals—A review of latest advancements. *Gels, 8*(3), 173. https://doi.org/10.3390/gels8030173

Holder, C. F., & Schaak, R. E. (2019). Tutorial on powder X-ray diffraction for characterizing nanoscale materials. *ACS Nano, 13*(7), 7359–7365. https://doi.org/10.1021/acsnano.9b05157

Hossain, N., Mobarak, M. H., Mimona, M. A., Islam, M. A., Hossain, A., Zohura, F. T., & Chowdhury, M. A. (2023). Advances and significances of nanoparticles in semiconductor applications – A review. *Results in Engineering, 19*, 101347. https://doi.org/10.1016/j.rineng.2023.101347

Hu, X., Zhang, Y., Ding, T., Liu, J., & Zhao, H. (2020). Multifunctional gold nanoparticles: A novel nanomaterial for various medical applications and biological activities. *Frontiers in Bioengineering and Biotechnology, 8*. https://doi.org/10.3389/fbioe.2020.00990

Joudeh, N., & Linke, D. (2022). Nanoparticle classification, physicochemical properties, characterization, and applications: A comprehensive review for biologists. *Journal of Nanobiotechnology, 20*(1), 262. https://doi.org/10.1186/s12951-022-01477-8

Khan, I., Saeed, K., & Khan, I. (2019). Nanoparticles: Properties, applications and toxicities. *Arabian Journal of Chemistry, 12*(7), 908–931. https://doi.org/10.1016/j.arabjc.2017.05.011

Köhler, J. M. (2021). Challenges for nanotechnology. *Encyclopedia, 1*(3), 618–631. https://doi.org/10.3390/encyclopedia1030051

Korin, E., Froumin, N., & Cohen, S. (2017). Surface analysis of nanocomplexes by X-ray photoelectron spectroscopy (XPS). *ACS Biomaterials Science & Engineering, 3*(6), 882–889. https://doi.org/10.1021/acsbiomaterials.7b00040

Kumah, E. A., Fopa, R. D., Harati, S., Boadu, P., Zohoori, F. V., & Pak, T. (2023). Human and environmental impacts of nanoparticles: A scoping review of the current literature. *BMC Public Health, 23*(1), 1059. https://doi.org/10.1186/s12889-023-15958-4

Liu, M., Zhao, F., Zhu, D., Duan, H., Lv, Y., Li, L., & Gan, L. (2018). Ultramicroporous carbon nanoparticles derived from metal–organic framework nanoparticles for high-performance supercapacitors. *Materials Chemistry and Physics, 211*, 234–241. https://doi.org/10.1016/j.matchemphys.2018.02.030

Liu, C., Zhou, H., & Zhou, J. (2021). The applications of nanotechnology in crop production. *Molecules, 26*(23), 7070. https://doi.org/10.3390/molecules26237070

Long, C. M., Nascarella, M. A., & Valberg, P. A. (2013). Carbon black vs. black carbon and other airborne materials containing elemental carbon: Physical and chemical distinctions. *Environmental Pollution, 181*, 271–286. https://doi.org/10.1016/j.envpol.2013.06.009

López-Lorente, Á. I., & Mizaikoff, B. (2016). Recent advances on the characterization of nanoparticles using infrared spectroscopy. *TrAC Trends in Analytical Chemistry*, Recent advances and trends in Analytical Nanoscience and Nanotechnology, *84*, 97–106. https://doi.org/10.1016/j.trac.2016.01.012

Meirzadeh, E., Evans, A. M., Rezaee, M., Milich, M., Dionne, C. J., Darlington, T. P., Bao, S. T., et al. (2023). A few-layer covalent network of fullerenes. *Nature, 613*(7942), 71–76. https://doi.org/10.1038/s41586-022-05401-w

Mobeen, H., Safdar, M., Fatima, A., Afzal, S., Zaman, H., & Mehdi, Z. (2022). Emerging applications of nanotechnology in context to immunology: A comprehensive review. *Frontiers in Bioengineering and Biotechnology, 10*. https://doi.org/10.3389/fbioe.2022.1024871

Mochalin, V. N., Shenderova, O., Ho, D., & Gogotsi, Y. (2012). The properties and applications of nanodiamonds. *Nature Nanotechnology, 7*(1), 11–23. https://doi.org/10.1038/nnano.2011.209

Mohammad, Z. H., Ahmad, F., Ibrahim, S. A., & Zaidi, S. (2022). Application of nanotechnology in different aspects of the food industry. *Discover Food, 2*(1), 12. https://doi.org/10.1007/s44187-022-00013-9

Moreno-Vega, A.-I., Gómez-Quintero, T., Nuñez-Anita, R.-E., Acosta-Torres, L.-S., & Castaño, V. (2012). Polymeric and ceramic nanoparticles in biomedical applications. *Journal of Nanotechnology, 2012*, e936041. https://doi.org/10.1155/2012/936041

Mourdikoudis, S., Pallares, R. M., & Thanh, N. T. K. (2018). Characterization techniques for nanoparticles: Comparison and complementarity upon studying nanoparticle properties. *Nanoscale, 10*(27), 12871–12934. https://doi.org/10.1039/C8NR02278J

Musielak, E., Feliczak-Guzik, A., & Nowak, I. (2022). Synthesis and potential applications of lipid nanoparticles in medicine. *Materials, 15*(2), 682. https://doi.org/10.3390/ma15020682

Nascimento, M. A., Cruz, J. C., Rodrigues, G. D., de Oliveira, A. F., & Lopes, R. P. (2018). Synthesis of polymetallic nanoparticles from spent lithium-ion batteries and application in the removal of reactive blue 4 dye. *Journal of Cleaner Production, 202*, 264–272. https://doi.org/10.1016/j.jclepro.2018.08.118

Nayl, A. A., Abd-Elhamid, A. I., Aly, A. A., & Bräse, S. (2022). Recent progress in the applications of silica-based nanoparticles. *RSC Advances, 12*(22), 13706–13726. https://doi.org/10.1039/D2RA01587K

Neme, K., Nafady, A., Uddin, S., & Tola, Y. B. (2021). Application of nanotechnology in agriculture, postharvest loss reduction and food processing: Food security implication and challenges. *Heliyon, 7*(12). https://doi.org/10.1016/j.heliyon.2021.e08539

Ng, K. K., & Zheng, G. (2015). Molecular interactions in organic nanoparticles for phototheranostic applications. *Chemical Reviews, 115*(19), 11012–11042. https://doi.org/10.1021/acs.chemrev.5b00140

Oh, W.-K., Yoon, H., & Jang, J. (2010). Size control of magnetic carbon nanoparticles for drug delivery. *Biomaterials, 31*(6), 1342–1348. https://doi.org/10.1016/j.biomaterials.2009.10.018

Pan, K., & Zhong, Q. (2016). Organic nanoparticles in foods: Fabrication, characterization, and utilization. *Annual Review of Food Science and Technology, 7*, 245–266. https://doi.org/10.1146/annurev-food-041715-033215

Patel, D. K., Kim, H.-B., Dutta, S. D., Ganguly, K., & Lim, K.-T. (2020). Carbon nanotubes-based nanomaterials and their agricultural and biotechnological applications. *Materials, 13*(7), 1679. https://doi.org/10.3390/ma13071679

Pokrajac, L., Abbas, A., Chrzanowski, W., Dias, G. M., Eggleton, B. J., Maguire, S., Maine, E., et al. (2021). Nanotechnology for a sustainable future: Addressing global challenges with the International Network4Sustainable Nanotechnology. *ACS Nano, 15*(12), 18608–18623. https://doi.org/10.1021/acsnano.1c10919

Raha, S., & Ahmaruzzaman, M. (2022). ZnO nanostructured materials and their potential applications: Progress, challenges and perspectives. *Nanoscale Advances, 4*(8), 1868–1925. https://doi.org/10.1039/D1NA00880C

Ramos, A. P., Cruz, M. A. E., Tovani, C. B., & Ciancaglini, P. (2017). Biomedical applications of nanotechnology. *Biophysical Reviews, 9*(2), 79–89. https://doi.org/10.1007/s12551-016-0246-2

Salem, S. S. (2023). A mini review on green nanotechnology and its development in biological effects. *Archives of Microbiology, 205*(4), 128. https://doi.org/10.1007/s00203-023-03467-2

Shafique, M., & Luo, X. (2019). Nanotechnology in transportation vehicles: An overview of its applications, environmental, health and safety concerns. *Materials, 12*(15), 2493. https://doi.org/10.3390/ma12152493

Sharma, A., Manocha, D., Arora, A., Kumar, A., Srivastava, C. M., Rawat, V., Kim, H., Verma, M., & Rao, G. K. (2022). Chapter 14 – Applications of green nanomaterials in electronic and electrical industries. In U. Shanker, C. M. Hussain, & M. Rani (Eds.), *Green nanomaterials for industrial applications* (Micro and nano technologies) (pp. 397–421). Elsevier. https://doi.org/10.1016/B978-0-12-823296-5.00016-2

Silva, D. F., Melo, A. L. P., Uchôa, A. F. C., Pereira, G. M. A., Alves, A. E. F., Vasconcellos, M. C., Xavier-Júnior, F. H., & Passos, M. F. (2023). Biomedical approach of nanotechnology and biological risks: A mini-review. *International Journal of Molecular Sciences, 24*(23), 16719. https://doi.org/10.3390/ijms242316719

Singh, K. R. B., Nayak, V., Sarkar, T., & Singh, R. P. (2020). Cerium oxide nanoparticles: Properties, biosynthesis and biomedical application. *RSC Advances, 10*(45), 27194–27214. https://doi.org/10.1039/D0RA04736H

Smith, D. J. (2015). *Characterization of nanomaterials using transmission electron microscopy*. https://doi.org/10.1039/9781782621867-00001

Toshima, N., & Yonezawa, T. (1998). Bimetallic nanoparticles—Novel materials for chemical and physical applications. *New Journal of Chemistry, 22*(11), 1179–1201. https://doi.org/10.1039/A805753B

Vladár, A. E., & Hodoroaba, V.-D. (2020). Chapter 2.1.1 – Characterization of nanoparticles by scanning electron microscopy. In V.-D. Hodoroaba, W. E. S. Unger, & A. G. Shard (Eds.), *Characterization of nanoparticles* (Micro and nano technologies) (pp. 7–27). Elsevier. https://doi.org/10.1016/B978-0-12-814182-3.00002-X

Younis, S. A., El-Fawal, E. M., & Serp, P. (2018). Nano-wastes and the environment: Potential challenges and opportunities of nano-waste management paradigm for greener nanotechnologies. In C. M. Hussain (Ed.), *Handbook of environmental materials management* (pp. 1–72). Springer International Publishing. https://doi.org/10.1007/978-3-319-58538-3_53-1

Zahra, Z., Habib, Z., Hyun, S., & Sajid, M. (2022). Nanowaste: Another future waste, its sources, release mechanism, and removal strategies in the environment. *Sustainability, 14*(4), 2041. https://doi.org/10.3390/su14042041

Zhang, X.-F., Liu, Z.-G., Shen, W., & Gurunathan, S. (2016). Silver nanoparticles: Synthesis, characterization, properties, applications, and therapeutic approaches. *International Journal of Molecular Sciences, 17*(9), 1534. https://doi.org/10.3390/ijms17091534

Nanoengineered Bioinspired Structures and Functions

P. Divyadharshini, D. Jhoshitha, Prathiksha Srikanthan, Pooja Pillai, Angeline Julius, and Suresh Malakondaiah

Abstract

Nanoengineering is a field of engineering that utilizes nanomaterials to design, build, and utilize nanoscale engines, machines, and structures, advancing various multiscale methods. Naturally available bioinspired structural materials are composed of hard and soft phases in complex hierarchical architectures, ranging from nanoscale to macroscale, built at ambient temperature. Various biological systems have sophisticated architectures for impact resistance and energy absorption. The findings could lead to data-driven methodologies in the low-cost manufacturing technologies for efficient bioinspired structures and in the development of engineering structural design. Top-down and bottom-up methods are used to fabricate functional nanostructures, with top-down involving photolithography and bottom-up requiring less tooling. Chemical synthesis tools have produced nanoscale structures for medical applications, while researchers explore alternative patterning technologies for next-generation applications. Researchers explore electron beam lithography and nanoimprint technology for next-generation integrated gadgets, while DNA nanotechnology improves surface properties in biosensing, materials science, and cell biology. 3D printing advances in macroscale manufacturing, enabling shape-independent products and structures. It also offers speed, less waste, and economic viability in nanotechnology, with techniques like nanotribological printing. The application of nanoscience and nanotechnology in engineering establishes a direct connection between industrial and everyday life and academic research in these fields. Consequently,

P. Divyadharshini · D. Jhoshitha · P. Srikanthan · P. Pillai
Department of Industrial Biotechnology, Bharath Institute of Higher Education and Research, Chennai, Tamil Nadu, India

A. Julius · S. Malakondaiah (✉)
Center for Material Engineering and Regenerative Medicine, Bharath Institute of Higher Education and Research, Chennai, Tamil Nadu, India

D. K. Gupta et al. (eds.), *Nano-Ecosystems*,
https://doi.org/10.1007/978-981-95-5033-3_3

a plethora of nanomaterials, nanodevices, and nanosystems for diverse technical applications have been created and employed for the welfare of humankind. New bioinspired nanoplatforms hold potential in nanomedicine, including cancer therapy, antimicrobial, immunotherapy, biosensing, and diagnosis, with various applications currently underway.

Keywords

Nanoengineering · Bioinspired structural material · Chemical synthesis · Nanoimprint · Biosensing · Nanotribological printing

1 Introduction to Nanoengineering: Bioinspired Nanostructures and Functions

Nanoengineering is a sector of study that brings together students from various fields of engineering and deals with the synthesis, fabrication, and utilization of material and devices that are in a range of 1–100 nm. There have been recent strides made in the field of nanoengineering that can be considered significant. For instance, in medicine, nanomaterials are being used to create advanced drug delivery methods that can deliver medicine to specific locations in the body in order to reduce the general adverse health effects of medication (Cheng et al., 2023). Also, the approaches like nanolithography and self-assembly are revolutionizing the material depositing technologies that afford the control over minute properties of a material and can build up nanostructures of distinct forms (Jambhulkar et al., 2024). Biomimetic nanocomposites in their simplest form refer to the kind of nanomaterials synthesized that are synthesized by borrowing concepts from biosphere with enhanced characteristics and performances compared to their conventional inorganic counterparts. They imitate some properties of the biological structures and activities—such as the use of bone on the cellular level, or the lotus effect on a macro scale, and thus it serves as a source for new areas of research. Significant advancements have been recorded in the synthesis of bioinspired nanomaterial systems for selective drug delivery and tissue engineering with drastic recovery abilities and minimized side effects (Zhang et al., 2023c). In addition, these materials are being considered for use in the energy applications, environmental management, and production of a wide range of services, with high added value defining the prospects and versatility of such materials.

The material and manufacturing advancements of the last few years have demonstrated a very promising line in terms of the creation of a new generation of materials for use in biomechanics and bioinspired technologies. For example, in their latest article, Zhang et al. (2023c) reported large-scale synthesis of ultrathin 2D nanosheets bearing nacre-like nesting structure for electromagnetic interference shielding applications. Peptide nanomaterials that are able to self-assemble have been identified as potential biomedical materials for uses like drugs and tissue development (La Manna et al., 2021). Also, there has been advanced progress in the development of nanoscale materials in the field of catalysis through the use of nanozymes that are structures with enzymatic properties on the nanoscale. Zhang et al. (2024) described the intricate features of the nanozyme field, giving information about the functions of the devices in industries and the medical field such as self-assembly and biomimetic catalysis.

Computer modeling proves useful in the optimization of bioinspired materials for use in engineering applications; this is because architects can test or model the complicated biological structures and the behavior of the same under different conditions. The most frequently applied procedures are molecular dynamics (MD), finite element analysis (FEA), and density function theory (DFT); others include computational chemistry, computational fluid dynamics, and Monte Carlo modeling.

MD simulations have been used by researchers in a bid to model how peptides assemble themselves to form nanofibers and in the process get a glimpse of the fibers' stability and performance (Nana Cao et al., 2024). FEA facilitates the ability to model the mechanical behavior on a bioinspired composite and make some prognosis of strength and also the toughness. In recent years, the bioinspired materials received a lot of interest in various fields, more particularly in medicine, energy, and environmental control. Using bioinspired material in the medical field is indeed advantageous, especially when it comes to drug delivery, tissue engineering, and diagnostic tools. Peptide nanomaterials have been found to offer potential abilities in designing drug delivery system for treatment that has the capacity to respond to specific cells to release the substances for treatment in an appropriate measure.

2 Principle of Bioinspired Nanostructures

Synthetic biology hybrid materials for biomimetics are a rapidly growing area in materials science and engineering that deals with the creation of new materials inspired by the properties of biological materials.

2.1 Hierarchical Structures

Nature has made it into a special science to do the best of hierarchical designs in structures, and it can be seen in materials like bones, shells, fruit, and plant tissues. Hierarchical structures include a process whereby several building blocks are assembled layer by layer at various length scales ranging from the nanoscale to the macroscale in order to deliver enhanced properties, including strength, toughness, and light weight. Biological engineering and one of the best examples of hierarchical structure is bone, an element containing stratum. Bone tissue type consists of three basic components: type I collagen fibers forming woven structures that are embedded in a mineralized ground substance made up of hydroxyapatite crystals. This structural arrangement gives bone the required tensile, compressive, and impact strength and elasticity to absorb and resist loads exerted on it (Wei et al., 2023).

Knowing and replicating this hierarchy arrangement is important for the further development of the concept of nanomaterials based on nature, especially in terms of mechanical properties. Scientists hope to form hybrid materials with well-defined interfaces between reinforcement and matrix phases to achieve bioinspired design and function. The hierarchical construction of bone can be mimicked in order for researchers to attain improved mechanical properties in composites, which was seen to have interest in aerospace, automobiles, and biomedicine industries. The new advancement of nanocomposite materials has shown potential of reproducing biological structures in reduced scales. Through further designing the nanoscale reinforcements and ease of polymer matrices, researchers have noticed increased properties like modulus of elasticity, strength, and toughness in the polymers, which is comparable to the natural materials like bone hierarchical structures (Chen et al., 2021).

2.2 Self-Assembly

Self-assembly is an intriguing process in which molecules spontaneously form complex structure through noncovalent forces, that is, hydrogen bonding, hydrophobic forces, and van der Wall forces. This self-assembling method is inherent, meaning that no external guidance is required to form a given structure of desired packed geometry, thus recalling nature-like building process. Self-assembly is the process of utilizing molecular and nano-level structures for constructing new materials that self-organize and arrange themselves in specific forms. Proteins are excellent examples of the self-assembly phenomena in nature. An additional dimension of protein architecture is achieved through folding and assembly, which can create new functional units, like enzymes, receptors, and structural elements of cells and tissues. This self-assembly process is crucial since it is necessary for the correct operation of biochemical networks, which illustrates the importance of the self-assembly process in the fabrication of nanoscale bioinspired systems (Katiyar et al., 2021). For instance, nanotechnology has improved in the fabrication of DNA-based nanostructures to make sophisticated structures composed of DNA linkers that self-assemble due to specific algorithms. Thus, through the base-pairing interactions of DNA strands, investigators can induce a highly selective and programmable assembly of functional units and matter at the nanometer scale for the fabrication of nanoscale appliances and systems, including drug delivery vehicles and molecular detectors and sensors (Francisco Lossada et al., 2020).

2.3 Functional Surfaces

Eco-engineering takes the form of biomimetic nanostructures, where the focus is on copying the surface activities of the natural models to obtain materials with predesigned adhesion, wettability, or catalytic coefficients. For example, adhesive technology developed from gecko feet, which helps the gecko lizard to stick on flat as well as the vertical surfaces. It is specifically grasshopper or gecko that naturally have feet bearing millions of hair-like structures called setae that are made of smaller structures known as spatulae that adhere to the surfaces through van der Waals forces (Wang et al., 2020). This mechanism has been replicated to develop disruptive and highly sticky adhesives that are also easily reversible for applications in robotics, biomedical engineering, and devices. The transportability of water droplets found in lotus leaves' superhydrophobicity has encouraged the creation of surfaces that can be cleaned automatically. Small structures that are visible on the surface of the leaves allow the creation of air gaps and minimize the area covered by the water drops (Wenlong Su et al., 2024). This principle has been used for the creation of coatings of textiles, glass, and other surfaces for anti-icing, water shedding, and decision (Collins & Safiuddin, 2022).

Moreover, the catalytic functionality of natural enzymes, which are capable of catalyzing biochemical reactions with high selectivity and efficiency, has led to the emergence of soft nanozymes. Such enzyme-templated nanomaterials are carefully designed to target applications in environmental cleanup technologies, biosensors, and industrial catalysis, with important studies showing their stability and multifunctionality (Zhang et al., 2024). Preparation of biomimetics nanostructures helps engineers to create innovative solutions that are associated with improved performance and function in different areas.

3 Recent Advancements of Bioinspired Nanomaterials and Their Applications

Iron-based nanomaterials can be used as catalysts in renewable biofuel production, potentially reducing production costs due to their low amount. These nanomaterials can be prepared via a green route, which is nontoxic and sustainable (Srivastava et al., 2021). The goal is to enhance biofuel production processes for sustainable commercialization by converting waste into value-added technology. Green methods like microorganisms, plant extracts, and waste biomass are being used to synthesize iron-based nanomaterials, which have potential applications in thermochemical and biochemical energy production (Rauf et al., 2024). The global challenge of increasing importance of safe drinking water has led to the development of ecologically sustainable nanomaterials with unique features like efficiency, earth abundance, stability, renewability, and inexpensive production techniques (Dutta et al., 2022). Carbon nanoparticles have a lot of promise for the environment and energy, but more environmentally acceptable synthesis methods based on affordable, renewable, and natural waste materials are required. The main areas of research include improved carbonaceous nanomaterials (NMs), techniques, and innovative nanosorbents for treating groundwater, wastewater, drinking water, and removing ions from the environment. Organic molecules and heavy hydrocarbons are needed to meet the increasing demand for carbon-based nanostructures in developing fields such as biomedicine, sensing, and gas storage (Cuong et al., 2022). The treatment of wastewater is essential for both environmental preservation and public health. Wastewater treatment can benefit from greener nanotechnology and sustainable nanoparticles; carbon-based nanoparticles have advantages in terms of infrastructure, resource usage, and safety.

In medicine, super paramagnetic iron oxide nanoparticles (SPIONs) are employed for both diagnostic and treatment. Targeted medication administration is one of the new study directions made possible by their guided assembly into nanochains. Natural magnetic nanochains are thought to have a role in the navigation of bacteria and animals. Magnetic nanochains have a great deal of potential for use in biological applications. In order to make the most of these nanostructures, new protocols that effectively isolate magnetosome chains while protecting 1D assemblies and magnetosome membranes must be created. It would also be ideal to develop novel methods of assembling bioinspired magnetic nanochains. These methods ought to enable easy production scale-up, a large selection of materials for joining nanoparticles, and accurate nanochain diameters. An important evolutionary accomplishment of magnetotactic bacteria is the creation of magnetic assemblages. This magneto strategy is frequently employed in cancer therapy and cardiovascular disease (Farzin et al., 2020).

The passage of a Jeffrey fluid in a limited asymmetric channel contains gold nanoparticles. The fluid has no magnetic properties and conducts electricity. Effects of viscous dissipation and thermal radiation are taken into account. The series findings are derived using a perturbation approach, which involves graphing versus the temperature, pressure, velocity, and trapping parameters (Bhatti, 2021). The behavior of gold nanoparticles is suspended in an extrinsic homogeneous magnetic field-induced Jeffrey fluid, which is irrotational, incompressible, and electrically conducting. Additionally taken into account are the effects of thermal radiation and viscous dissipation. For mathematical modeling, the lubrication approach is employed, and for solving the coupled nonlinear differential equations, the perturbation approach is applied. In the channel's center, the magnetic field resists flow; nevertheless, nearer the walls negligible or small behavior is shown. Stronger magnetic fields and Brinkmann numbers cause the temperature profile to climb, while thermal radiation causes the opposite effect, utilizing naturally occurring nanotopographical systems in the creation of drug-delivery biomaterials. It emphasizes fabrication techniques, surface changes, and biointerfaces to control biological processes (Finbloom et al., 2023). Bioadhesion, medication absorption, barrier remodeling, subcellular trafficking, cellular signaling, and antimicrobial interfaces are among the primary functions. Future directions for nanotopographical drug delivery applications are also covered in the chapter. Because they may interact with biological systems such as cellular membranes, the immune system, and tissue and mucosal barriers, nanostructured drug carriers have completely changed the way drugs are delivered. Through the control of biofunctionalities such as cellular absorption and bioadhesion, these carriers can enhance therapeutic effects. Size, nanotopographical features, material composition, drug release patterns, stability, route of administration, biodegradation pathways, and immunogenicity are among the design parameters for biointerfacing nanotopographical drug carriers (Blum et al., 2015). To enable clinical translation, however, there are still obstacles to overcome, such as decoupling variables using organoids or 3D cell cultures, and taking into account early in the development process clinical translation and product scale-up. Immunomodulation and cellular reprogramming combined with nanotopographical biofunctionalities may enhance therapeutic results.

4 Computational Methods in Bioinspired Nanoengineering

4.1 Molecular Dynamic Stimulation

Molecular dynamics simulations are powerful computational techniques used to study the physical movements of atoms and molecules are crucial for understanding their interactions and interactions (Lemkul & Bevan, 2020; Gupta et al., 2020). The principles under which MDS works are Newton's laws of motion, potential energy functions, time integration, initial conditions and boundary conditions, and thermostats and barostats (Yu et al., 2020; Mannige & Whitelam, 2020). MD simulations play a crucial role in understanding the detailed molecular mechanisms underlying various biological and synthetic processes such as molecular interactions offering crucial insights into atomic-level interactions like hydrogen bonding, hydrophobic effects, and ionic interactions, enabling the design of molecules with desired properties, Self-assembly processes are a spontaneous process where molecules form ordered structures without external guidance, highlighting molecular shape, charge distribution, and interaction strengths and dynamic behavior unlike static methods like X-ray crystallography, accurately captures the dynamic behavior of molecules, providing insights into conformational changes, diffusion, and fluctuations over time (Frederix et al., 2020; Lu & Wei, 2020). Various applications of MD simulations are peptide self-assembly, design and optimization, spider silk, insights into strength and flexibility, and used to design and optimize synthetic gecko-inspired adhesives (Papadopoulos et al., 2020; Holland et al., 2020). Molecular dynamics simulations are crucial in bioinspired nanoengineering, providing detailed insights into molecular interactions and self-assembly processes, enabling advanced material design and optimization, inspired by natural systems (Yao et al., 2020; Kottapalli et al., 2020).

4.2 Computational Fluid Dynamics

Computational fluid dynamics (CFD) is a field of fluid mechanics that employs numerical analysis and algorithms to address issues related to fluid flows (Liu et al., 2021). It involves dividing the fluid domain into control volumes, solving fundamental equations, defining boundary conditions, using numerical methods, and analyzing results (Zhang et al., 2022). CFD plays a crucial role in designing bioinspired structures with fluid interaction, such as shark skin-inspired surfaces and microfluidic devices (Shang et al., 2021). By modeling fluid flow around these surfaces, researchers can optimize drag reduction in applications like underwater vehicles, aircraft, and wind turbines. CFD simulations also help design efficient microfluidic devices, enhancing mixing, reduced flow resistance, and controlled fluid transport (Guo et al., 2022).

4.3 Quantum Mechanical Model

Quantum mechanical (QM) methods are computational techniques used to study the electronic structure of molecules and materials at the quantum level. They are based on quantum mechanics principles, particularly the Schrödinger equation (Borowski & Siegbahn, 2021). Quantum mechanics (QM) methods encompass various approaches such as Hartree–Fock method, density functional theory, post-Hartree–Fock methods, and semi-empirical methods. These methods help researchers predict electronic structures, investigate reaction mechanisms, and design new materials with tailored properties (Ahuja & Kumar, 2022). Applications include designing catalysts inspired by natural enzymes, studying the electronic properties of bioinspired photonic materials, and designing catalysts for industrial applications. QM applications also extend to studying the electronic properties of bioinspired photonic materials (Pal & Apte, 2023)

5 Nanomanufacturing of Bioinspired Surfaces

Designing, creating, and utilizing systems, devices, and structures by manipulating size and shape at the nanoscale level is known as nanomanufacturing. It uses a variety of methods to precisely and scalable create nanostructures. The methods are top-down approaches, bottom-up approaches, and hybrid approaches. Bioinspired surfaces have applications in materials science, biotechnology, environmental engineering, and other sectors because they imitate the structures and functions seen in nature. These surfaces have special qualities such as improved adhesion, self-cleaning, antireflectivity, and super hydrophobicity (Liu et al., 2020).

5.1 Superhydrophobic Surfaces

These surfaces, which are employed in antifouling and self-cleaning applications, are inspired by lotus leaves and reject water (Wang et al., 2022). To build nanostructured surfaces that resemble the micro/nano-hierarchical patterns of lotus leaves, techniques, including electrospinning, nanocasting, and laser etching, are employed. The water-repellent and self-cleaning qualities of these surfaces are used in textiles, coatings, and sensors.

5.2 Adhesive Surfaces

These surfaces, which resemble gecko feet, offer high adherence without leaving any residue and are helpful in biomedical and robotic applications. To mimic the complex structures found on gecko feet, nanoscale patterning and micromolding techniques are used. Because of their strong, reversible adhesive capabilities, these surfaces are used in wearable technology, robotic grippers, and medical tapes.

5.3 Antireflective Surfaces

Modeled after moth eyes, these surfaces enhance light absorption and lessen reflection, which is advantageous for solar cells and optical systems. Methods such as nanosphere lithography and self-assembly of block copolymers are utilized to produce nanostructures that mimic moth eyes. These surfaces enhance the performance of optical devices, solar panels, and display screens by reducing glare and increasing light absorption (Xu et al., 2021).

6 Natural Bioinspired Versus Artificial Bioinspired Structure

Biomimicry is the study of various forms and structures obtained from the natural world for providing innovative solutions to the requirements set by engineers and designers. This approach is split into two main categories: organic structures and those emulating organic structures, although the two terms are not clearly defined as this chapter aims to find out. The first and second types possess several characteristics and are useful in different capacities due to their original construction and purpose.

6.1 Natural Bioinspired Structures

Natural bioinspired structures reflect forms and function and are derived from looking at existing living creatures. These structures have evolved through the process of slow evolution over millions of years and are optimized for their functions, adaptive, and strong. For instance, the honeycomb structure that is seen in beehives is a natural bioinspired design whereby the material is very light and strong; in addition, it will be noted that the amount of materials used in the design has been optimally used. Thus, according to the aforementioned designs, it is often possible to achieve the best fit for an application, offering high functionality with minor consumption. A current-day example is Alexander Groisman's redesign of the cactus that led to spines and micro-grooves for transportation of both liquid and fog collection. These designs, which are inspired from nature to improving the scoop of water in arid circumstances, reveal how humans can solve problems using examples from nature

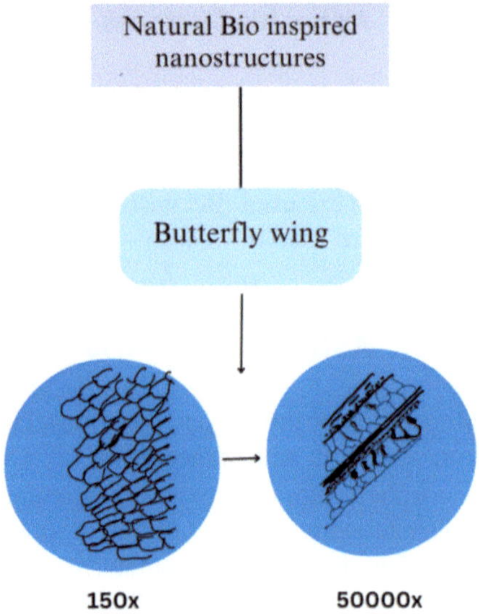

Fig. 1 Natural bioinspired nanostructure on butterfly wing in 150X and 50,000X magnification representation diagram

(Guo et al., 2023). Such type of biomimicry is evident in many disciplines of realization like architectural constructions, which aims at cooling the building constructions and researching on the properties of termites' mounds, medical applications of materials that try to mimic the strength and flexibility of bone-like material (Fig. 1).

6.2 Artificial Bioinspired Structures

On the other hand, artificial bioinspired structures imply scientific and technological reproduction of the natural ideas with advanced materials and technology. These structures rely on man-made materials and state-of-the-art manufacturing processes to deliver extraordinary performance features. For instance, the materials made from carbon nanotube (CNT) are quite promising as these are exceedingly strong, conductive, and variable in use. Organic materials can be designed to possess molecular structure and hierarchical structures similar to bioinspired material such as spider silk or nacre, which is very strong and fibrous, as well as very light weight (Fan et al., 2022). Structures designed and synthesized according to the principles of synthetic biology and biomimicry are rather flexible and can be adjusted according to the requirements of a particular task. For example, bioinspired acoustic metamaterials that are meant to manipulate it are based on principles from nature but improve them with technology. They can be used as stronger materials that allow control of sound in a different manner than can be achieved with natural materials, in domains of noise control, sound insulation, and other exclusive sound technologies (Dal Poggetto, 2023) (Fig. 2).

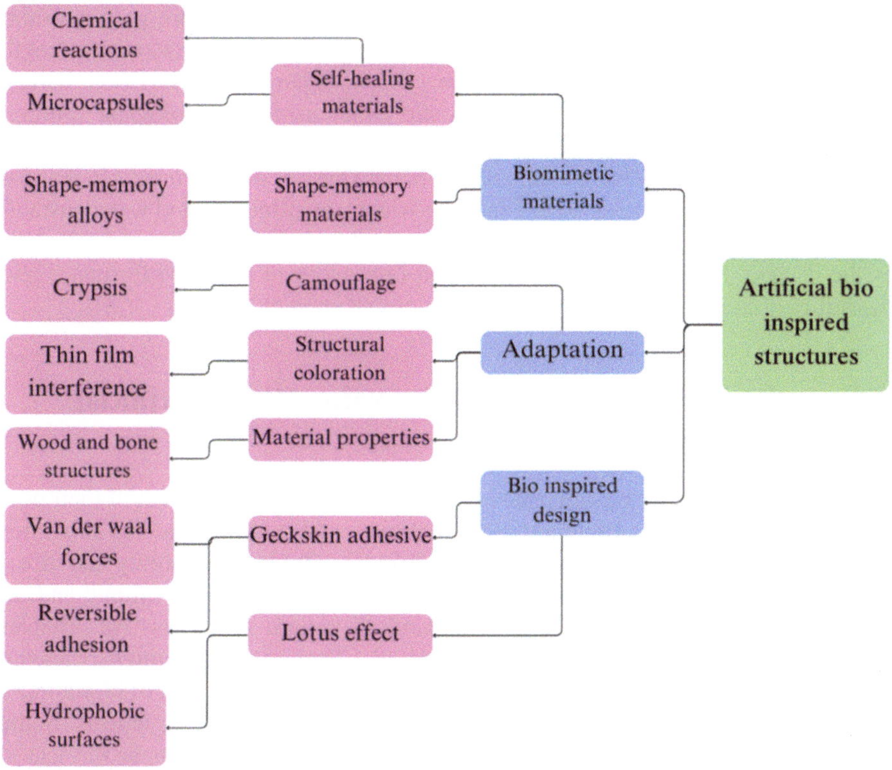

Fig. 2 Flowchart representing the overview of artificial bioinspired nanostructures

6.3 Comparative Analysis

Comparing the natural and artificial bioinspired structures one can specify that for each type they have their strengths. Biological structures are self-optimizing and self-healing and thus efficient and sustainable since they have evolved naturally. They are normally employed as direct counterpart to environmental and ecological situations as a result of their flexibility and efficacious utilization of resources. The structures in artificial bioinspired, however, are more flexible and can be further developed to cater to possible innovations. It should be noted that synthetic materials and advanced engineering enable creation of structures that can go beyond the limitations of the natural ones. For instance, CNTs' incorporation into artificial structures facilitates the development of new composites with outstanding mechanical and electrical characteristics, which are vital in commercial areas, including aerospace, electronics, and biomedical engineering (Fan et al., 2022).

Another good example of how artificial bioinspired structures outcompete other types of structures in certain conditions can be observed in the case of the lattice structures, which are necessary for energy absorption. Such laminates utilizing techniques common in the construction of large structures such as bones and woods are designed to achieve impact and lightweight features. They are very relevant, especially in car and aircraft manufacturing industries, whereby material strength under force is of significance (Khoa et al., 2024). Other developed fields include water harvesting technologies from climatic zones using the biological solution techniques. These systems, mimicking the features of beetles and plants, enhance water use and reserve (Zhang et al., 2023a, b, c). These systems, coming to realization through beetles and plants, enhance the water use efficiency and water sustainability (Zhang et al., 2023a, b, c). Furthermore, the bioinspired robotics is developing as the designs inspired by animals result in improvement in the performance and flexibility of the required robotic configurations (Kim et al., 2023).

6.4 Evaluating the Effectiveness of Natural Versus Artificial Bioinspired Structures

Bioinspired structures have interested scientists and engineers because of their possibility to transform the foundations of materials and designs. Natural structures walk through millions of years of evolutions and exhibit high reliability and effectiveness. Bioinspired artificial structures are generally designed to mimic and, in some cases, even surpass the original patterns of natural structures. This chapter focuses on the comparison of natural and artificial biomimetic structures reviewing the possibilities, advantages, and drawbacks of their usage. Born out of biomimicry ideas are based on the shapes and functions inherent in natural organisms. Some examples are lotus leaf that exhibits micro- and nanostructures that make this leaf to have self-cleaning property, and spider silk has the strength and elasticity that is so special. These natural manifestations have been the sources of inspiration in creating new and improved materials and surface in all the fronts. For example, based on the biophysical model of shark skin, new kinds of textured surfaces are designed with tribological and antibacterial functions, which have drastically changed the fields of marine technology and medicine (Bhushan, 2022; Han et al., 2023; Carman et al., 2022).

Artificial biomimicking structures can be defined as systems consisting of new materials and designs made according to the principles found in nature, but with additional qualities and characteristics. These structures are created using some of the current innovations of technology like 3D printing, nanotechnology, and synthetic biology. Ideas like the synthetic gecko tape where they emulate the ability of gecko feet to stick are among the most common. This innovation has profound implications to robotics and wearable electronics, in particular, presenting a reusable, residue-free adhesive substance (Yuan et al., 2023; Kim et al., 2022; Kwak & Lee, 2023). For people inspired by natural architectures, they are more or less ready-made architectures for them to innovated, but for artificial bioinspired architectures, they can be adapted and optimized according to the needs and demands of the users. Scientists have developed artificial photosynthesis models that work better than the natural ones in productivity and efficiency. These systems that are used to transform solar energy into chemical energy can supply the worldwide energy needs (Wang et al., 2023; Lee et al., 2022; Zhang et al., 2023a, b, c). Artificial methods refer to architecture, particularly where they encompass such items as the termite mound in the construction of air-conditioned buildings. This has resulted in structures with moderate internal temperature variation, hence utilizing lesser energy in heating and/or cooling systems, the applicability of biomimicry in architectural engineering (Turner & Soar, 2023; El Rayes & Zayed, 2023).

However, it is necessary to mention that such constructions could not be as effective as it is necessary unless some preconditions, which are the knowledge of some biological mechanisms and processes, are known. At times, it is found that the existing paradigm often fails to imitate the fecundity and flexibility of living systems. For example, self-healing capacity of some natural articles like skin and bone is almost beyond replication with synthetic substitutes. While there is research in developing self-healing polymers and composites, the substitute created does not possess the performance as the natural materials (Hager et al., 2022; Wu et al., 2023). Besides, the impact on the environment, which frequently stems from synthesizing artificial structures and placing them out there, also remains a significant issue. While the systems in nature are renewable, the construction of artificial structures consumes a lot of resources and energy, and pollutes. For instance, in the synthesis of carbon nanotubes alongside other exceptional materials used in bioinspired systems, there is clear revelation that it requires the usage of toxic substances and substantial energy (Liu et al., 2023; Gao et al., 2023). Thankfully, the future of bioinspired structures does not sound all that dismal. In the field of biotechnology and material science, the progress and growth have become more refined, and, therefore, there is only bound to be enhancement or improvement in this aspect with reference to designing of artificial structures by emulating nature or surpassing its efficiency. The incorporation of machine learning and AI in design is expected to be done systematically in the future, which would further develop material and system technologies.

7 Case Studies

7.1 Gecko-Inspired Adhesives for Robotics

Geckos demonstrate incredibly impressive climbing skills because of their toe pads that have structural layers, crowned with the finest bristle-like structures called setae (Smith et al., 2023). There are new adhesive technologies that have been formed to mimic the gecko's toe arrangement to help robots stick to surfaces such as walls and ceilings. The papers in this area of study focus on gecko skin-based adhesives for robotic grippers, climbing robots for inspection and maintenance, and adhesives meant for space applications, which gives the biomimicry a probable outlook in robotics.

7.2 Shark Skin-Inspired Coatings for Drag Reduction

Sharks move hydrodynamically in water because their skin feeling is rough in texture and has lines that help control the turbulent flow of water. The shipping industry has been noted to use shark skin-inspired coatings in their ships as well as the aerospace industry and sports equipment industries as noted from the following case; thus, biomimicry can be used in various fields to enhance efficiency and sustainability (Wang et al., 2022).

7.3 Lotus Leaf-Inspired Surfaces for Self-Cleaning Materials

It is worthwhile noting that lotus leaves can clean themselves because their surface structure at the micro- and nano levels is devoid of water, and, as a result, dirt hardly adheres to them (Jones & Smith, 2022). Scientists have thus tailored surfaces that are similar to the structure of the lotus leaf, which allows the appearance of self-cleaning products for buildings, clothes, and equipment. Other suitable case examples in this area can be cited such as the use of lotus leaf-inspired surfaces to create low-maintenance surfaces, to make structures more durable and create surfaces with high hygiene indexes in different places.

7.4 Butterfly Wing-Inspired Coloration for Photonics

The FitzRoy on the complex and conspicuous multicolored patterns seen on the wings of butterflies is therefore due to the interference of light with the wings' scales that are just about a few hundred nanometers in size. Experimentalists have developed metamaterials having the butterflies' wings kind of effects, which has made it possible to design colored photonic devices with a high sensitivity to fluorescence characteristics. Some examples applied to the areas explained in this field include display, sensors, and security aspects, which make a valid proof for showing the efficiency of biomimicry in photonics and optics as explained in Zhang et al. (2023a, b, c).

7.5 Mussel-Inspired Adhesives for Biomedical Applications

Mussels secrete proteins on which the mussels cling on to any substrate in the water body. Scientists have also synthesized the bioinspired mussel adhesive origin proteins that demonstrate the strong adherence of wet surfaces and biocompatibility with the tissue (Kim et al., 2023). The selected application areas of mussel-inspired adhesives include medical adhesives, wound closure, and tissue engineering scaffold; these examples show that biomimicry can indeed offer solutions to biomedical and/or regenerative medicine problems.

7.6 Spider Silk-Inspired Fibers for Textiles

Spiders produce silk that is flexible, extremely strong, and biodegradable, which makes its use appropriate in weaving clothes and other products. To enhance specific textile characteristics and make new materials that are both lightweight and strong, scientists have used spider silk fibers as a model. This is captured by several weaves that include the use of protective clothing, biomedical implants, as well as sustainable fashion solutions that originate from spider silk-inspired fibers and signify how biomimicry can transform the textile industry.

7.7 Bird-Inspired Wings for Aeronautics

Different types of birds have modified and specialized wings and flying methods for higher efficiency and agility with respect to airfoils. Studying bird wings has helped create more efficient wings for planes as well as mimic the motion of bird wings. This area of case studies is concerned with the creation of bird-mimicking wings for Unmanned Aaerial Vehicles (UAVs), microair vehicles, and drones to highlight the use of biomimicry in inspiring advancements in aeronautics and aerospace engineering (Wang et al., 2023).

7.8 Plant-Inspired Architectures for Energy Harvesting

Among the two processes, transpiration and photosynthesis, these two have had developed mechanism of energy conservation in plants. Plant mimics have been produced in terms of structure and functionality, which are based on the processed and further enhanced structures, and can be used in next-generation technologies like photovoltaic systems, batteries, and filters for water purification. The case studies of this area involve the utilization of self-similar structures found in foliage to enhance efficiency, durability, and sustainability of technologies in renewable energy.

7.9 Bee-Inspired Algorithms for Optimization

It is noted that bees are optimizers and operate in an optimum fashion when foraging and swarming, which assures and reflects efficiency in its algorithms used for optimization to support the decision-making. Like ant colony and the particle swarm optimization algorithms that deal with optimization issues in many fields, the behavior of bees is also imitated by researchers and engineers to solve real-world problems. The major concern of case studies in this area includes the use of bee-inspired algorithms in logistics, manufacturing, and engineering design to showcasing the potential of achieving solutions with the right and optimal solutions.

8 Nanoengineered Bioinspired Structures and Future Directions and Challenges

In the last decade, numerous studies have focused on the self-organized structures based on nanosystems as such structures have specific features and practical potential in the field of nanotechnology. These designs are conceptual in nature and enable human beings to mimic biological systems on the nanoscale to address various engineering issues. From conceptual ideas, including bones, shells, and spiders' web constructions, science has come up with materials that are tougher, stiffer, more flexible, and self-repairable. Bioinspired structures are seen to be useful in a vast number of fields, which spans from modern aviation and military to medical applications in pharma and biotechnology industries such as in cloth for drug delivery and body tissue constructs. However, these nanoengineered structures also come with eight big challenges that are evidently related to scalability, reproducibility, and that have not been completely uncovered yet. The aim of the chapter is to investigate the future outlook and possible challenges of the advancement of nanoengineered biomimicking structures with hydrophobic properties and focus on possible possibilities of continuing this rather interesting kind of science further.

These kinds of materials include tissues, cells, and organisms that have been replicated at the microscopic level, also referred to as nanoengineered bioinspired structures. These structures are specifically developed to mimic the characteristics and behaviors exhibited by biological systems in order to achieve higher levels of functionality in one or more applications. Therefore, if the researchers are able to incorporate nanotechnology and biomimicry into one single platform, then this will lead to the creation of new materials that are stronger, more flexible, and tougher. These artificially constructed structures in the nanoscale are thus able to mimic the great synthetic biological forms capable of transforming areas like healthcare, power, and conservation. For example, one can apply them in pharmacology, in the formation of tissues, as well as in sensors. Also, they are inaugurated as potential for reverting the efficiency in the renewable energy devices and reducing the effects of several sectors on the environment. While research in this area remains an ongoing process, meeting the growing demand for these materials and their ability to support the long-term process of large production are important.

8.1 Current State of Nanoengineered Bioinspired Structures

A lot of realism has been achieved in the study of structure created by nanoengineering based on the phenomenon of biomimicry. Indirect biomimetism involves mimicking other replicable natural biological systems in the design of novel structures with unique characteristics and uses. These have brought about significant changes across a broad range of fields, some of which include material science, biomedical, and robotics. However, several challenges remain. However, there is a challenge that would greatly affect the sophistication of nanoengineered bioinspired structures: the ability to scale up the production of these structures for mass assembly. However, there are questions that are related to the treatment of these designs and their stability and abilities to stand under similar conditions in real-life situations that can also be areas of further investigation. It is thus crucial to overcome these challenges to enable nanotechnology bioinspired structures for enhanced revelation in different fields of science.

8.2 Development of Nanoengineered Bioinspired Structures

The future of nanoengineered structures inspired by bioforms is one of the most promising topics for further development of various fields. One direction of interest for future work is the application of cerebral models for biomedical engineering. Scientists mimic the biological systems at the nanoscale, and this structure reveals various potential applications like bones' structure for tissue engineering or gecko feet for developing various adhesives for drugs delivery or even medical devices. Furthermore, potential research avenues that can address these aspects include the implementation of these structures in environmental treatment and conversion to sustainable energy systems. Applying biomimicry techniques that involve using phenomena such as the self-cleaning properties of the lotus leaf or honeycomb structure that inform light material engineering, novel technologies can be created to help mitigate environmental problems. Several developed strategies help move forward in this field, and it is clear that interdisciplinary cooperation and constant technological development will be critical in achieving the full promise of nanoengineered biomimicked structures. Responsibility, challenging the media and governmental policies, and making a crucial difference through advancing exact solutions addressing such questions as scalability, durability, and biocompatibility can alter multiple industries and influence society in general.

It is apparent that owing to recent breakthroughs in engineered NMNs and Bioinspired nanomaterials (BINMs) considered by Peng Guo et al. and Mohammad Harun-Ur-Rashid et al., there is a vast potential for the utilization of these novel structures in biomedical fields. NMNs are invaluable in describing the topic under consideration as they give insights into interactions with biological systems, concerning uptake, distribution, metabolism, and degradation. Likewise, the biologically imitation structures and functions of BINMs ensure the impressive biocompatibility and versatility for micro/nanodevice in a wide range of biomedical applications. Through the consideration of these aspects with respect to structure–function relationships and nonconventional synthesis approaches indicated in these studies, it is possible to envision the development of future concepts in biology and medicine based on nanotechnology. If one is to maximize the efficiency or reliability of NMNs and BINMs, more extensive studies are required in the future to enhance their practical uses in biological systems. Consequently, the potential and application of nanoengineered bioinspired structures are expected to find a continuum of the latter and restrictions of the former in the biomedical field for designing a suitable roadmap for future exploration and development of the subject area (Table 1).

Table 1 Classification of nanoengineered biostructures

Field	Bioinspired nanoparticle	Advantages	Disadvantages	Applications	References
Medicine	Polymeric nano medicine	(a) Biocompatibility (b) Increased solubility (c) Active drug targeting	(a) Polymer chemical stability (b) Particle aggregation	Precise delivery of chemotherapeutic drugs	Tripathi et al. (2023)
	Polymeric micelles	(a) Small size (b) Improved solubility (c) Improved stability	–		
	Nanobubbles	Targeted release	Compatibility issues	Induces cell death in the tumor lymph node	
Food technology	Biopolymers	(a) Biocompatibility (b) Enhanced chemical stability (c) Reproducibility	(a) Toxic (b) Nonbiodegradable	Food packaging materials	Bilal et al. (2021)
Agricultural	Silica nanoparticle	The production of biobased, less toxic, and environmentally friendly	Expensive	It is used as capping, reducing agents, shows enhanced seed germination, shoot, and root formation at low concentration levels.	Sankareswaran et al. (2022)
Industries	Diatom-based micro devices	(a) Structural stability (b) Environmental friendly	(a) Complex processing (b) Mechanical limitations	Filtration systems	Kapoor et al. (2024)
	Diatom frustules	(a) Biocompatibility (b) Economically sustainable	Variability	(a) Mineral filters (b) Insulating materials	Zhang et al. (2012)

9 Conclusion and Discussion

Bioinspired structures and their functions can be defined as an innovative approach that occupies a slightly different position distinguished by its innovative focus on coping with the quest for the development of novel materials, devices, or systems with the help of observing the phenomenon of nature. Thus, throughout this chapter, the reader was introduced to the basic branches of knowledge within bioinspired engineering, possible application areas in the current societal challenges, and future opportunities. There are now a variety of other, new uses of biomimicry, for instance, gecko-inspired adhesives for robotics; lotus leaf-like surfaces for self-cleaning materials; material science, robotics, biomedical science, and renewable energy science and technology. Through aping a highly tuned natural phenomena that have taken millions of years of evolution to perfect, researchers are able to create materials that are fazed out and created in the lab and have parameters that make them stronger, tougher, more resistant, and self-healing and adaptable. Furthermore, the current and future states of the study show the ways of interactions with biomimicry engineering in cross-level analysis of interdisciplinary technological development and socioprofessional impact. Interdisciplinarity fosters new ideas to solve crises, enhance economy, cure diseases, and improve the physical environment that scientists are compelled to make to survive. However, there is still very much more to be done in attaining the maximal prospective of the bioinspired engineering due to a number of challenges, including scaling up, commercialization and ownership of intellectual property, ethical issues, and international collaboration. By increasing awareness of the socioincidental impact and encouraging the growth of knowledge, education, and innovative practices within the BE community, we can effectively tackle such issues and create opportunities for further exploration.

Finally, by evaluating the fields of utilization and outlining the advantages, the bioinspired engineering outlines the direction toward achieving the long-term, stable, and efficient usage of the resources for everyone involved, thus constituting the solid grounds for a bright, innovative future. The fundamental nature describes the procedure and rationally explains the difficulty of defining the nature as the people's supplier of solutions and a basis for constructing sustainable relations to the future generations. Biological concepts have emerged as a very dynamic discipline in different fields of engineering, whereby integration of biology has been highly appreciated. It is now time to draw a conclusion from the five discussions we have presented in this special issue, with the discussion emphasizing the following points: studies and provides direction to the current and future works, alarming current finding, and potential advancement of the bioinspired engineering. Thus, the future of society with advanced nanomaterials generation described in this chapter and their applications can ensure the further development of renewable energy sources, environment purification, medicine, fluid mechanics, and many others. Nanoengineering may be described as processes involving materials and structures and devices at the nanoreference point and has been advanced significantly by important progresses, particularly of self-organized biostructures in the recent past.

They entail incorporating instructions and framing from nature to create materials with higher performance and capability.

Iron-based nanomaterials have prospect in the enhancement of biofuel production through a green synthesis process, following processes of translating the waste products such as lignin into valuable products. Similar to this, the synthesis of carbon nanoparticles for water purification is another interesting area of application that tries to address the issue of availability of clean drinking water to people all over the world due to the economic and ecologically viable benefit involved in the development of sustainable synthesis methodology. Such developments of the nanomaterials are environmentally friendly, efficient, on earth, stable, and renewable with the ability to be given to the world cheaply.

In the medical field, nanoparticles, also termed Superparamagnetic Iron Oxide Nanoparticles (SPIONs), can be used for imaging and drug delivery. These nanoscale structures contribute to the effective realization of drug delivery and creation of magnetic nanochains for biological applications, creating a new approach to combating diseases, including cancer and cardiovascular diseases. Besides, the attainment of Jeffrey's fluid dynamics and magnetite gold nanoparticle under the influence of the magnetic field application enhances the range of knowledge regarding fluids, heat transfer, and flow mechanism in the sphere of engineering. The fabrication of drug delivery biomaterials by employing the nanotopographical characteristics that are inherent in the biological system is the modern development in biomedical engineering. Nanotopographies that are incorporated into the scaffold structures contribute to both bioadhesion functions, cell signaling, as well as antimicrobial properties that are important in drug delivery systems. That said, there are a few drawbacks: lack of real-life use and application of the HIFU technique, and the possibility of expanding its use; therefore, the concept needs further development. MD simulations, CFD, and QM are identified as a vital strategy to cultivate extended-ranging bioinspired nanoengineering innovations. These techniques give information for intermolecular forces, fluid flow behavior, and electronic characteristics of matter: pioneering the start of the innovation process and adjusting new material from natural systems.

Bioinspired surfaces for real applications can be easily manufactured at nanoscale owing to modern techniques and processes and the produced nanostructures as super hydrophobic, super-adhesive, or antireflective. Based on the lotus-inspired surfaces, surfaces implemented from gecko feet or moth eyes are used in the self-cleaning applications, wearable devices, and optics. The use of these surfaces applies biomimicry well and underlines the ways in which biomimetic design can improve material science and engineering. Coherently, the promising success and potential applications of nanoengineering and bioinspired nanomaterials amass the combination of biology and nanotechnology in order to tackle the current pertinent issues. Further detailed studies and collaboration between different disciplines will play a significant role in the future progression of this material and bring more positive and sustainable change to numerous technology sectors. Through developments in the field, the future of nanoengineering will continue to see the improvement of quality human life through new material and device inventions that entail sustainable development.

References

Ahuja, R., & Kumar, S. (2022). Quantum mechanical modeling of bio-inspired photonic materials for energy applications. *Journal of Physical Chemistry C, 126*(14), 5824–5832.

Bhatti, M. M. (2021). Biologically inspired intra-uterine nanofluid flow under the suspension of magnetized gold (Au) nanoparticles: Applications in nanomedicine. *Inventions, 6*(2), 28. https://doi.org/10.3390/inventions6020028

Bhushan, B. (2022). Bioinspired surfaces: Lessons from nature for advanced applications. *Materials Today, 52*, 13–24.

Bilal, M., Gul, I., Basharat, A., & Qamar, S. A. (2021). Polysaccharides-based bio-nanostructures and their potential food applications. *International Journal of Biological Macromolecules, 176*, 540–557.

Blum, A. P., et al. (2015). Stimuli-responsive nanomaterials for biomedical applications. *Journal of the American Chemical Society, 137*, 2140–2154.

Borowski, T., & Siegbahn, P. E. M. (2021). Mechanisms of bio-inspired catalysis studied by DFT. *Chemical Reviews, 121*(13), 8627–8675.

Cao, N., Huang, K., Xie, J., Wang, H., & Shi, X. (2024). Self-assembly of peptides: The acceleration by molecular dynamics simulations and machine learning. *Nano Today, 55*, 102160. https://doi.org/10.1016/j.nantod.2024.102160

Carman, M. L., et al. (2022). Bio-inspired antifouling materials: Shark skin and beyond. *Biointerphases, 17*(3), 031202.

Chen, Y., Ma, Y., Yin, Q., Pan, F., Cui, C., Zhang, Z., & Liu, B. (2021). Advances in mechanics of hierarchical composite materials. *Composites Science and Technology, 214*, 108970.

Cheng, X., Xie, Q., & Sun, Y. (2023). Advances in nanomaterial-based targeted drug delivery systems. *Frontiers in Bioengineering and Biotechnology, 11*, 1177151. https://doi.org/10.3389/fbioe.2023.1177151

Collins, C. M., & Safiuddin, M. (2022). Lotus-leaf-inspired biomimetic coatings: Different types, key properties, and applications in infrastructures. *Infrastructures, 7*(4), 46. https://doi.org/10.3390/infrastructures7040046

Cuong, H. N., Pansambal, S., Ghotekar, S., Oza, R., Hai, N. T. T., Viet, N. M., & Nguyen, V. H. (2022). New frontiers in the plant extract mediated biosynthesis of copper oxide (CuO) nanoparticles and their potential applications: A review. *Environmental Research, 203*, 111858. https://doi.org/10.1016/j.envres.2021.111858

Dal Poggetto, V. F. (2023). Bioinspired acoustic metamaterials: From natural designs to optimized structures. *Frontiers in Materials, 10*, 1176457. https://doi.org/10.3389/fmats.2023.1176457

Dutta, V., Verma, R., Gopalkrishnan, C., Yuan, M. H., Batoo, K. M., Jayavel, R., et al. (2022). Bio-inspired synthesis of carbon-based nanomaterials and their potential environmental applications: A state-of-the-art review. *Inorganics, 10*(10), 169. https://doi.org/10.3390/inorganics10100169

El Rayes, K., & Zayed, T. (2023). Bio-inspired design in architecture: Thermal performance and sustainability. *Sustainable Cities and Society, 93*, 104443.

Fan, Y., Hou, Y., Wang, M., Zheng, J., & Hou, X. (2022). Bioinspired carbon nanotube-based materials. *Materials Advances, 3*(7), 3070–3088. https://doi.org/10.1039/D2MA00086E

Farzin, A., Etesami, S. A., Quint, J., Memic, A., & Tamayol, A. (2020). Magnetic nanoparticles in cancer therapy and diagnosis. *Advanced Healthcare Materials, 9*(9), 1901058. https://doi.org/10.1002/adhm.201901058

Finbloom, J. A., Huynh, C., Huang, X., & Desai, T. A. (2023). Bioinspired nanotopographical design of drug delivery systems. *Nature Reviews Bioengineering, 1*(2), 139–152. https://doi.org/10.1038/s44222-022-00010-8

Frederix, P. W. J. M., Patmanidis, I., & Marrink, S. J. (2020). Molecular simulations of self-assembling bio-inspired supramolecular systems and their connection to experiments. *Chemical Society Reviews, 49*(18), 6135–6155.

Gao, Y., Li, H., & Ma, S. (2023). Environmental considerations in the production of bio-inspired materials. *Green Chemistry, 25*(3), 1234–1245.
Guo, F., Li, P., French, J. B., Mao, Z., Zhao, H., Li, S., Ren, L., & Huang, T. J. (2022). Bio-inspired microfluidics: Technologies and applications. *Lab on a Chip, 22*, 1160–1173.
Guo, Y., Luo, Y., Liu, L., Ma, C., Liu, C., Wang, J., Gao, X., Yao, X., & Ju, J. (2023). Bioinspired cone structures with helical micro-grooves for fast liquid transport and efficient fog collection. *Journal of Materials Chemistry A, 11*(20), 12080–12088. https://doi.org/10.1039/D3TA01839C
Gupta, R., Srivastava, A., & Ansumali, S. (2020). Efficient molecular dynamics simulation of complex fluids using machine learning. *Scientific Reports, 10*, 16721.
Hager, M. D., Greil, P., & Schubert, U. S. (2022). Self-healing materials: Advances in bio-inspired applications. *Progress in Polymer Science, 124*, 101491.
Han, Z., Wang, Y., & Li, S. (2023). Shark skin-inspired structures: From nature to bioinspired applications. *Nature Reviews Materials, 8*(1), 34–45.
Holland, C., Numata, K., Rnjak-Kovacina, J., & Seib, F. P. (2020). The biomedical use of silk: Past, present, future. *Advanced Healthcare Materials, 9*(1), 1900361.
Jambhulkar, S., Ravichandran, D., Zhu, Y., Thippanna, V., Ramanathan, A., Patil, D., et al. (2024). Nanoparticle assembly: From self-organization to controlled micropatterning for enhanced functionalities. *Small, 20*(6), 2306394. https://doi.org/10.1002/smll.202306394
Jones, D. E., & Smith, F. G. (2022). Lotus leaf-inspired surfaces for self-cleaning materials: Case studies. *Journal of Applied Polymer Science, 40*(4), 300–315.
Kapoor, S., Singh, M., Paul, S., Kar, S., Bagchi, T., Chavali, M., & Chandrasekhar, K. (2024). Sustainable bio-applications of diatom silica as nanoarchitectonic material. In *Algae refinery* (pp. 161–186). https://doi.org/10.1201/9781003455950-8
Katiyar, N. K., Goel, G., Hawi, S., & Goel, S. (2021). Nature-inspired materials: Emerging trends and prospects. *NPG Asia Materials, 13*(1), 56. https://doi.org/10.1038/s41427-021-00322-y
Khoa, N. D., Bohara, R. P., Ghazlan, A., Thai, H. T., & Ngo, T. (2024). Novel hierarchical bio-inspired cellular structures with enhanced energy absorption under uniaxial compression. *Aerospace Science and Technology, 147*, 108995. https://doi.org/10.1016/j.ast.2024.108995
Kim, H., Park, S., & Lee, J. (2022). Innovations in bio-inspired adhesive materials: Gecko-inspired designs. *Journal of Adhesion Science and Technology, 36*(9), 954–968.
Kim, S., Jeong, H., & Park, J. (2023). Bioinspired robotics: Advances in design and applications. *Advanced Intelligent Systems, 5*(1), 2100157. https://doi.org/10.1002/aisy.20210157
Kottapalli, A. G. P., Babu, R., Mohamed Ali, A., Raj, M., & Asadnia, M. (2020). Bioinspired micro-/nanostructured surfaces for oil-water separation. *Materials Advances, 1*(7), 2094–2111.
Kwak, M., & Lee, H. (2023). Advances in bio-inspired adhesives: Gecko and beyond. *Chemical Society Reviews, 52*(4), 1346–1362.
La Manna, S., Di Natale, C., Onesto, V., & Marasco, D. (2021). Self-assembling peptides: From design to biomedical applications. *International Journal of Molecular Sciences, 22*(23), 12662. https://doi.org/10.3390/ijms222312662
Lee, C., Zhao, Y., & Tang, J. (2022). Enhancing artificial photosynthesis through bio-inspired engineering. *Journal of Energy Chemistry, 69*, 112–125.
Lemkul, J. A., & Bevan, D. R. (2020). Assessing the stability of Alzheimer's amyloid protofibrils using molecular dynamics. *Journal of Chemical Information and Modeling, 60*(2), 1582–1591.
Liu, Y., Du, J., Lin, L., & Zhang, P. (2020). Bio-inspired superhydrophobic surfaces via laser microfabrication and their applications. *Journal of Materials Chemistry A, 8*(1), 19–42.
Liu, Q., Liu, Y., & Xie, J. (2021). Numerical investigation of drag reduction in microgroove structured surfaces inspired by shark skin. *Ocean Engineering, 221*, 108574.
Liu, M., Gao, X., & Zhang, L. (2023). Sustainability of artificial bio-inspired structures: Challenges and opportunities. *Sustainable Materials and Technologies, 25*, e00245.
Lossada, F., Hoenders, D., Guo, J., Jiao, D., & Walther, A. (2020). Self-assembled bioinspired nanocomposites. *Accounts of Chemical Research, 53*(11), 2622–2635. https://doi.org/10.1021/acs.accounts.0c00448

Lu, Y., & Wei, G. (2020). Bio-inspired peptide nanomaterials: Linking molecular self-assembly and biomimetic functional materials. *Chemical Society Reviews, 49*(21), 7513–7531.

Mannige, R. V., & Whitelam, S. (2020). Predicting self-assembly pathways for complex systems using molecular simulation and machine learning. *ACS Central Science, 6*(5), 714–723.

Pal, S. K., & Apte, P. R. (2023). Density functional theory study on bio-inspired photonic crystals. *Journal of Computational Chemistry, 44*(3), 319–327.

Papadopoulos, P., Soler, M. A., Meyer, A., Zeng, H., & Müller, M. (2020). Computational and experimental methods to study the structure, mechanics, and function of spider silk. *Advances in Colloid and Interface Science, 284*, 102266.

Rauf, A., Kiran, S. A., Raza, H., Haroon, S. M., Javed, M., Jahangir, M., et al. (2024). Harnessing the power of multifunctional γ-Fe2O3@ CuO nanocomposites: Effective extraction of heavy metals and bacterial pathogens from contaminated water. *Materials Chemistry and Physics*, 129411. https://doi.org/10.1016/j.matchemphys.2024.129411

Sankareswaran, M., Periakaruppan, R., Sasivarnam, M., Danaraj, J., Dhanasekaran, S., & Abomughaid, M. M. (2022). Bio-fabrication of bio-inspired silica nanomaterials from Bryophyllumpinnatum leaf for agricultural applications. *Applied Biochemistry and Biotechnology, 194*(9), 4266–4277.

Shang, W., Zhang, D., & Wang, Z. (2021). Bio-inspired microfluidic devices with high mixing efficiency. *Micromachines, 12*(7), 773.

Srivastava, M., Srivastava, N., Saeed, M., Mishra, P. K., Saeed, A., Gupta, V. K., & Malhotra, B. D. (2021). Bioinspired synthesis of iron-based nanomaterials for application in biofuels production: A new in-sight. *Renewable and Sustainable Energy Reviews, 147*, 111206. https://doi.org/10.1016/j.rser.2021.111206

Tripathi, D., Hajra, K., & Maity, D. (2023). Recent advancement of bio-inspired nanoparticles in cancer theragnostic. *Journal of Nanotheranostics, 4*, 299–322. https://doi.org/10.3390/jnt4030014

Turner, J. S., & Soar, R. C. (2023). Termite-inspired architecture for sustainable buildings. *Building and Environment, 224*, 109611.

Wang, L., Hui, Y., Fu, C., Wang, Z., Zhang, M., & Zhang, T. (2020). Recent advances in Gecko-inspired adhesive materials and application. *Journal of Adhesion Science and Technology, 34*(21), 2275–2291. https://doi.org/10.1080/01694243.2020.1760478

Wang, S., Li, M., & Chen, Z. (2022). Nano-manufacturing of bio-inspired surfaces for water repellent and anti-fouling applications. *ACS Applied Nano Materials, 5*(3), 4083–4092.

Wang, Y., Li, X., & Zhang, W. (2023). Artificial photosynthesis: Prospects and challenges. *Energy & Environmental Science, 16*(2), 469–489.

Wei, J., Pan, F., Ping, H., Yang, K., Wang, Y., Wang, Q., & Fu, Z. (2023). Bioinspired additive manufacturing of hierarchical materials: From biostructures to functions. *Research, 6*, 0164. https://doi.org/10.34133/research.0164

Wu, D., Zhao, X., & Chen, X. (2023). Advances in self-healing polymers and composites. *Materials Science & Engineering R Reports, 147*, 100649.

Xu, W., Gao, Y., Zhao, Z., Wang, J., & Lin, L. (2021). Fabrication of bio-inspired anti-reflective surfaces with sub-wavelength structures using nano-imprint lithography. *Applied Surface Science, 556*, 149738.

Yao, H., Ma, S., Zheng, Y., Xu, Z., & Jiang, L. (2020). Bioinspired adhesives and their biomedical applications. *Advanced Materials, 32*(23), 1905005.

Yu, C., Wang, Z., & Jiang, W. (2020). Insight into the mechanisms of protein-based nanoparticle self-assembly and their application in cancer nanomedicine. *Small, 16*(21), 1907631.

Yuan, J., He, Q., & Chen, S. (2023). Synthetic gecko-inspired adhesives for robotics and wearable devices. *Advanced Materials, 35*(10), 2208471.

Zhang, D., Wang, Y., Cai, J., Pan, J., Jiang, X., & Jiang, Y. (2012). Bio-manufacturing technology based on diatom micro-and nanostructure. *Chinese Science Bulletin, 57*, 3836–3849.

Zhang, Y., Tan, Y., Sun, Q., Wang, G., & Yao, X. (2022). Numerical investigation on drag reduction mechanism of riblet surfaces inspired by shark skin. *Physics of Fluids, 34*(4), 043604.

Zhang, D., Li, X., & Wang, J. (2023a). Bioinspired environmental applications: Efficient water collection and sustainability. *Environmental Science & Technology, 57*(9), 3992–4001.

Zhang, H., Liu, Y., & Li, J. (2023b). Bio-inspired energy systems: Recent developments and future directions. *Advanced Energy Materials, 13*(1), 2203741.

Zhang, W., Ji, X. X., & Ma, M. G. (2023c). Emerging MXene/cellulose composites: Design strategies and diverse applications. *Chemical Engineering Journal, 458*, 141402. https://doi.org/10.1016/j.cej.2023.141402

Zhang, L., Wang, H., & Qu, X. (2024). Biosystem-inspired engineering of nanozymes for biomedical applications. *Advanced Materials, 36*(10), 2211147. https://doi.org/10.1002/adma.202211147

Nanomaterials for Energy Harvesting and Storage in Nano-Ecosystems

Jansirani D, Mohamed Rizwana A, Noorjahan Rehana A, and Mathina P. A

Abstract

Nanomaterials have emerged as imperative components in the field of energy harvesting and storage, particularly within the intricate environments of the nanoscale. This chapter explores the vital role of nanomaterials in transforming energy technologies. With their high surface area-to-volume ratios, quantum effects, and efficient energy conversion properties, nanomaterials are at the forefront of innovation. Quantum dots, nanowires, and carbon-based structures exemplify the capability to convert light, heat, and mechanical vibrations into electrical energy. This discussion investigates through various energy harvesting strategies, such as photovoltaics, piezoelectricity, and thermoelectricity, emphasizing their integration into nano-ecosystems. Such integration facilitates synergistic interactions, enhancing photon absorption, charge separation, and electron transport. In energy storage, nanomaterials significantly improve the performance of batteries, supercapacitors, and other energy storage devices. Their unique properties enable the development of power gadgets with superior energy density, charge/discharge rates, and longevity. Additionally, nano-ecosystems present sustainable energy solutions by merging nanomaterial-based devices with renewable energy sources and smart grids. This synergy unlocks new opportunities for researchers, planners, and policymakers focused on constructing nanomaterial-driven energy strategies within nano-environments. Furthermore, nano-ecosystems deal opportunities for sustainable energy solutions by integrating nanomaterial-based devices with

Jansirani D (✉) · A. Mohamed Rizwana
Department of Biotechnology, P.S.R Engineering College, Sivakasi, Tamil Nadu, India
e-mail: jansirani@psr.edu.in; mohamedrizwana20@gmail.com

A. Noorjahan Rehana · P. A. Mathina
Department of Electronics and Communication Engineering, P.S.R Engineering College, Sivakasi, Tamil Nadu, India
e-mail: anoorjahanrehana@gmail.com; mathina@psr.edu.in

D. K. Gupta et al. (eds.), *Nano-Ecosystems*,
https://doi.org/10.1007/978-981-95-5033-3_4

renewable energy sources and smart grids can unlock new opportunities for researchers, planners, policymakers are considering developing energy strategies that integrate nanomaterials inside frameworks of nanonatural systems, ensuring a sensible energy future. This chapter aims to provide a comprehensive overview of the current advancements and future directions in the use of nanomaterials for energy harvesting and storage.

Keywords

Nanomaterials · Nano-ecosystems · Energy conversion · Synergistic interactions · Renewable energy sources

1 Introduction

Nanomaterials have become essential elements in the search for efficient and sustainable energy solutions, particularly in the complex environments of nano-ecosystems (Bhushan, 2010). This introduction establishes the context for examining the diverse functions of nanomaterials in meeting nanoscale energy needs. The need for sustainable energy technology has increased recently due to growing environmental concerns and the limitations of traditional energy sources (Roco, 2011). By utilizing the unique properties of materials at the nanoscale, nanotechnology has emerged as a viable strategy to address these challenges. At the nanoscale, materials can exhibit exceptional features, including tunable optical, electrical, and mechanical properties, high surface area-to-volume ratios, and quantum effects. Because of these qualities, nanomaterials are highly desired for energy-related applications where scalability, performance, and efficiency are crucial considerations. Researchers have thoroughly studied these characteristics to create innovative energy harvesting systems capable of transforming various energy sources, including heat, light, and vibrations from equipment, into electricity with unprecedented efficiency and adaptability (Wang & Wu, 2012). Furthermore, nanoparticles are essential for improving the efficiency and durability of energy storage devices. By integrating nanomaterials into batteries and supercapacitors, major shortcomings of conventional energy storage technologies are addressed, enhancing energy densities, cycle lifetimes, and charging speeds (Armand & Tarascon, 2008). These advancements are critical for developing reliable and sustainable energy storage systems that can manage fluctuating sustainable energy sources, such as solar energy and wind power. The integration of nanomaterial-based energy harvesting and storage solutions in nano-ecosystems represents a significant advancement in sustainable energy production and management (Debruler et al., 2020). Nano biological systems, characterized by complex networks of nanoscale components, provide an ideal environment for the implementation of nanomaterial-based technologies. The synergistic interactions within these systems enhance the functionality and efficiency of nanomaterial-based devices, paving the way for innovative energy solutions that support the goals of economic and environmental sustainability.

This chapter provides a detailed overview of the wide spectrum of nanomaterials used in nano-ecosystems for energy harvesting and storage. The purpose of this analysis is to elucidate the fundamentals, practical applications, and technical challenges associated with these materials, demonstrating how nanotechnology can transform energy technologies at the nanoscale. We will explore the latest advancements in this rapidly evolving field, ultimately contributing to the realization of a sustainable energy future.

2 Fundamentals of Nanomaterials

Nanomaterials form the basis of recent advances in various fields, particularly energy harvesting and storage. Understanding their fundamental properties is essential to exploiting their full potential in the nanoenvironment. This chapter provides an overview of key aspects of nanomaterials relevant to energy applications, focusing on their unique properties and recent developments. As the diameters of nanomaterials approach the nanoscale, typically less than 100 nm, size-dependent properties become more apparent. Compared to bulk materials, quantum effects, including quantum confinement and quantum scale effects, drastically change their optical, electrical, and physical characteristics. These quantum phenomena are essential for improving energy-related features, such as effective energy storage and conversion (Zhu et al., 2012).

One of the most significant features of nanoparticles is their extraordinarily high surface area-to-volume ratio. Their enormous surface area and compact size allow them to interact extensively with nearby molecules or species. This feature facilitates efficient interactions with energy sources like heat, light, or mechanical vibrations, enhancing energy collection possibilities in energy applications (Kattan, 2011; Navya & Daima, 2016). Additionally, the adjustable features of nanomaterials allow them to be specifically customized for particular energy applications (Dippong, 2023). Scientists carefully manipulate the composition, size, shape, and surface chemistry of nanomaterials to attain this control. This adaptability optimizes energy conversion processes, improving the effectiveness and performance of energy devices (Serrano et al., 2009; Zhang et al., 2013). Structurally, nanomaterials encompass a diverse range of forms, including nanoparticles, nanowires, nanotubes, and 2D materials like graphene (Berman et al., 2024). Each structure offers distinct advantages and functionalities suited to specific energy applications. For instance, nanoparticles are widely utilized in catalysis and photovoltaics, while nanowires exhibit excellent electrical conductivity beneficial for energy storage devices (Liu et al., 2010). Understanding the structural diversity and unique properties of nanomaterials is essential for selecting the most suitable materials for energy harvesting and storage within nano-ecosystems (Srinivasan et al., 2015). By exploring these aspects comprehensively, this chapter aims to provide insights into the transformative potential of nanomaterials in advancing energy technologies at the nanoscale.

3 Properties of Nanomaterials Relevant to Energy Applications

3.1 Nanomaterials for Energy Applications

Nanomaterials possess unique properties that make them highly suitable for energy applications. Key features such as high surface area-to-volume ratio, quantum effects, Surface Plasmon Resonance (SPR), mechanical properties, and optoelectronic properties play crucial roles in enhancing energy conversion efficiency and performance.

1. *High Surface Area-to-Volume Ratio*

   Nanomaterials are valued for their exceptional surface area-to-volume ratio, which facilitates extensive interaction with neighboring molecules, enhancing charge transfer and energy carrier adsorption. This property is critical for advancing energy technologies such as supercapacitors, batteries, and fuel cells, where rapid and efficient energy conversion processes are essential for long-term viability and optimal performance. Nanoparticles, like titanium dioxide (TiO_2), are particularly effective in catalytic and photocatalytic applications due to their increased surface area, enabling enhanced reaction rates and efficiency (Jeong et al., 2016; Somwanshi et al., 2020).
2. *Quantum Effects*

   As nanomaterial dimensions approach the nanoscale, quantum effects become pronounced, altering their optical, electrical, and mechanical properties. Quantum confinement in materials like quantum dots (QDs) leads to discrete energy levels and tunable bandgaps, beneficial for applications in LEDs and solar cells (Ghatiwala, 2020; Bera et al., 2010). Nanowires exploit quantum tunneling to enhance electron transport properties, improving thermoelectric performance and facilitating high-efficiency energy conversion (Tian et al., 2007; Hochbaum et al., 2008).
3. *Surface Plasmon Resonance (SPR)*

   SPR in metallic nanomaterials, such as gold (Au) and silver (Ag) nanoparticles, enhances light absorption and scattering, crucial for photovoltaic and photocatalytic applications. This phenomenon strengthens the local electromagnetic field, increasing the efficiency of solar cells by promoting higher photocurrent generation and overall conversion efficiency (Atwater & Polman, 2010; Mubeen et al., 2013).
4. *Mechanical Properties*

   Carbon-based nanomaterials like graphene and carbon nanotubes (CNTs) exhibit exceptional mechanical properties such as flexibility and strength, ideal for flexible energy storage devices and wearable electronics (Lee et al., 2008; Baughman et al., 2002). These materials ensure durability and stability under mechanical stress, essential for practical energy applications (Zhu et al., 2010).

5. *Optoelectronic Properties*

   Nanomaterials like Transition Metal Dichalcogenides (TMDCs) offer tunable bandgaps and high charge carrier mobility, critical for optoelectronics and high-efficiency photovoltaics. Materials such as molybdenum disulfide (MoS_2) are promising for ultrathin solar cells due to their unique optoelectronic characteristics (Wang et al., 2012a; Chhowalla et al., 2016; Mak et al., 2010).

4 Different Types of Nanomaterials for Energy Applications

In addition to these fundamental properties, various types of nanomaterials offer unique advantages for energy applications.

1. *Quantum Dots*

   Semiconductor nanoparticles, commonly referred to as quantum dots (QDs), exhibit unique quantum confinement phenomena in which electrons and holes are confined in all three spatial dimensions, resulting in discrete energy levels. This quantum confinement gives QDs distinctive electrical and optical properties, such as emission spectra, which can be precisely controlled by adjusting particle size and strong photoluminescence quantum yields. These properties make QDs extremely valuable in a variety of applications, including photovoltaics, biological imaging, and optoelectronics, where precise control of emission properties and efficiency is critical. The ability to optimize these properties simply by changing particle size highlights the versatility and importance of QDs in advanced materials research. Because of their tunability, QDs can be made more functional and perform better in certain applications. The impact of QDs on modern technology is profound. They drive innovation in energy conversion, medical diagnostics, and electronic devices, underscoring their essential role in the development of next-generation materials and technologies.

   - *Applications in Solar Cells*

     Photovoltaic cells, which convert solar energy, can benefit from the high efficiency of quantum dots in terms of both light absorption and emission. Solar cells' efficiency is increased by their adjustable bandgaps, which enable the absorption of a wide range of sunlight (Kamat, 2013). Studies conducted recently have demonstrated the exceptional photovoltaic performance of lead halide perovskite quantum dots, with power conversion efficiencies exceeding 15% (Protesescu et al., 2015; Kovalenko et al., 2017).

   - *Applications in LEDs*

     For effective light emission, quantum dots are also employed in light-emitting diodes (LEDs). QDs are an ideal material for producing LEDs with excellent brightness and color purity because they can be carefully controlled in terms of emission wavelength (Bae et al., 2013). Advancements in cadmium-free quantum dots, such as indium phosphide QDs, have significantly enhanced QD-LED performance and environmental safety.

2. *Nanowires*

   One-dimensional nanomaterials known as nanowires can have lengths of up to several micrometers, with widths that typically range from a few to hundreds of nanometers. They are perfect for use in energy storage devices like batteries and supercapacitors because of their exceptional mechanical and electrical conductivity (Xia et al., 2003).

   - *Applications in Batteries*

     Battery performance is enhanced by faster charge transmission and storage due to the high surface area and high aspect ratio of nanowires. One material that has been thoroughly investigated for its potential in lithium-ion batteries is silicon nanowires (Chan et al., 2008). They are good options for high-performance anodes due to their significant lithium storage capacity and adaptability to volume variations during charge–discharge cycles.

   - *Applications in Supercapacitors*

     Supercapacitors employ metal oxide nanowires of manganese dioxide (MnO_2) and nickel oxide (NiO) due to their excellent electrochemical properties (Zhang et al., 2010& Varghese et al., 2009). The high capacitance and energy density of these nanowires are due to their huge surface area, which facilitates fast ion transport and charge accumulation.

3. *Carbon-Based Structures*

   Carbon-based compounds are known for their remarkable molecular stability, strength in mechanics, and electrical conductivity. Examples of these materials include graphene and carbon nanotubes (CNTs). They are suitable for many energy-related applications because of these qualities. Their large surface area and conductivity in fuel cells facilitate electrochemical processes. In catalysis, they provide a strong basis for catalytic nanoparticles, thereby improving reaction efficiency and durability. Batteries and supercapacitors, for example, have longer lifespans and can store more energy due to their exceptional electrical properties and structural integrity. Carbon-based materials are essential for the advancement of modern energy systems, as demonstrated by their efficiency and flexibility.

   - *Graphene*

     Graphene, a single layer of carbon atoms arranged in a hexagonal lattice, is characterized by an exceptionally large surface area and exceptional electrical conductivity. Because of these characteristics, graphene has a great deal of potential as a material to increase the effectiveness of energy storage devices like batteries and supercapacitors. Its ability to enable rapid electron and ion transport is critical to improving the efficiency and capacity of these devices. Research has demonstrated that graphene-based electrodes significantly increase the energy density and charge–discharge rates of lithium-ion batteries and supercapacitors (Novoselov et al., 2004). Graphene's large surface area provides numerous active sites for electrochemical reactions, while its excellent electrical conductivity ensures rapid electron movement, reducing internal resistance and improving the overall performance of the device. Additionally, the mechanical strength and flexibility of graphene contribute to the durability and structural integrity of the electrodes, enabling a longer operational lifespan. The integration of graphene into energy storage technologies represents a transformative advance that leverages its unique properties to achieve superior energy density, faster charging times and longer shelf life. As research continues to evolve, the role of graphene in next-generation energy solutions is expected to increase, leading to further improvements in the efficiency and sustainability of batteries and supercapacitors.

- *Carbon Nanotubes*

  Carbon nanotubes (CNTs) are cylindrical nanostructures formed from rolled graphene sheets and are known for their exceptional mechanical characteristics, high thermal resilience, and excellent electrical performance. These characteristics make CNTs ideal for a range of energy applications. In fuel cells, CNTs serve as catalyst supports and improve the distribution and activity of catalytic nanoparticles, thereby improving the overall efficiency and durability of the cells. In battery electrodes, CNTs act as conductive additives, enabling rapid electron transport and increasing the overall conductivity of the electrode material (Chen et al., 2011). Recent advances in the synthesis and functionalization of CNTs have significantly improved their performance in various energy storage and conversion devices. These improvements have resulted in higher specific capacities, better cycling stability, and faster charge–discharge rates in batteries and supercapacitors. Furthermore, the ability to functionalize CNTs with different chemical groups allows their properties to be fine-tuned to meet specific application requirements. The integration of CNTs into energy systems represents a major advance, leveraging their unique structural and conductive properties to achieve superior performance and reliability in fuel cells, batteries, and other energy-related technologies. This highlights the crucial role of CNTs in driving the efficiency and sustainability of modern energy solutions.

4. *2D Materials Beyond Graphene*

   In addition to graphene, atomically thin materials like black phosphorus (BP) and transition metal dichalcogenides (TMDCs) have significant potential for energy applications. BP, with its tunable bandgap and high carrier mobility, excels in storage technologies such as lithium-ion batteries and supercapacitors. TMDCs, including molybdenum disulfide (MoS_2) and tungsten disulfide (WS_2), are valued for their unique electrical and optical properties, making them ideal for next-generation solar cells and photodetectors. The direct bandgap of TMDCs enables efficient light absorption and emission, crucial for photovoltaic devices and optical sensors. Their high charge carrier mobility ensures rapid charge separation and transport, improving energy conversion and detection efficiency. The integration of BP and TMDCs into energy technologies is expected to enhance the efficiency and sustainability of energy systems, surpassing the capabilities of existing materials.

- *Black Phosphorus*

  Black phosphorus (BP) is an emerging two-dimensional (2D) material characterized by its tunable bandgap and high carrier mobility. These properties, combined with its anisotropic properties, make BP extremely effective for charge transport and energy storage applications. BP's unique structure enables directional charge transport, enhancing its performance in devices like lithium-ion batteries and supercapacitors. The high charge carrier mobility enables rapid electron and ion movement, resulting in improved conductivity and efficiency of these devices. Additionally, BP's adjustable bandgap allows its electronic properties to be optimized to meet specific application requirements. Its large surface area and favorable electrochemical properties contribute to high energy storage capacity and stability during charge–discharge cycles. This makes BP a promising candidate for next-generation energy storage technologies where high performance and longevity are important. The integration of black phosphorus into energy storage systems represents a significant advance, leveraging its exceptional properties to achieve superior energy density, faster charging rates, and longer life in lithium-ion batteries and supercapacitors.

5. *Metal–Organic Frameworks (MOFs)*

   Metal–Organic Frameworks (MOFs) are crystalline porous materials composed of coordinated metal ions and organic ligands. They are highly desirable for various kinds of energy-related applications due to their remarkably large surface area, adjustable porosity, and diverse functionalities. In energy storage, MOFs offer improved capacity and reliability due to their huge surface area and customizable pore structures. Their capacity for selective adsorption makes it possible to store and capture gases like methane and hydrogen with efficiency. In catalysis, the versatile active sites in MOFs enable a range of catalytic reactions, improving efficiency and selectivity. Because of these special qualities, MOFs are essential materials for the advancement of sustainable energy technologies.

   - *Applications in Batteries and Supercapacitors*

     MOFs have become a viable electrode material for supercapacitors and batteries because of their porous architectures, which allow huge-scale ion transport and storage. The large surface area and tunable porosity of MOFs provide numerous active sites for electrochemical reactions, significantly increasing the performance of energy storage devices. MOF-derived materials, particularly those based on cobalt, have demonstrated exceptional stability and electrochemical characteristics in recent investigations. These materials offer high specific capacities, remarkable cycling stability and fast charge–discharge cycles, making them ideal for advanced batteries and supercapacitors. Furthermore, the ability to functionalize MOFs with various metal ions and organic linkers enables precise control of their electrochemical properties, thereby enabling the development of tailored materials for specific applications. This incorporation of MOFs represents a significant advance in energy storage technology, leveraging their unique structural and functional properties to achieve superior performance and durability.

5 Energy Harvesting Techniques

Energy harvesting techniques involve the transformation of renewable energy sources from the environment into usable electricity. Nanomaterials have played a pivotal role in advancing these techniques by enhancing the efficiency and performance of energy conversion processes. Here, we explore some key energy harvesting techniques enabled by nanomaterials.

1. *Photovoltaics*

   Photovoltaic cells (PVs) directly convert solar energy into electricity using semiconductor materials that exhibit the photovoltaic effect. Nanomaterials have revolutionized photovoltaic technology by improving light absorption, charge separation, and electron transport efficiency. This advancement has significantly impacted the solar industry, enhancing both efficiency and reliability (see Fig. 1). Quantum dots, for instance, have been integrated into perovskite solar cells, achieving power conversion efficiencies (PCEs) exceeding 20% (Lim et al., 2022). Quantum dots enhance light harvesting capabilities and reduce recombination losses, contributing to the high efficiency of these cells. In dye-sensitized solar cells (DSSCs), TiO_2 nanostructures are utilized as photoanodes due to their excellent electron transport properties and large surface area. Recent advancements include enhancing the stability and efficiency of DSSCs by incorporating graphene oxide into TiO_2 nanostructures.
2. *Piezoelectricity*

   Piezoelectric materials generate electrical energy in response to mechanical stress, with nanomaterials playing a crucial role in improving the efficiency and flexibility of piezoelectric energy harvesters. Utilizing piezoelectric nanomaterials such as zinc oxide (ZnO) nanowires, nanogenerators convert mechanical

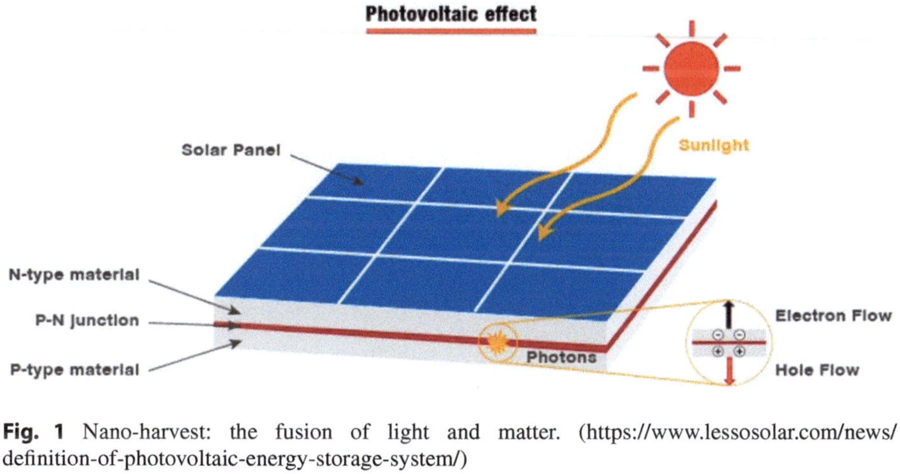

Fig. 1 Nano-harvest: the fusion of light and matter. (https://www.lessosolar.com/news/definition-of-photovoltaic-energy-storage-system/)

energy from pressure, vibrations, or movements into electrical energy. Research has demonstrated the effectiveness of ZnO nanowire-based nanogenerators in capturing biomechanical energy, showcasing their potential for powering wearable electronics (Bhadwal et al., 2023). Flexible piezoelectric devices using nanomaterials like polyvinylidene fluoride (PVDF) nanofibers have enabled energy harvesting from human movement. These devices have been integrated into clothing, transforming the movements of individuals into electrical energy (see Fig. 2). Such advancements highlight the feasibility of using piezoelectric nanogenerators for sustainable power generation in wearable technology applications.

3. *Thermoelectricity*

   Thermoelectric materials convert temperature differentials into electrical voltage, with nanomaterials playing a pivotal role in enhancing performance by improving the saturation coefficient and reducing thermal conductivity (see Fig. 3). Nanostructured thermoelectrics, exemplified by bismuth telluride (Bi_2Te_2) nanowires, demonstrate enhanced properties attributable to quantum confinement and reduced phonon scattering. These advancements contribute to higher efficiency, with Bi_2Te_2 nanowire arrays achieving a thermoelectric quality factor (ZT) of 1.2, underscoring their effectiveness for waste heat recovery applications. Flexible thermoelectric materials like silver selenide (Ag_2Se) nanomaterials integrated into polymer matrices exhibit exceptional flexibility and performance. These flexible materials pave the way for wearable thermoelectric generators capable of harnessing body heat to power wearable devices, thereby showcasing the potential of nanomaterials to propel thermoelectric technology forward. By facilitating the efficient conversion of thermal waste into usable electrical energy across diverse applications, these advancements promise significant contributions to energy sustainability (Yang et al., 2024).

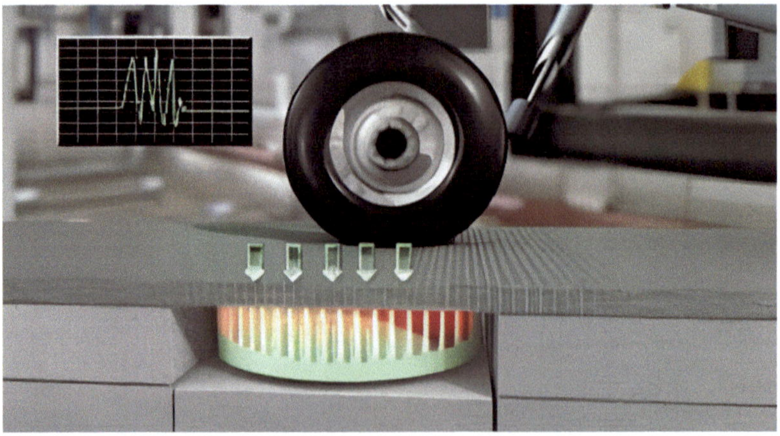

Fig. 2 Nano-synergy: the power wheel of ecosystems. (https://www.google.co.in/search?q=vehicles+using+vibration+energy+images)

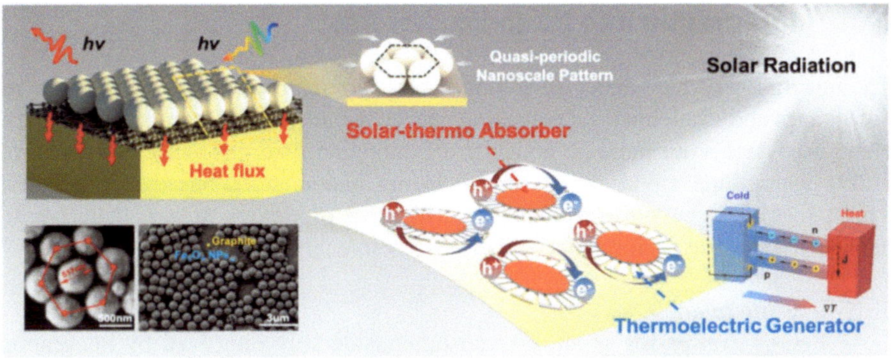

Fig. 3 Solar-thermo and thermoelectric energy conversion process. (https://phys.org/news/2023-02-nanoparticles-self-assemble-harvest-solar-energy.html)

4. *Triboelectric Energy Harvesting*

   Triboelectric nanogenerators (TENGs) harness the triboelectric effect to convert mechanical energy from material separation and contact into electrical energy. Nanomaterials, such as nanowires and nanoporous structures, have been instrumental in enhancing the performance of high-efficiency TENGs, thereby increasing their effectiveness. For instance, a silicon nanowire-based TENG prototype demonstrated a remarkable output power density, positioning it ideally for capturing energy from ambient vibrations. TENGs have found application in diverse automated systems, including wearable electronics and environmental monitoring. A flexible TENG integrated into a shoe sole exemplified electricity generation from walking movements, underscoring its potential for wearable energy harvesting devices (Chen & Wang, 2022).
5. *Photocatalytic Water Splitting*

   Photocatalytic water splitting utilizes light to catalyze the separation of water molecules into hydrogen and oxygen, offering a sustainable method for hydrogen production. Nanomaterials integrated with photocatalysts have significantly improved the efficiency of this process. Nanostructured photocatalysts, such as titanium dioxide (TiO_2), enhanced with heavy metals like platinum (Pt), have demonstrated enhanced photocatalytic activity. Pt-doped TiO_2 nanotubes, for example, have shown promising results in producing sustainable hydrogen energy under sunlight exposure (Xiang et al., 2012). Additionally, 2D materials such as molybdenum disulfide (MoS_2) have been investigated for their photocatalytic capabilities in water splitting. The large surface area and efficient charge transfer properties of MoS_2 nanosheets contribute to its effective photocatalytic activity. These advancements underscore the potential of nanomaterials in advancing photocatalytic water splitting technologies, paving the way for efficient and sustainable hydrogen production.

6 Integration into Nano-Ecosystems

The integration of nanomaterials into nano-ecosystems presents a dual prospect of advantages and challenges for bolstering energy harvesting and storage capabilities. Nano-ecosystems are intricate networks of interconnected nanoscale entities, providing an optimal environment for advancing nanomaterial-based technologies. This integration seeks to leverage the distinctive properties of nanomaterials to achieve heightened performance and efficiency in energy systems.

1. *Synergistic Interactions*

   Energy conversion processes can be considerably improved by integrating nanomaterials into nano-ecosystems, which promotes synergistic interactions. For example, combining several nanomaterials can result in composite systems with enhanced characteristics. Research has shown that hybrid nanocomposites with enhanced photocatalytic performance are produced when the strong photoactivity of titanium dioxide (TiO_2) combines with the high conductivity of graphene. In order to maximize the performance of energy-harvesting devices within nano-ecosystems, these synergistic interactions are essential. TiO_2's photoactivity guarantees efficient light absorption and conversion, while graphene's high conductivity promotes effective charge transport (Zhong et al., 2021). As a result, a composite material that possesses these qualities works better than any of its constituent parts. These advances highlight the critical need to explore and develop hybrid, nanomaterials as they hold promise for improving the efficiency of energy conversion devices. All things considered, enhancing the functionality of energy-harvesting devices by the addition of nanomaterials to nano-ecosystems is a promising approach.
2. *Nanostructured Interfaces*

   The efficient integration of nanomaterials into energy devices depends heavily on the design of nanostructured interfaces. Efficient transfer of power and decrease recombination losses are made possible by these well-designed interfaces. For example, improved cycling stability and higher lithium storage capacity have been demonstrated with lithium-ion batteries with nanostructured silicon anodes containing porous silicon nanowires. This demonstrates the critical role that nanostructured interfaces play in improving the longevity and efficiency of energy storage applications and highlights the significance of these interfaces in developing reliable, high-efficiency energy solutions.
3. *Scalability and Fabrication Techniques*

   Scalable and economical fabrication methods are necessary for the proper incorporation of nanomaterials into nano-ecosystems. Mass-producing high-quality nanomaterials has become more feasible due to scientific advancements in nanofabrication techniques, such as layer-by-layer deposition and vapor-phase deposition. These innovative methods have significantly streamlined the production process, enabling the large-scale synthesis of nanomaterials with meticulous control over their attributes and quality. This advancement ensures the consistent production of high-quality materials, which is crucial for their effective integration into various technological applications. The production process has been greatly expedited by these novel techniques, making it possible to synthesize nanomaterials on a massive scale while precisely controlling their qualities and characteristics.

For instance, scalable CVD techniques have been established to produce large-area graphene films, which are applicable in various energy technologies, including transparent conductive electrodes and flexible electronics. These advancements in fabrication techniques are crucial for the widespread adoption of nanomaterial-based energy solutions (Biswas et al., 2011). By ensuring scalability and cost-effectiveness, these methods enable the practical implementation of nanomaterials in diverse applications, thereby driving progress in energy technologies. The ability to produce high-quality nanomaterials at scale ensures their integration into energy devices, promoting enhanced performance and efficiency in energy.

4. *Multifunctional Nanodevices*

   The introduction of nanomaterials into nano-ecosystems enables the creation of multifunctional nanodevices that can perform multiple tasks at the same time. These advanced devices can harvest, store, and manage energy within the same device, significantly increasing their efficiency and utility. For example, multifunctional nanodevices have been developed that combine triboelectric nanogenerators with supercapacitors, allowing the simultaneous harvesting of mechanical energy and the storage of electricity. This dual functionality provides a self-sufficient power source ideal for wearable electronics and other applications (Barri et al., 2021). The invention of such integrated devices highlights the potential of nanomaterials to revolutionize energy solutions by combining multiple functions in a compact and efficient system, ultimately contributing to more sustainable and versatile energy technologies.

5. *Smart Grids and IoT*

   Smart grid and Internet of Things (IoT) performance can be greatly improved by incorporating nanomaterials into nano-ecosystems. Nanomaterials enhance the efficiency of power conversion along with storage appliances within smart grids, enabling greater dependability and effectiveness in energy management. For instance, the utilization of nanostructured materials in supercapacitors has shown substantial improvements in energy density and power performance. Incorporating carbon nanotubes into supercapacitor electrodes, for example, enhances these properties, making them highly suitable for integration into smart grid systems. These advancements enable better energy storage and distribution, supporting the seamless operation of IoT devices. Overall, the integration of nanomaterials into smart grid technologies represents a pivotal step toward more efficient and resilient energy management solutions.

6. *Environmental Impact and Sustainability*

   The integration of nanomaterials into nano-ecosystems has profound implications for environmental sustainability. Nanomaterials enhance the productivity of sustainable energy devices, thereby lessening dependence on fossil fuels and lowering atmospheric carbon levels. For instance, nanostructured photocatalysts have demonstrated their utility in photovoltaic water splitting for hydrogen production, achieving high rates of hydrogen generation with minimal environmental impact. This illustrates the potential of nanomaterials to enable sustainable energy solutions. By increasing the efficiency of renewable energy systems and advancing cleaner, more efficient energy sources, nanomaterials play a critical role in global initiatives aimed at addressing climate change and promoting environmental sustainability. Their ability to enhance the performance and reliability of renewable energy technologies underscores their significance in the transition to a more sustainable and environmentally friendly energy landscape.

7. *Energy Harvesting from Wearable Devices*

   The integration of nanomaterials into wearable devices is a rapidly growing area of interest within nano-ecosystems. Wearable energy harvesting devices can harvest energy from body movement, heat and ambient light, providing a continuous source of energy for health monitoring and other applications. For example, portable piezoelectric nanogenerators using PVDF nanofibers can convert biomechanical energy into electrical energy (Sobianin et al., 2022). The development shows that the integration of nanoparticles into wearable technology could lead to the production of renewable energy. These innovations enable the development of self-powered wearable devices, reducing the need for external power sources and improving the functionality and convenience of health monitoring systems. The integration of nanomaterials into wearables represents a significant step toward more sustainable and efficient energy solutions in personal technology.

7 Energy Storage Systems

Energy storage devices are essential for optimizing the application of sustainable energy sources, enhancing grid, stability and balancing supply and demand for energy. Nanomaterials play a critical role in advancing the capabilities of various battery storage devices, including batteries, supercapacitors, and fuel cells. The special contributions that nanomaterials provide to these systems are discussed in this section, along with how they affect important performance indicators like longevity, charge/discharge rates, power output, and energy capacity (Table 1). These devices for storing energy may function far more effectively due to the incorporation of nanomaterials, offering enhanced energy and power capacities, quicker storage and discharge rates, and extended lifespans. Ensuring a stable and efficient energy supply and optimizing the integration of renewable energy into the grid require these improvements.

1. *Lithium-Ion Batteries (LIBs)*

   In energy storage systems, lithium-ion cells are widely employed, particularly in electric vehicles and portable electronics, where the efficiency of these cells is critical. These cells are favored for their substantial energy storage capacity, extended operational lifespan, and reliability, making them integral to the perfor-

Table 1 Energy storage systems

Energy storage system	Nanomaterial application	Advantages
Lithium-ion batteries	Silicon nanoparticles embedded in carbon matrices to enhance capacity and cycling stability	Higher energy density and longer lifespan
Supercapacitors	Graphene and carbon nanotubes in electrode materials for increased power density and capacitance	Rapid charge/discharge cycles and durability
Sodium-ion batteries	Titanium dioxide nanospheres as anodes to improve sodium-ion storage capability	Cost-effective alternative with high efficiency
Fuel cells	Platinum nanoparticles on graphene for enhanced catalytic activity in proton exchange membranes	Efficient conversion of chemical energy to electricity
Solid-state batteries	Lithium lanthanum zirconium oxide nanowires in solid electrolytes for improved ionic conductivity	Enhanced safety and mass of energy

mance and advancement of modern electric transportation and compact consumer devices. Their superior characteristics ensure that they remain a cornerstone in the development of efficient and durable energy solutions. Nanomaterials play a crucial role in improving LIBs. For example, silicon nanoparticles embedded in a carbon matrix have been shown to significantly improve anode capacity and cycling stability. The nanostructured design increases surface area and conductivity, allowing for more efficient lithium-ion diffusion and electron transport. These advances highlight the capacity of nanomaterials for addressing key challenges in lithium-ion battery (LIB) technology, such as enhancing energy storage capacity and durability, thereby contributing to the advancement of sustainable energy solutions. By improving these critical aspects, nanomaterials play a pivotal role in the evolution of more efficient and long-lasting energy storage systems, which are essential for the broader adoption of renewable energy technologies and the reduction of reliance on nonrenewable resources.

2. *Supercapacitors*

   Supercapacitors are valued for their large density of power and fast charge/discharge capabilities, making them ideal for functions that require rapid bursts of energy. Supercapacitor electrodes often utilize nanomaterials due to their exceptional electrical conductivity and extensive surface area, which are critical for their performance. Research has shown that hybrid electrodes combining graphene with manganese oxide (MnO_2) nanosheets exhibit impressive specific capacity and cycling stability. The synergy between graphene's conductivity and MnO_2's pseudocapacitive behavior improves overall performance and highlights the role of nanomaterials in advancing supercapacitor technology. These advancements facilitate a wider integration of supercapacitors into energy storage systems for a range of consumer and industrial applications while also enhancing their efficiency and dependability.

3. *Sodium-Ion Batteries (SIBs)*

   Sodium-ion batteries (SIBs) are becoming increasingly important as a cost-effective alternative to lithium-ion batteries, particularly for applications requiring tremendous power retention. Nanomaterials play a critical role in addressing key challenges associated with SIBs, such as low energy levels and inadequate cycling reliability. For example, titanium dioxide (TiO_2) nanospheres were investigated as anodes in SIBs and showed excellent sodium-ion storage ability. This performance corresponds to the huge surface area and unique nanostructure of the nanospheres, which improve the effectiveness of sodium-ion diffusion and mitigate volume expansion problems during cycling. Such advances highlight the ability of nanomaterials to improve the performance and durability of sodium-ion batteries, thereby contributing to the advancement of more sustainable and efficient energy storage solutions.
4. *Fuel Cells*

   Integral to electrochemical processes, fuel cells facilitate the conversion of stored chemical energy into electrical power. By efficiently transforming the stored energy in fuels into usable electrical power, fuel cells play a crucial role in advancing sustainable energy technologies and reducing dependence on traditional fossil fuels. This capability underscores their importance for the development of cleaner, more efficient energy systems. By facilitating advancements in these technologies, nanomaterials are integral to creating sustainable energy solutions that reduce environmental impact and improve overall energy efficiency. These devices play a crucial role in efficient energy conversion, providing a sustainable and reliable power source for various applications. Nanomaterials, particularly noble metal nanoparticles, are essential for enhancing the endurance and catalytic activity of fuel cell electrodes. For example, platinum (Pt) nanoparticles atop nitrogen-doped graphene were explored for proton exchange membrane fuel cells (PEMFCs). For the oxygen reduction process (ORR), which is essential to fuel cell performance, the Pt nanoparticles exhibit strong catalytic activity (Long et al., 2013). Meanwhile, nitrogen-doped graphene contributes to the excellent conductivity and stability of the electrode structure, improving overall performance and longevity. These advances highlight the role of nanomaterials in optimizing fuel cell technologies and promise more efficient and sustainable energy conversion solutions for diverse applications, from wearable electronics to automobiles and industrial sectors.
5. *Solid-State Batteries*

   Solid-state batteries, employing solid electrolytes instead of liquid ones, promise enhanced safety and electrical energy. For example, nanowires comprising zirconium, lanthanum, and oxygen, integrated within a polymer framework, have demonstrated substantial enhancements in both ionic conductivity and mechanical robustness of polymer electrolytes. These advancements contribute to enhanced performance of solid-state batteries by facilitating rapid ion diffusion and reducing structural deterioration. Such advancements underscore the transformative impact of nanomaterials on revolutionizing advanced energy storage systems, positioning them as promising alternatives for next-generation uses, including portable electronics and electric mobility. By enhancing efficiency, capacity, and durability, nanomaterials facilitate substantial progress in energy storage, ultimately improving the performance and reliability of modern energy-dependent technologies. The unique properties of nanomaterials enhance the efficiency, capacity, and longevity of these batteries, paving the way for more sustainable and high-performance energy storage technologies.

6. *Redox Flow Batteries*

   Redox flow batteries (RFBs) are progressively preferred for massive energy storage because of their scalability and extended cycle life. Nanomaterials play a pivotal role in enhancing the electrochemical characteristics of RFB electrodes. Their unique properties, such as extensive surface area and superior conductivity, contribute significantly to improving the efficiency and reliability of redox flow energy systems. These attributes enhance the effectiveness of energy retention and utilization, thereby advancing the performance and reliability of these systems. This enhancement is crucial for advancing the performance and reliability of these storage solutions. Additionally, nitrogen-doped carbon nanotubes (NCNTs) have been explored as electrode materials in vanadium redox flow systems (VRFBs). These NCNTs offer a huge surface area and exhibit excellent catalytic activity during redox reactions, which significantly improves energy efficiency and maintains capacity over time in batteries. Their unique properties contribute to the overall performance and longevity of the energy storage systems, making them an invaluable component in battery technology (Wang et al., 2012a, b). Such advancements underscore the abilities of nanomaterials to address key challenges in RFB technology, including enhancing energy storage capabilities and prolonging battery lifespan, thereby contributing to the advancement of sustainable and efficient energy storage solutions for grid-scale applications.

8 Sustainable Energy Solutions

Sustainable energy solutions are critical to addressing global problems concerning energy demand and environmental impacts. Nanomaterials are an essential component of many advanced technologies used to effectively utilize sustainable energy sources. These materials are crucial for enhancing the effectiveness and performance of devices for converting and storing energy owing to their unique properties, which include expanded surface area, improved electrical transport, and customized catalytic activity. These unique attributes contribute significantly to advancing more efficient and reliable energy solutions (Lu et al., 2021). Applications for nanomaterial-based solutions are numerous and include electric vehicle batteries that use nanocomposites to maximize charge storage capacity, solar cells that utilize nanostructured photovoltaic materials for enhanced light absorption, and nanocatalysts that enable the effective conversion of biomass or carbon dioxide into renewable fuels. By leveraging these properties, nanoparticles make a significant contribution to the advancement of sustainable energy technologies and promise solutions that are both efficient and environmentally friendly, ensuring a more sustainable energy future.

1. *Photovoltaics*

   The use of solar energy to generate electricity is based primarily on photovoltaic (PV) technology. Perovskite solar cells provide evidence that nanomaterials can improve solar cell performance and efficiency. Perovskite nanocrystals are used in perovskite solar cells to increase productivity. For example, researchers have recently shown that incorporating carbon quantum dots into perovskite films enhances charge transport and light absorption, leading to increased energy conversion efficiency. These developments show how nanomaterials can be employed to improve photovoltaic technology's performance and make solar energy collecting systems more robust and effective.
2. *Photocatalysis*

   Photocatalysis uses light to accelerate chemical reactions, especially water splitting and pollutant degradation. Nanomaterials such as titanium dioxide (TiO_2) nanoparticles are preferred because of their exceptional photocatalytic activity. These nanoparticles are highly effective in promoting chemical reactions under light, making them ideal for various applications in environmental remediation and energy conversion technologies. Their efficiency in harnessing light energy and facilitating catalytic processes underscores their importance in advancing sustainable technological solutions. Research has demonstrated that applying TiO_2 with noble metal nanoparticles such as platinum significantly increases its efficiency in photocatalytic processes, especially in hydrogen production through water splitting. The integration of nanomaterials into photocatalytic technologies for environmental remediation and renewable energy applications is emphasized in this approach, offering a sustainable method to generate clean fuels.
3. *Piezoelectric Energy Harvesting*

   Piezoelectric nanoparticles are very useful for powering small, portable electronic devices because they can convert mechanical energy derived from motion and vibrations into electrical power. These nanoparticles efficiently capture kinetic energy, rendering them suitable for energy harvesting in diverse compact and wearable technologies, thereby contributing to advancements in autonomous electronic devices. For instance, zinc oxide (ZnO) nanowires have been investigated in flexible piezoelectric nanogenerators. These nanowires exhibit excellent piezoelectric properties, making them ideal for converting mechanical energy into electrical energy in flexible and wearable electronic devices.

   This innovative approach highlights the potential of ZnO nanowires in developing advanced, sustainable energy harvesting technologies. These nanowires exhibit excellent piezoelectric properties and enable efficient energy harvesting from everyday activities. Some research highlights the ability of piezoelectric nanomaterials to contribute to the growth of self-sufficient electronic devices and provide sustainable energy solutions that harness ambient mechanical energy for practical applications.

4. *Thermoelectric Energy Harvesting*

   Thermoelectric materials have the unique ability to convert temperature differentials directly into electrical energy. Nanomaterials serve an essential part in enhancing thermoelectric performance by improving the Seebeck coefficient and reducing thermal conductivity. For example, research on bismuth telluride (Bi_2Te_3) nanoplates has shown that nanostructuring can significantly enhance the thermoelectric figure of merit. This improvement makes nanomaterial-enhanced thermoelectric materials more efficient for applications such as waste heat recovery, offering promising solutions for converting otherwise wasted heat into useful electrical power.

5. *Hydrogen Storage*

   Hydrogen is recognized as a clean fuel, yet efficient storage remains a significant challenge. Nanomaterials such as MOFs provide solutions by offering large surface areas and adjustable pore sizes, facilitating efficient hydrogen storage. For instance, research on MOF-5 has highlighted its capability to achieve high hydrogen uptake capacity as a result of its extensive surface and porous structure. These attributes enable substantial hydrogen adsorption, positioning MOF-5 as a promising material for hydrogen storage applications. These properties allow for substantial hydrogen adsorption, making MOF-5 a promising material for hydrogen storage applications. These properties make MOFs promising candidates for sustainable hydrogen storage solutions, addressing the demand for effective storage methods to promote the widespread adoption of hydrogen as a clean energy source.

6. *Bioenergy*

   Bioenergy production, including biofuels and biogas, can benefit significantly from nanomaterials. For example, nanocatalysts are essential for enhancing the efficiency of systems that convert biomass. Their unique properties significantly improve reaction rates and selectivity, making them crucial for optimizing biomass conversion processes. Research on nickel nanoparticles as catalysts in hydrothermal liquefaction of biomass has shown promising results, leading to higher yields of bio-oil, an alternate source of energy. These discoveries illustrate how nanomaterials can enhance bioenergy technologies by providing more sustainable and effective methods for converting biomass production into valuable energy products. This contributes to reducing dependency on fossil fuels and diversifying renewable energy sources (Markandan & Chai, 2022).

7. *Smart Grids*

   The incorporation of nanomaterials into smart grid technologies presents a substantial opportunity to enhance energy distribution and management. In smart grids, nanomaterials like graphene can greatly enhance sensor and energy storage performance. Fast energy delivery and storage are made possible by graphene-based supercapacitors, as demonstrated by research, and this is essential for effectively balancing supply and demand in smart grid applications. Based on these advancements, future commercial and urban infrastructures can implement more efficient and eco-friendly energy management strategies, utilizing nanomaterials to optimize the efficiency and reliability of smart grid systems.

9 Conclusion

In conclusion, nanomaterial integration in energy systems holds transformative potential across energy harvesting, storage, and sustainability. The MOFs enhance batteries and supercapacitors with improved ion diffusion and cobalt-based MOFs showing excellent stability. Photovoltaics, piezoelectricity, thermoelectricity, triboelectricity, and photocatalytic water splitting benefit from nanomaterials like quantum dots, zinc oxide nanowires, and bismuth telluride nanowires, boosting efficiency. Nano-ecosystems foster synergies, optimized interfaces, scalable fabrication, and smart grid applications. Nanomaterials also enhance sustainability by improving energy device efficiency, reducing fossil fuel reliance, and enabling cleaner solutions in supercapacitors, batteries, and fuel cells with graphene, TiO_2 nanospheres, and platinum nanoparticles. They advance photovoltaics, photocatalysis, piezoelectrics, thermoelectrics, MOFs for hydrogen storage, and nanocatalysts for bioenergy, addressing global energy challenges and promoting a more efficient, sustainable energy landscape.

References

Armand, M., & Tarascon, J. M. (2008). Building better batteries. *Nature, 451*, 652–657.

Atwater, H. A., & Polman, A. (2010). Plasmonics for improved photovoltaic devices. *Nature Materials, 9*, 205–213.

Bae, W. K., Park, Y.-S., Lim, J., Lee, D., Padilha, L. A., McDaniel, H., Robel, I., Lee, C., Pietryga, J. M., & Klimov, V. I. (2013). Controlling the influence of Auger recombination on the performance of quantum-dot light-emitting diodes. *Nature Communications, 4*(1), 1–7.

Barri, K., Jiao, P., Zhang, Q., Chen, J., Wang, Z. L., & Alavi, A. H. (2021). Multifunctional meta-tribomaterial nanogenerators for energy harvesting and active sensing. *Nano Energy, 86*, 106074–106079.

Baughman, R. H., Zakhidov, A. A., & Deheer, W. A. (2002). Carbon nanotubes—The route toward applications. *Science, 297*, 787–792.

Bera, D., Qian, L., Tseng, T.-K., & Holloway, P. H. (2010). Quantum dots and their multimodal applications: A review. *Materials, 3*(4), 2260–2345.

Berman, D., Jastrzebska, A., Papi, M., & Rosenkranz, A. (2024). Introduction to 2D materials and their applications. *RSC Advances, 14*(24), 17234–17235.

Bhadwal, N., Mrad, R. B., & Behdinan, K. (2023). Review of zinc oxide piezoelectric nanogenerators: Piezoelectric properties, composite structures and power output. *Sensors, 23*(8), 3859–3882.

Bhushan, B. (2010). Self-assembled monolayers for nanotribology and surface protection. In *Nanotribology and nanomechanics II* (Vol. 39, pp. 1309–1346). https://doi.org/10.1007/978-3-642-02525-9_39

Biswas, A., Bayer, I. S., Biris, A. S., Wang, T., Dervishi, E., & Faupel, F. (2011). Advances in top–down and bottom–up surface nanofabrication: Techniques, applications & future prospects. *Advances in Colloid and Interface Science, 170*(2), 1–88.

Chan, C. K., Peng, H., Liu, G., McIlwrath, K., Zhang, X. F., Huggins, R. A., & Cui, Y. (2008). High-performance lithium battery anodes using silicon nanowires. *Nature Nanotechnology, 3*, 31–35.

Chen, B., & Wang, Z. L. (2022). Toward a new era of sustainable energy: Advanced triboelectric nanogenerator for harvesting high entropy energy. *Small, 18*(43), 2107034–2107050.

Chen, Z., Ren, W., Gao, L., Liu, B., Pei, S., & Cheng, H.-M. (2011). Three-dimensional flexible and conductive interconnected graphene networks grown by chemical vapour deposition. *Nature Materials, 10*, 424–428.

Chhowalla, M., Jena, D., & Zhang, H. (2016). Two-dimensional semiconductors for transistors. *Nature Reviews Materials, 1*, 16052.

Debruler, C., Wu, W., Cox, K., Vanness, B., & Liu, T. L. (2020). Integrated saltwater desalination and energy storage through a pH neutral aqueous organic redox flow battery. *Advanced Functional Materials, 30*(24), 200038–2000385.

Dippong, T. (2023). Synthesis, physicochemical characterization and applications of advanced nanomaterials. *Materials, 16*(4), 1674.

Ghatiwala, A. (2020). Tuning the optical properties of quantum dots to increase the efficiency of solar cells. *International Journal of Engineering Research, 9*(9), 108–119.

Hochbaum, A. I., Chen, R., Delgado, R. D., Liang, W., Garnett, E. C., Najarian, M., Majumdar, A., & Yang, P. (2008). Enhanced thermoelectric performance of rough silicon nanowires. *Nature, 451*, 163–167.

Jeong, S., Kim, M. W., Jo, Y.-R., Leem, Y.-C., Hong, W.-K., Kim, B.-J., & Park, S.-J. (2016). High-performance photoresponsivity and electrical transport of laterally-grown ZnO/ZnS core/shell nanowires by the piezotronic and piezo-phototronic effect. *Nano Energy, 30*, 208–216.

Kamat, P. V. (2013). Quantum dot solar cells. The next big thing in photovoltaics. *The Journal of Physical Chemistry Letters, 4*(6), 908–918.

Kattan, P. I. (2011). Ratio of surface area to volume in nanotechnology and nanoscience. *Basic Nanomechanics Series*, 1–32.

Kovalenko, M. V., Protesescu, L., & Bodnarchuk, M. I. (2017). Properties and potential optoelectronic applications of lead halide perovskite nanocrystals. *Science, 358*, 6364745–6364750.

Lee, C., Wei, X., Kysar, J. W., & Hone, J. (2008). Measurement of the elastic properties and intrinsic strength of monolayer graphene. *Science, 321*, 385–388.

Lim, S., Lee, D. H., Choi, H., Choi, Y., Lee, D. G., Cho, S. B., Ko, S., Choi, J., Kim, Y., & Park, T. (2022). High-performance perovskite quantum dot solar cells enabled by incorporation with dimensionally engineered organic semiconductor. *Nano-Micro Letters, 14*, 204–218.

Liu, C., Li, F., Ma, L.-P., & Cheng, H.-M. (2010). Advanced materials for energy storage. *Advanced Materials, 22*(8), 28–62.

Long, N. V., Yang, Y., Thi, C. M., Van Minh, N., Cao, Y., & Nogami, M. (2013). The development of mixture, alloy, and core-shell nanocatalysts with nanomaterial supports for energy conversion in low-temperature fuel cells. *Nano Energy, 2*(5), 636–676.

Lu, H., Tournet, J., Dastafkan, K., Liu, Y., Ng, Y. H., Karuturi, S. K., Zhao, C., & Yin, Z. (2021). Noble-metal-free multicomponent nanointegration for sustainable energy conversion. *Chemical Reviews, 121*(17), 10271–10366.

Mak, K. F., Lee, C., Hone, J., Shan, J., & Heinz, T. F. (2010). Atomically thin MoS2: A new direct-gap semiconductor. *Physical Review Letters, 105*(13), 1–15.

Markandan, K., & Chai, W. S. (2022). Perspectives on nanomaterials and nanotechnology for sustainable bioenergy generation. *Materials, 15*(21), 7769–7789.

Mubeen, S., Lee, J., Singh, N., Krämer, S., Stucky, G. D., & Moskovits, M. (2013). An autonomous photosynthetic device in which all charge carriers derive from surface plasmons. *Nature Nanotechnology, 8*, 247–251.

Navya, P. N., & Daima, H. K. (2016). Rational engineering of physicochemical properties of nanomaterials for biomedical applications with nanotoxicological perspectives. *Nano Convergence, 3*, 1–34.

Novoselov, K. S., Geim, A. K., Morozov, S. V., Jiang, D., Zhang, Y., Dubonos, S. V., Grigorieva, I. V., & Firsov, A. A. (2004). Electric field effect in atomically thin carbon films. *Science, 306*(5696), 666–669.

Protesescu, L., Yakunin, S., Bodnarchuk, M. I., Krieg, F., Caputo, R., Hendon, C. H., Yang, R. X., Walsh, A., & Kovalenko, M. V. (2015). Nanocrystals of cesium lead halide perovskites (CsPbX3, X = Cl, Br, and I): Novel optoelectronic materials showing bright emission with wide color gamut. *Nano Letters, 15*(6), 3692–3696.

Roco, M. C. (2011). The long view of nanotechnology development: The National Nanotechnology Initiative at 10 years. *Journal of Nanoparticle Research, 13*, 427–445.

Serrano, E., Rus, G., & Garcia-Martinez, J. (2009). Nanotechnology for sustainable energy. *Renewable and Sustainable Energy Reviews, 13*(9), 2373–2384.

Sobianin, I., Psoma, S. D., & Tourlidakis, A. (2022). Recent advances in energy harvesting from the human body for biomedical applications. *Energies, 15*(21), 7959–7983.

Somwanshi, S. B., Somvanshi, S. B., & Kharat, P. B. (2020). Nanocatalyst: A brief review on synthesis to applications. *Journal of Physics, 1644*, 1–8.

Srinivasan, S., Kannan, A. M., Kothurkar, N. K., Khalil, Y. F., & Kuravi, S. (2015). Nanomaterials for energy and environmental applications. *Journal of Nanomaterials*, 979026. https://doi.org/10.1155/2015/979026

Tian, B., Zheng, X., Kempa, T. J., Fang, Y., Nanfang, Y., Guihua, Y., Huang, J., & Lieber, C. M. (2007). Coaxial silicon nanowires as solar cells and nanoelectronic power sources. *Nature, 449*, 885–889.

Varghese, O. K., Paulose, M., LaTempa, T. J., & Grimes, C. A. (2009). High-rate solar photocatalytic conversion of CO2 and water vapor to hydrocarbon fuels. *Nano Letters, 9*(2), 731–737.

Wang, Z. L., & Wu, W. (2012). Nanotechnology-enabled energy harvesting for self-powered micro-nanosystems. *A Journal of German Chemical Society, 51*(47), 11700–11721.

Wang, Q. H., Kalantar-Zadeh, K., Kis, A., Coleman, J. N., & Strano, M. S. (2012a). Electronics and optoelectronics of two-dimensional transition metal dichalcogenides. *Nature Nanotechnology, 7*, 699–712.

Wang, S., Zhao, X., Cochell, T., & Manthiram, A. (2012b). Nitrogen-doped carbon nanotube/graphite felts as advanced electrode materials for vanadium redox flow batteries. *The Journal of Physical Chemistry Letters, 3*(16), 2160–2167.

Xia, Y., Yang, P., Sun, Y., Wu, Y., Mayers, B., Gates, B., Yin, Y., Kim, F., & Yan, H. (2003). One-dimensional nanostructures: Synthesis, characterization, and applications. *Advanced Materials, 15*(5), 353–389.

Xiang, Q., Jiaguo, Y., & Jaroniec, M. (2012). Synergetic effect of MoS2 and graphene as cocatalysts for enhanced photocatalytic H2 production activity of TiO2 nanoparticles. *Journal of the American Chemical Society, 134*(15), 6575–6578.

Yang, D., Shi, X.-L., Li, M., Nisar, M., Mansoor, A., Chen, S., Chen, Y., Li, F., Ma, H., Liang, G. X., Zhang, X., Liu, W., Fan, P., Zheng, Z., & Chen, Z.-G. (2024). Flexible power generators by Ag2Se thin films with record-high thermoelectric performance. *Nature Communications, 15*, 923–934.

Zhang, H., Xia, X., Li, W., Zeng, J., Dai, Y., Yang, D., & Xia, Y. (2010). Facile synthesis of five-fold twinned, starfish-like rhodium nanocrystals by eliminating oxidative etching with a chloride-free precursor. *A Journal of German Chemical Society, 49*(31), 5296–5300.

Zhang, Q., Uchaker, E., Candelaria, S. L., & Cao, G. (2013). Nanomaterials for energy conversion and storage. *Chemical Society Reviews, 42*(7), 3127–3136.

Zhong, Y. L., Basu, S. N., & Sun, Z. (2021). Nanomaterials and composites for energy conversion and storage. *The Journal of the Minerals, Metals & Materials Society (TMS), 73*, 2752–2753.

Zhu, Y., Murali, S., Cai, W., Li, X., Suk, J. W., Potts, J. R., & Ruoff, R. S. (2010). Graphene and graphene oxide: Synthesis, properties, and applications. *Advanced Materials, 22*(35), 3906–3924.

Zhu, Y., Murali, S., Cai, W., Li, X., Suk, J. W., Potts, J. R., & Ruoff, R. S. (2012). Graphene and graphene oxide: Synthesis, properties, and applications. *Advanced Materials, 22*(35), 3906–3924.

Biomimetic Nanodevices for Environmental Sensing and Monitoring

M. Abiyoga and P. Saravanakumari

Abstract

Biomimicry involves the action of natural elements, structures, and processes to develop novel devices with unique features and functionalities similar to biomolecules. The combination of biomimicry with nanotechnology has led to the development of nanodevices, including biosensors. The application of nanomaterials in biosensors has become an important topic as they provide accurate, fast, and reliable results in various fields, including agriculture, remediation, food safety, and medicine. Biosensors can detect analytes in low concentrations due to their sensitivity at the nanoscale level. They play a commendable role in environmental diagnostics and targeted treatments of critical conditions. In agriculture, they ensure the monitoring and maintenance of biotic and abiotic factors for crop cultivation. They also assess pollutant levels and optimize detoxification in the environment. The operation of these devices is very convenient as they are portable, low-cost, and can be customized based on prevailing conditions. These devices have a significant role in diagnosis of plant infections, customized medication, heavy metal toxicity analysis, and other contaminant determination and clean-up. They have also proven to be the best tool for hazard analysis and environmental monitoring. Furthermore, these devices offer potential for future exploration and advancements. This review briefly discusses biomimetic nanodevices, their types, and advancements in sensing and monitoring techniques in various fields with their applications.

Keywords

Biomolecules · Detoxifications · Hazard analysis · Nanodevices · Nanosensors

---

M. Abiyoga · P. Saravanakumari (✉)
Department of Microbiology, Rathnavel Subramaniam (RVS) College of Arts and Science, Coimbatore, Tamil Nadu, India

D. K. Gupta et al. (eds.), *Nano-Ecosystems*,
https://doi.org/10.1007/978-981-95-5033-3_5

1 Introduction

Transformation toward the use of natural compounds has become very essential in the current scenario. The organic revolution has ceased the use of chemically synthesized materials, which tremendously affected the livelihood of the food chain. Magnificent usage of chemical pesticides, fertilizers, and plastics affected the environment and human health terrifically. The number of cancer cases is increasing, and about 20, 000 million individuals are affected every year. The most common is breast cancer (12.4%) in women and prostate cancer (7.3%) in men, and it is predicted the emergence of new cancer cases will be around 35 million in 2050 (Bray et al., 2024). These complications in human health are associated with disorders in the function of biomolecules at the cellular level. The roles of these biomolecules are irreplaceable. But the advancements in technology have brought down alternate molecules/particles like the naturally existing agents with no side effects and specific action similar to natural biomolecules were identified (Yu et al., 2016; Felgueiras, 2021). These compounds are similar in process or structure or function and are used for addressing environmental or health monitoring and solutions to environmental issues.

One of the major issues in the application of any artificial chemical excessively or continuous usage results in accumulation and causes biomagnification upon entry into a human or animal or plant system that ends up in genotypic and phenotypic changes in the system. Commonly used chemical substances also become potential hazardous materials while assessing the long-term impacts of these pollutants on air, soil, and water. Therefore, it affects plants, animals, and human health (Pathak et al., 2022). These consequences end in an unmanageable level of slow pile-up to a toxic concentration that deteriorates the environment. Regular testing of pollutant levels and continuous monitoring and improving current strategies to protect environmental health is very essential. Hence, usage of any compound at minimum or nanoscale level is preferred in the environment. Moreover, nanotechnology involves green synthesis of biocompatible and bioactive nanodevices. Plant-based synthesis of nanodevices is more stable, recyclable, and less hazardous as they are capped and stabilized by phytochemicals. Such stable nanodevices are used for clean-up of mixed pollutants such as heavy metals and persistent organic pollutants on an average of 82.26% without affecting living organisms in the environment (Roy et al., 2021; Abuzeid et al., 2023). So, nanoparticles are gaining attention to constructing systems/devices mimicking biological systems or biomolecules to use safely without any alarming side effects. The usage of nanoparticles is well regulated based on the hazard severity and exposure potential in the environment by in silico and in vitro methods for safe release in the environment and disposal, so are considered safe for use (Gupta & Xie, 2018). The nanodevices are constructed of nanoparticles that mimic biological material and perform its specific function like monitoring tasks. These devices at nanoscale level are precise and specific, with the utmost sensitivity used for sensing, monitoring, delivery of deficit molecules, and treatment of prevailing weak conditions. Commercially, different nanoparticles are produced and used as nanosensors, nanopesticide, and nanofertilizers with multiple beneficial roles in the environment that wires organic farming.

2 Environmental Issues

There are a huge number of environmental issues raised in various ways in day-to-day life that need special attention and solved specifically. These issues include global warming due to combustion of fossil fuels, emission of greenhouse gases, food waste due to fast deterioration of fruits and vegetables (60 million tons per year in America), loss of biodiversity due to accumulation of toxic chemicals (microplastics—14 million tons/year), loss of indigenous beneficial microbes, water pollution, air pollution, changes in environmental conditions like pH, oxygen concentration, temperature, humidity, acidity, alkalinity of the soil, etc.. These factors need to be monitored regularly to estimate excess accumulation of toxic compounds in the environment, which is a severe threat to the crop cultivation in agriculture and leads to loss of biological community. The environment is under serious threat due to the accumulation of exacerbated amount of pollutants that exert extreme stress on plants, loss of microbial diversity, and affect the food web, which ultimately affect the food chain and biodiversity loss. Severe climate changes cause drastic effects on food production. In response to climate change, biodiversity collapse is increasingly driving the emergence and spread of infectious diseases, food loss, and waste accumulation, causing health at the greatest risk with direct and indirect health consequences (Williams et al., 2023). This is the need of the hour to be regularly monitored and rectified to overcome environmental issues.

3 Nanodevices

Nanodevices or nanomachines are nanoparticles with specific functions. Biological molecules are systematically arranged, and the assembly of atoms is brought forward by magnetic waves or mechanical assembly in which design or functionality is similar to existing biological systems. Simply modifying the atomic-level mobility, the magnetic effects, melting point, and interaction with molecules can be changed without altering the configuration. They are referred to as biomimetics. These devices are inserted in place of living cells to play various roles and safeguard the environment. Additionally, these devices are energy efficient, generated from renewable, easily available raw materials, made-to-order, and highly compatible (Harun-Ur-Rashid et al., 2023).

The nanodevices are used as batteries to store and release energy in a stabilized way, fuel efficiency of motors, catalysts, pollution control or emission control, membrane filters, solar cells, molecular engines, diagnosis and drug delivery, orthopedic implants, thermal conversion units, photodetectors, food production, food preservation, and drug delivery in human diseases. Also used in nanophotonic devices, biosensing devices, chemical sensing LASER devices, insulations, and in bioelectronics, nanorobots constructed from nanodevices are used to perform precise surgery (especially in the brain or eyes).

4 Construction of Nanodevices

Each atom has a specific force field called quantum mechanics, which is calculated using ab initio method in static conditions. Molecular dynamics is used to measure the specific molecular interaction in motion. Using these calculations, a molecular mimic is developed with nanoparticles. These nanodevices are molecular modeled/customized using the concept of artificial neural networks using the genetic algorithms, MATLAB, fuzzy systems, Furiousatoms, Aionics.io, Ascalaph designer, CAD, Cadnano, CoNTub, CORBA, CP2K, nextnano, Ninithi, nanoxplorer, SAMSON, etc., like applications to receive biological simulations cost efficiently at faster rate in a super computer (high-performance computer) or clusters of computers or grid computers (dispersed computers integrated by a network). Upon receiving the molecular properties, the nanodevices designed from an atom-for-atom assembly of structures or atoms are added one after another upon a scaffold to get a molding or by achieving an etching process. From atom-by-atom addition, done using spinning, supercritical fluid synthesis, laser pyrolysis, chemical vapor deposition, molecular condensation, chemical reduction, and green synthesis, size reduction from larger particle to nanoscale level done by mechanical milling, laser ablation, sputtering, electro-explosion, etc. (Baig et al., 2021).

Different types of nanodevices are developed to solve environmental issues. Nanosensors are used to detect pollutants in the environment and ensure the safety and quality of food products. Advanced sensors of nanophotonic devices are used for environmental monitoring, performing minimally invasive surgery, or assembling nanoscale components for manufacturing and sensing. First, synthesized nanodevices are simple nanoparticles used for specific functions. In 1914, Bridgman produced black phosphorus used for semiconductor properties with a direct band-gap of 0.35 eV and charge carrier mobilities of 10,000 cm^2/V, which enables photothermal conversion efficiency, thereby used in solar cell, nanolasers, tissue engineering, and other biomedical applications (Ghazi et al., 2023). In soft contact lenses, a polymeric nanodevice made of pullulan and polycaprolactone bound by high-affinity binding with antibiotic, ciprofloxacin, incorporated for controlled and sustained release to prevent infections caused by *Pseudomonas aeruginosa* and *Staphylococcus aureus* was incorporated during the usage (Garhwal et al., 2012).

Green synthesis of organelles like nanodevices is developed at a high rate of synthesis of products under controlled conditions by synthetic biology. A nanoreactor mimicking 60 subunits of dodecahedron of a viral head self-assembled to express highly stable (thermal and pH) biocatalyst (+)-γ-lactamase contiguously from ketohydroxyglutarate aldolase gene was obtained from *Thermatoga maritime* (Li et al., 2018). The enzyme, (+)-γ-lactamase, is used in the medical industry to prepare viral drugs and synthesis of antibiotics to control multidrug-resistant pathogens. Nanodevices work at a high speed with reduced power, high sensitivity, low-cost, high-throughput detection, easier production, high-portability, nonvolatile, radiation-resistant, instruments used in transistorless computing, nanolithography, nanofabrication, and nanogenerators instruments to power-up devices in electronic instrument.

Now fields are applied in different fields efficiently for precise results. Nanodevices have the advantage of measuring the analyte in a single component assay and as a multicomponent assay. Thus, nanoparticles are used synergistically with potential applications in nanosensing, multimodal imaging, surface-enhanced Raman scattering, heterologous catalysis, and drug delivery. Highly portable, low-cost, and quantitatively accurately measuring DNA-based nanosensor was developed to detect the specific analyte such as environmental contaminants like uranium (9.1 nM sensitivity), diverse microbes detected based on their DNA sequence using aptamers/DNAzymes/aptazymes, and cocaine at 3.4 μM minimum concentration. When an analyte is detected, an indicator like invertase enzymes is released, which is proportional to the quantity of the analyte. The released invertase converts sucrose into glucose, which develops color and can be easily detected in colorimetry through color development by reaction with a detector reagent (Xiang & Lu, 2011). Even in the presence of interfering molecules, it precisely measures the least concentration of analyte by sensors. Such multicomponent nanodevices are formulated by means of co-precipitation, microemulsion, molecular interactions through bonds, sol–gel, and hydrothermal/solvothermal methods. For example, on natural polymers, other components are loaded to produce multicomponent nanodevices. Gluconic acid, folic acid, or glycan, dextran, chitosan, alginate, polyethylene glycol (PEG), polyacrylic acid, and metals such as silica or gold are used for multicomponent nanodevice formulation. For example, caffeine folic acid-chitosan–methotrexate is used in targeted breast cancer therapy without any side effects (Torres-Pérez et al., 2020; Díez et al., 2022; Qin et al., 2024). These molecules have high specificity, versatility, low immunogenicity, and mimic naturally existing molecules.

5 Classification of Nanodevices

Nanodevices are formulated according to the functional properties and availability of specific molecular components, based on the availability of the reaction platform, control mechanism, and engineering process that is provided (Baysal, 2020). Each nanoparticle is electrically charged, differs in semiconductor property, thermal and electrical conductivity, mechanical properties, and unique in its characteristics and entirely absent in properties of the bulk chemical part (Baig et al., 2021) (Table 1).

Table 1 Different methods of classification of nanodevices

Classification of nanodevices based on their properties		
Polymeric composition	Structures	Function
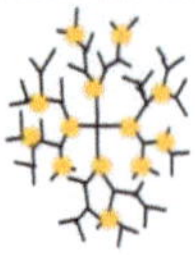 Dendrimers 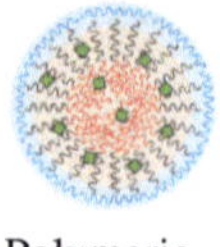 Polymeric micelles	Carbon nanotubes 	Nanoadsorbents 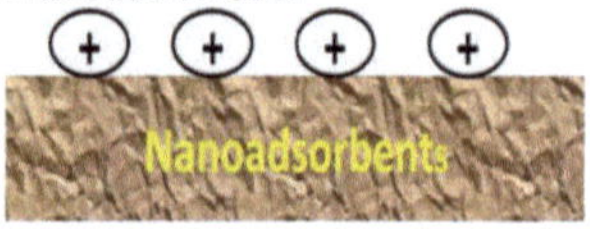
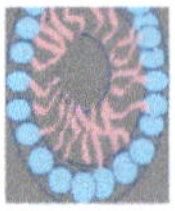 Niosomes  Nanocapsules	Dendrimers 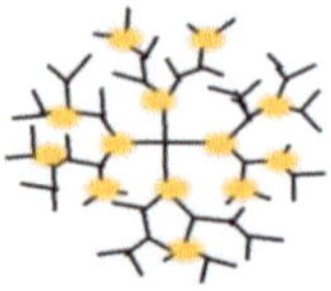	Nanofilteration 
 Nanospheres	Nanocrystals 	Nanophotocatalysis 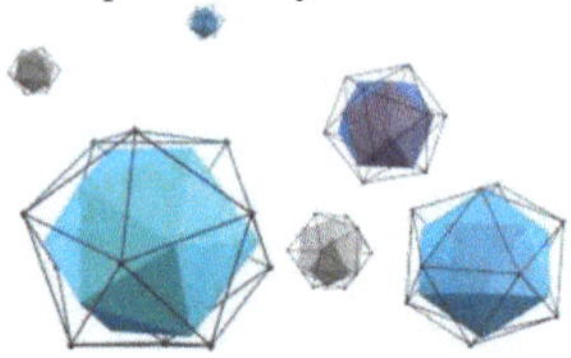
Lipid base	Nanoparticles 	Magnetic nanoparticles 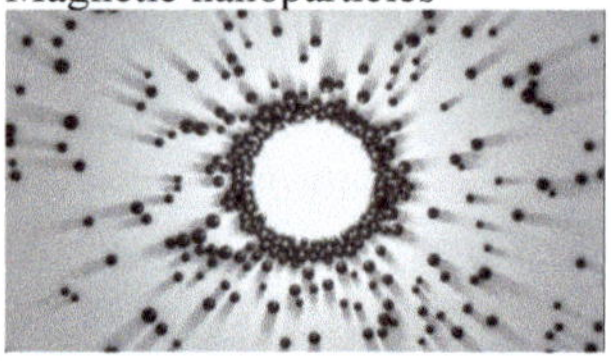
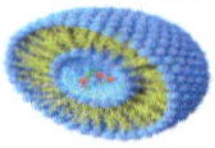 Liposomes 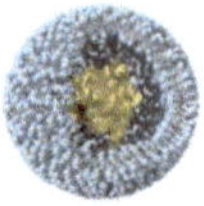 Lipid vesicles	Nanoshell 	Nanosensors 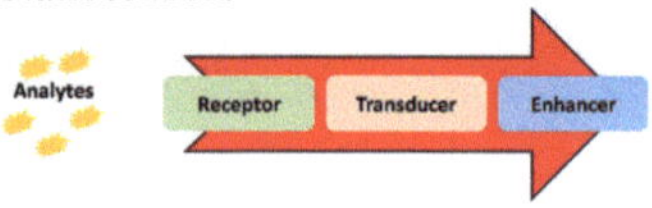
Metal based Metal alloids 	Nanowires 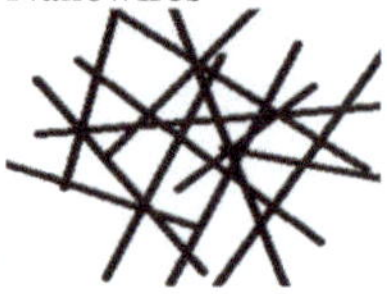	
Carbon nanotubes 	Quantum dots 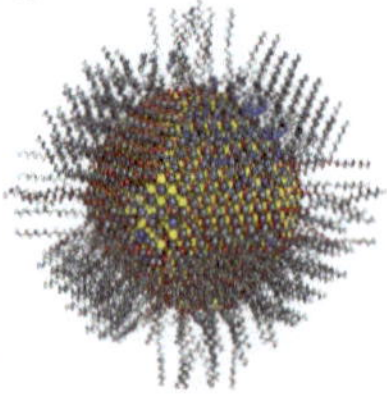	
Gold nanoshells 		
Ceramic based 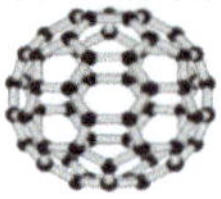 MnO_2,$MgNO_2$		

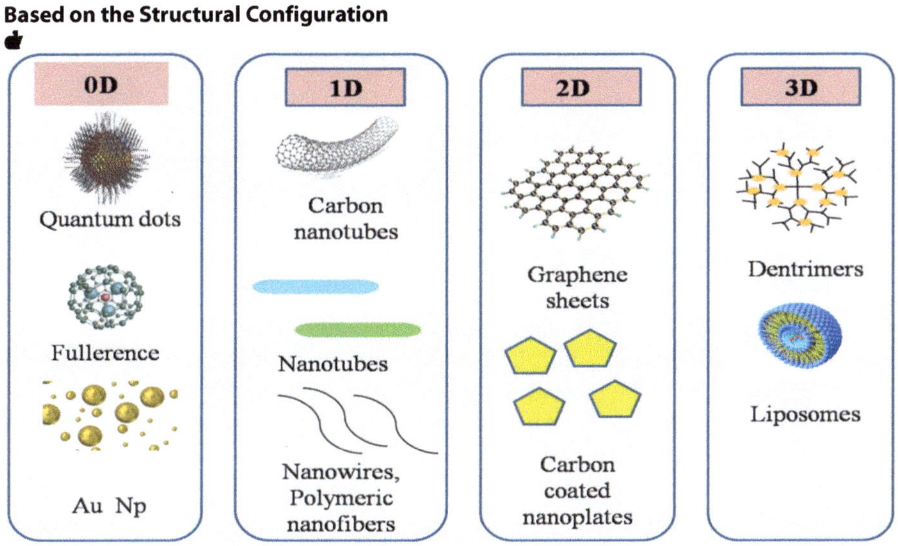

6 Nanodevices in Environmental Monitoring

Nanodevices have vast applications in environmental sensing and remediation, including detection of pollutants and their concentration in soil, water, and air, remediation of pollutants, prevention of pollution, energy production and conservation, and carbon capture. The compound to be detected is referred to as an analyte. Presence of an analyte in an environment is sensed by receptors present on the nanodevice, which react with the analyte; the concentration of end product formation depends on the quantity of the analyte. The end product acts as the indictor by emitting light or changing the color to detect the presence of an analyte visually (Fig. 1).

To reduce engine noise in Shinkansen bullet trains, the modified front end is as sharp as the streamlined beak of a kingfisher, which reduces noise, fuel economy, and efficient function (increased speed). This biomimicry inspired us to take up natural designs to construct efficient biosensor systems.

7 Detection of Soil Contaminants and Bioremediation

Nanodevices are emerging as a highly selective technology used to detect environmental pollutants in recent decades. Contamination of soil by varieties of chemicals from crop fields such as pesticides, herbicides, industrial discharges, plastics, and industrial dyes leads to the accumulation of pollutants in land, which leads to a threat to food security and serious environmental risks. Nanodevices are used to detect heavy metals, herbicides, dyes, and persistent organic pollutants, including pesticides in soil, especially in agricultural soil (Dhanapal et al., 2024). Carbon

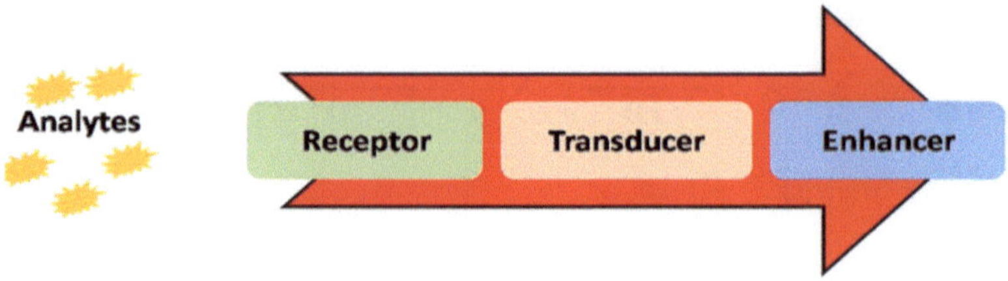

Fig. 1 Simple representation of biosensor device

nanotubes detect the metal contaminants in soil and oxidize. Similarly, quantum dots such as titanium oxide and ferric oxide immobilize the toxic cadmium and arsenic from soil. Even the presence of pollutants in ppm or ppb levels is also accurately detected. About 88.34% of hydrocarbons are removed from polluted environments, and organic dyes are adsorbed by gold or silver nanodevices. Polymeric nanodevices, sodium carboxy methyl cellulose, stabilized toxic zero valent iron into hexavalent nontoxic form.

Nanodevices have been developed as great analytical devices for the detection of the dispersion of agricultural chemicals and heavy metals in excess. The pollutants present in the soil were identified by nanodevices or nanosensors through its complementary biomolecules of the device due to their high selective properties that detect the soil pollutants accurately. Interest in these nanodevices is increased by their efficiency in the detection of pollutants without any high-end-equipped instruments from laboratories. It was possible to identify the prevalence of plant pathogens in crop cultivation soil using nanosensors containing ssDNA or RNA (Kapil et al., 2022). ssDNA, dsDNA, aptamers [single-stranded DNA/RNA loops with a high affinity to bind with specific biomolecule (analyte)], G-quadraplex DNA (guanine-rich four stranded DNA sequence, and complementary DNA) are used for the detection of heavy metals, pesticides, and toxic pollutants. These are especially devised to detect contamination by pesticide acetamiprid used to control aphids. Accumulation of acetamiprid in tea and celery leaf was identified by continuous monitoring using ssDNA apatamer conjugated with gold nanoparticle. Colorless ssDNA-Au nanoparticles disassociated with red fluorescence by the formation of acetamiprid salt. Development of red color is used to identify pesticide accumulation qualitatively and quantitatively (Yang et al., 2019).

Pesticides are used in crop fields to prevent insect infestation in crops and improve crop yield. Higher usage of pesticides increases the accumulation of pesticide residues in soil and water sources, which directly affects the human health and environment. Detection of pesticide residues in the environment has attracted tremendous attention in academic research, industries, and government. Various techniques are already available to recognize the presence of chemical pollutants, including liquid chromatography-mass spectrophotometry (LC-MS), gas chromatography (GC), enzyme-linked immune absorbent assay (ELISA), and high-performance liquid chromatography (HPLC). Due to the disadvantages of the high cost and requirement of specific chemicals, these methods are replaced by the usage of nanotechnology-based biosensors in pollutant analysis. Thiram was used as a pesticide and fungicide to prevent crop infection. Accumulation of this fungicide leads to serious environmental pollution. To detect the accumulation of this pollutant, Tween 80-capped Cu nanoparticles were tested by Nguyen Tuan Anh et al. in 2021. Tween 80-capped Cu nanoparticles were prepared using tween 80 as a stabilizer and sodium borohydride as a reducing agent. Thiram interacts with tween 80, creates sulfide bond cleaved from thiram, then binds to Cu nanoparticles and creates a response within 5 min. Therefore, the incidence of thiram is indicated. Gold, silver, and copper nanodevices used for the detection of pesticide, compared to other Au–Ag nanoparticles, and copper nanoparticles would be cost-effective, and multiple pesticide residues could be identified.

Titanium oxide, manganous oxide, or zinc oxide semiconductor nanodevices detect the presence of organic pollutants (herbicides or pesticides) and hydrocarbons by accelerated electrons and, in the presence of sunlight, oxidizes compounds into harmless end products such as carbon dioxide and water. Transgenic plants are developed using plasmids as carriers of target genes. Plasmids, like efficient hydroxy apatite (nanodevice) carrier molecules with reporter gene (green fluorescent protein), developed and competently transferred in plants like *Arabidopsis* sp. Transformed genes are ably expressed similar to plasmid, thereby useful in the development of environmental stress (drought, pesticide/herbicide)-resistant plant varieties (Izuegbunam et al., 2021).

Difference in pH is detected using silicon nanowire, where electrical conductivity is detected by nanoelectrode arrays, humidity indexed by surface acoustic wave (SAW) sensors referred to as cantilevers, accumulation of phosphate detected by nanoenzymatic sensors, nitrate concentration by doped polypyrrole nanowires, dissolved oxygen level detected by resistance in carbon nanotube, volatile organic compounds detected by SAW sensors, polymer-coated nanocantilevers, proteins accumulation detected by immune-functionalized cantilever and nanowires, and specific DNA detected by cDNA-coated nanoparticles. Nanoscale plant polymers commonly opt for environmental applications as microbial polymers have safety issues. These nanodevices mimic plant components, thereby easily penetrating through plasmodesmata into plant cells and enhancing the expression of heavy metal resistance genes. For example, cadmium toxicity resistance in plants is enhanced by the application of Fe3O4 nanoparticles (Kumar et al., 2021).

8 Detection of Water Contaminants and Bioremediation

Biomimetic snowflake-shaped magnetic nanostructures used for the detection and removal of heavy metals from soil and water were established to protect environmental pollution. Arsenic (As), cadmium (Cd), chromium (Cr), lead (Pb), mercury (Hg), and thallium (TL) were the most vital heavy metal components released from industrial processes. Accumulation of these metals leads to cancer and skin diseases in humans. Biomimetic nanotechnology solved this emerging difficulty using nanoscale adsorbent using mimicking snowflake-shaped ZnO@SiO2@Fe3O4/C nanostructures. The maximum adsorption capacities of the adsorbents for Pb, As, and organic pollutants are 94.3, 23.6, and 142.9 mg g^{-1}, respectively, which were significantly higher than the conventional adsorbents. On the other hand, the produced adsorbents were good at recycling capacities. Removal of heavy metals by this method is cheap, highly efficient, and stable (Zhang et al., 2014).

The chitosan nanocomposite obtained from crustacean shells with free amino groups and hydroxyl groups was used for the removal of uranium, lead, and heavy metals and organic pollutants by a high level of adsorption property as nanosorbent. In 2019, Rigo et al. reported cantilever nanobiosensors were used to remove heavy metals from rivers and drinking water (Fig. 1). Their work demonstrated functionalization of cantilever nanobiosensors with alkaline phosphatase was a feasible and promising method for the removal of heavy metal pollutants. Iron oxide ceramic nanomembrane (ferroxane membrane) traps the wastes in water like a filter to remove organic and inorganic pollutants in water. Similarly, nanofibers are used for the removal of pathogens from water. These membranes are also used for the removal of contaminants (arsenic and other heavy metals) from water by the reverse osmosis process. Nanodevices remove water pollutants by clarification, deaeration, decarbonation, sludge densification, or the high-density process used to treat acid mine drainage. These processes are used to treat water from mine drainage polluted with heavy metals, chlorinated solvents, halogenated chemicals, or pesticides.

Nanozerovalent iron has a superior adsorption ability, which transforms toxic contaminants into less noxious compounds. For example, hexavalent chromium is converted into triple-valent chromium, which is ferrous chromite. Recently, a nanomaterial from dried hydrogel made of poly vinyl alcohol cross-linked covalently with poly acrylic acid and N-hydroxysuccinimide ester prepared with hygroscopic or hydrophilic adhesive properties. Upon attachment with marine animals, it remained stable and sensed the multianimal sensing of biomechanics, collective behaviors, concurrent multispecies activity, and interspecific interactions with small to large animals (Duque Londono et al., 2024) (Table 2).

Release of waste effluents from industries, dye factories, and agricultural residues causes contamination in fresh water sources like lakes, river, and oceans. This factor would be a source of water pollution and water contamination. Recovering water from this polluted effluent was key research to find a solution to this problem. Separation and purification technology became the answer to this difficulty. Membrane-based water treatment, reverse osmosis (RO), forward osmosis (FO), membrane distillation, and electrodialysis (ED) are conventional methods used nowadays for purification, whereas compared to other methods, RO is the most commonly used. Polymer-based membranes used for filtration of water can withstand even at high pressure. For example, cellulose acetate (CA) membranes were the first type of membrane fabricated using cellulose and possess high salt rejection properties but are poor in water permeability.

To overcome this problem, biomimetic nanotechnology is applied to prepare new membrane systems for purification of water. One such technology is the biomimetic aquaporin membranes. Water reflux and salt rejection are the extraordinary features of this membrane. This membrane was fabricated using three different components:

Table 2 Nanobiosensors used to detect heavy metals (Manjceevan & Velauthamurty, 2022)

Heavy metals	Transduction method	Nanomaterial used	Sample	References
Hg^{2+}	Optical fluorescence emission	Quantum dots, gold nanoparticles	Buffer solution and river water	Li et al. (2011)
Hg^{2+}	Colorimetric detection	Gold nanoparticles	Aqueous solution	Xue et al. (2008)
Pb^{2+}	Colorimetric sensor	Cellulose acetate nanofibers	Water sample	Raj and Shankaran (2016)
Cd^{2+}	Electrochemical method	Multiwalled carbon nanotube	Environment samples	Sreekanth et al. (2021)
As^{3+}	Surface-enhanced Raman scattering	Silver nanoparticles	Real water samples	Li et al. (2011)
Pb^{2+}	Electrochemical	Gold nanoparticles	Water samples	Zhu et al. (2014)
Cu^{2+}	Fluorescent	Gold nanoparticles	Environmental and biological samples	Chen et al. (2009)
Pb^{2+}	Colorimetric method	Colorimetric	Aqueous solution	Slocik et al. (2008)
As^{3+}	Electrochemical method	Ag–Au alloy nanoparticles	Water from natural water courses, lakes, and wells	Yadav et al. (2020)
Hg^{2+}	Fluorescent	Graphene oxide	Water sample	Li et al. (2013)

aquaporins, amphiphilic housing, and porous solid substrate. Aquaporins, where membrane proteins act as water channels, porous solid substrate, enhance mechanical stability and are amphiphilic housing where aquaporins were reconstituted. Aquaporins are transmembranes that have a three-dimensional structure that were first discovered in red blood cells (RBCs) and also identified their presence in plants, animals, and microorganisms. Various conventional membranes were used, but the disadvantage over those was fouling due to the lack of isoporosity. Compared to other membranes, biomimetic membranes have uniform pore size, prevent blockage of pores, and are resistant to fouling (Fuwad et al., 2019). Carbon nanotubes, carbon aerogels, carbon graphenes, and other forms of carbon elements were found to be very effective in the treatment of wastewater from industries and had a high affinity toward heavy metals (Jiang et al., 2018).

For example, various nanobiosensors were developed to identify the presence of chemical pollutants and heavy metals from wastewater, including enzyme-based biosensors (EBBs), DNA-based biosensors (DBBs), immuno-based biosensors, cell-based biosensors (CBBs), and microbial-based biosensors (Singh et al., 2022). EBBs were developed as a massive technique for quantitative examination of analytes in food quality control, medicine, and drug industries (Ispas et al., 2012). Nucleic acid identification methods based on biosensors, DBBs, are being developed to test the inherited and contagious diseases like tuberculosis and cholera (Kavita, 2017). Immunosensors were divided based on the principle of typical and most advanced, which is further classified as electrochemical, optical, and microgravimetric immunosensors (Salouti & Khadivi Derakhshan, 2020). CBBs are biosensors where a whole part of the cell is combined with a transducer to identify like smart sensors (Turdean, 2011). Microbial biosensors are used for analyzing devices consisting of microorganisms to identify the particular analyte and the outcomes transferred in the form of biochemical, electrical, or physiological signals. For example, an algal biosensor for toxicity evaluation in water bodies such as *Cyanobacterium spirulina* is used as an algal bioreceptor (Salouti & Khadivi Derakhshan, 2020).

Methicillin-resistant *Staphylococcus aureus* is a predominant antibiotic-resistant pathogen. The mecA and "arg" genes were responsible for the antibiotic resistance. To identify the presence of this pathogen and its resistant genes, Riquelme 2017 and coworkers synthesized gold nanoparticles functionalization with oligonucleotides for monitoring of mecA and "arg." The results with synthesized nanoparticles showed higher detection of the "arg" gene was 70 ppm from environmental samples. Caffeine is one of the pharmaceutical products released abundantly into the environment for various reasons, including pharmaceutical wastewater, colas, tea, coffee beans, medicines, and soft drinks. Detection of caffeine in the environment is a tedious job for researchers, industries, and academicians. According to this, in 2018, Hu et al. developed a nanosensor of AgNPs doped by polymers for the detection of caffeine in wastewater detection, and the sensing limit was found to be 100 ng/l with a high precision.

A trace of chromium (Cr) is used commonly as a trace element for various purposes, including steel manufacturing, painting, tanning, welding, and as a catalyst too. High levels of usage resulted in an increase in the Cr level in the environment, which leads to poor effects on biological and ecological systems. Whereas Cr (VI) has carcinogenic and mutagenic effects on living organisms. In order to detect Cr in environmental water samples, in 2020, Zhang et al. synthesized carbon dot-based nanosensors. The synthesized nanosensors were able to detect Cr at 2.3 nM concentration at pH 6.

9 Detection of Air Pollution

Air-polluting nitrous oxide, carbon dioxide, and carbon monoxide are efficiently adsorbed by titanium oxide and indicate the pollution level in the air. Further, on exposure to UV- light, these toxins are converted into stable, nontoxic products like nitrates. Important toxic metals such as mercury are oxidized by silica–titania nanocomposite into less volatile mercury oxides. Dendrimers remove metal and metal ions by solubilizing them. Arsenic detects and removes iron oxide by binding by magnetic waves. A biosensor with 50 nm diameter, filled with palladium, molybdenum or tungsten, copper salts, and sodium chloride, is encapsulated inside a hydroscopic halide membrane. This membrane mimics a human olfactory sensory organ to sense the toxic airborne contaminants, such as Hg, CO, EtO, H2S, volatile organic materials, and hydrogen sulfide, which are often colorless and odorless.

Urease sol–gel matrix is used to detect mercury ion, copper ion, and cadmium ion. Acetyl choline esterase enzyme doped on carbon nanotubes are used to detect toxic paraoxon, tyrosinase cross-linked on poly vinyl alcohol are used to detect atrazine, catechol, etc., dispersed in air. An input and output ports sensor device like a diaphragm was developed to increase indoor air flow in sensor devices (Scharfenberg, 2023).

10 Microbial Fuel Cell (MFC) Biosensor

Microbial fuel cell biosensor is a kind of biosensor used in energy generation, detection, and treatment of heavy metals from water sources. Here, a biological organism like bacteria grows in anodes, utilizes organic pollutants, and converts chemical energy into electrical energy. The electrode operates as a cathode and accepts electrons for microbial utilization. MFC biosensors are simpler than conventional biosensors. Like other biosensor systems, MFC is able to operate on natural organisms (no genetically modified or mutated organisms), for example, *Enterobacter cloacae,* which is cost-effective. Compared to other methods of detection, the nanosensor-based MFC is easy to operate and reproducible. The MFC sensor needs more research to commercialize and work on it (Chansi & Basu, 2022).

11 Protein-Functionalized Microcantilever Sensors

Protein-functionalized microcantilever sensors were able to absorb heavy metals from a solution. In 2003, Cherian et al. demonstrated microcantilever sensors functionalization with protein and used to absorb heavy metals like Hg2+ and Zn2+. Modifying cantilever surfaces with other proteins can adsorb other metal ions (Fig. 2).

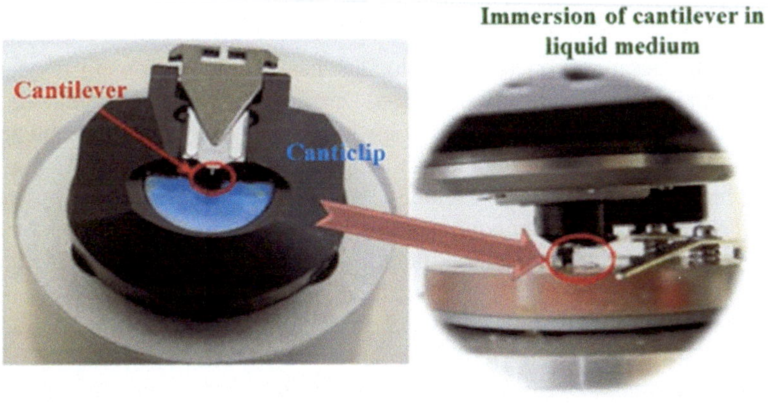

Fig. 2 Photograph of cantilever sensor immersed in a water sample. (Rigo et al., 2020)

12 Detection of Microorganism

The diversity of microorganisms in agricultural land directly affects the quality of food and water. Invasion of microorganisms into food and water through the environment is a primary source affecting human health. Various methods, including chemical, physical, and enzymatic, were employed to detect the presence of microorganisms. Introduction of nanotechnology to detect pathogens is a rapid, simple, and effective method. Using microorganisms as biosensors is an emerging technology used in various fields, like agriculture, the food industry, textiles, and the medical sector. Using antigen–antibody, ssDNA, and RNA, lock and key model enzymes as analyte was an extraordinary way to detect the presence of an unknown pathogen from the surroundings. Gold nanoparticles-based biosensors are considered appropriate biosensors because of their selectivity, low cost, and easy operating system. It involves two strong longitudinal surface plasma resonance (SPR) bands that are highly susceptible to changes in the environment. These biosensors are ideal for clinical applications and in the detection of microorganisms from the environment (Oluwaseun Adetunji et al., 2022).

Biomimetic SPR sensors proposed by Neslihan Idil et al., 2021 served as an effective tool to identify the presence of *Staphylococcus aureus* in food and clinical specimens. SPR sensors provide an appropriate method to detect pathogens. Cryogels in the biosensor act as pioneering material to capture *S. aureus,* which binds with cell wall protein A. SPR sensors could be used to identify other bacterial strains such as gram-negative pathogens with some modification in the sensor system. SPR sensors were mainly used to identify *Campylobacter jejuni, Helicobacter pylori, Salmonella typhimurium, Salmonella paratyphi,* and *E. coli.* They were responsible for causing severe infections, like peptic ulcer, paratyphoid, lung infections, and wound infections, urinary tract infections, and localized and systemic infections.

13 Determination of Analytes in Air

Air pollution caused by chemical pollutants has become the main concern for human health and environmental pollution. Detection of chemical pollutants needs to be addressed to ensure the quality of air for a healthy environment and society. Various conventional methods are available to identify the pollutants, such as chromatographic techniques. Those techniques were really time-consuming, require expensive reagents and highly equipped instruments, and well-trained personalities are essential. To overcome such difficulties, selective and sensitive methods of detection would be needed. One such method is biosensors. Biosensors are able to identify the analytes by using simple and rapid methods. On the other hand, using nanotechnology to rectify the issues would be a smart and advanced approach (Mohamed & Awad, 2022). Air quality monitoring nanodevices measuring lead contaminants are used to sense the traces of DNA accidentally falling from animal, plant, and fungal species precisely.

Soy protein isolate (SPI)–bacterial cellulose (BC) composite film is used as an air filter. SPI-BC composite films are novel and biobased air filters for the determination of chemical pollutants, including particulate matter (PM). Functional groups of SPI play an important role in capturing PM since it has electrostatic attraction and dipole interaction. Bacterial cellulose from *Acetobacter xylinum* is purified abundantly with high purity and size uniformity. These composite films have efficiency in the removal of PM, about 99.94% and 99.95% corresponding to $PM_{2.5}$ and $PM_{10.}$

Classical porous materials like aerogels were good at air filtration; continuous three-dimensional networks, flexible density, high specific surface area, and high porosity were the key features of aerogels to use as air filters. Plant polysaccharide aerogels, including cellulose aerogels, starch aerogels, and sodium alginate aerogels, are used for biosafety and are environmentally friendly. Some disadvantages of plant polysaccharide aerogels like poor mechanical and hydrophobic properties limit the usage of these materials. Konjac glucomannan (KMG) was a high molecular weight and water-soluble polysaccharide material used to prepare aerogels with a high specific surface area (García-Mayagoitia et al., 2020). Bielectronic nose is a biosensor combined with the olfactory system of mammals. It can differentiate thousands of molecules at low concentration. It is a fast, cheap, and reliable method of detecting chemical pollutants from the environment. By incorporating gas sensor technology, a bielectronic nose is used to monitor air pollution and smoke distribution in the environment. Bielectronic nose technology is also used to identify the accumulation of carbon monoxide (CO) and carbon dioxide (CO_2) (Gözde et al., 2022). Biomimetic nanoleaf with a high electrocatlyst function was developed to enhance oxygen evolution reaction (OER) and reduce the exploitation of renewable energy. These oxygen-evolving systems are developed by coating NiCO hydroxide on CuO nanowires to form nanosheets. CuO accelerates the stable OER with 9.3-fold increased conductivity (Chen et al., 2019). Thus, pollutants in air can be detected accurately using nanodevices.

14 Conclusion

The environmental use of nanodevices for sensing and monitoring provoking variation in environmental conditions has been the most important factor in advancing the frontiers of modern nanotechnology. More than 60 years of experience with nanoparticles has advanced the knowledge of nanoparticles in the ecosystem. Modern, safe techniques with nanoparticles have far fewer environmental effects compared to their benefits. Green nanotechnology can still reduce the negative effects of nanoparticles as biosensors. The use of nanodevices in sensing adverse effects of environmental pollutants is really essential for saving the ecosystem from deleterious effects. Moreover, these devices are relatively cost-effective and very sensitive. The benefits of biomimetic biosensors overshadow the risks of other diagnostic procedures. In order to limit the negative ill effects of nanoparticles, optimization of minimum usage of nanoparticles in sensing and remediation is necessary to achieve safety in environmental interventions. With emphasis on sustaining the health of the environment, the need for nanodevices is realized and is in great demand in the commercial sector. Implementation of nanosensors in the environment prevents over-exploitation. Vast applications of nanodevices need periodic inspection to ensure that all regulatory requirements are met and safe usage in the environment.

References

Abuzeid, H. M., Julien, C. M., Zhu, L., & Hashem, A. M. (2023). Green synthesis of nanoparticles and their energy storage, environmental, and biomedical applications. *Crystals, 13*(11), 1576. https://doi.org/10.3390/cryst13111576

Anh, N. T., Dinh, N. X., Van Tuan, H., Thuan, T. H., Tung, L. M., Le, V. P., & Le, A. T. (2021). Cost-effective tween 80-capped copper nanoparticles for ultrasensitive colorimetric detection of Thiram pesticide in environmental water samples. *Journal of Nanomaterials, 2021*(1), 5513401. https://doi.org/10.1155/2021/5513401

Baig, N., Kammakakam, I., & Falath, W. (2021). Nanomaterials: A review of synthesis methods, properties, recent progress, and challenges. *Materials Advances, 2*(6), 1821–1871.

Baysal, G. (2020). *The components of functional nanosystems and nanostructures*. IntechOpen. https://doi.org/10.5772/intechopen.92027

Bray, F., Laversanne, M., Sung, H., Ferlay, J., Siegel, R. L., Soerjomataram, I., & Jemal, A. (2024). Global cancer statistics 2022: GLOBOCAN estimates of incidence and mortality worldwide for 36 cancers in 185 countries. *CA: A Cancer Journal for Clinicians, 74*(3), 229–263. https://doi.org/10.3322/caac.21834

Chansi, C., R., Punya, & Basu, T. (2022). Utility of nano biosensors for heavy metal contamination detection in the environment. In *Nanobiosensors for environmental monitoring: Fundamentals and application* (pp. 299–338). Springer International Publishing. https://link.springer.com/chapter/10.1007/978-3-031-16106-3_16

Chen, W., Tu, X., & Guo, X. (2009). Fluorescent gold nanoparticles-based fluorescence sensor for Cu2+ ions. *Chemical Communications, 13*, 1736–1738. https://doi.org/10.1039/B820145E

Chen, B., Zhang, Z., Kim, S., Baek, M., Kim, D., & Yong, K. (2019). A biomimetic nanoleaf electrocatalyst for robust oxygen evolution reaction. *Applied Catalysis B: Environmental, 259*, 118017.

Cherian, S., Gupta, R. K., Mullin, B. C., & Thundat, T. (2003). Detection of heavy metal ions using protein-functionalized microcantilever sensors. *Biosensors and Bioelectronics, 19*(5), 411–416. https://doi.org/10.1016/S0956-5663(03)00226-4

Dhanapal, A. R., Thiruvengadam, M., Vairavanathan, J., Venkidasamy, B., Easwaran, M., & Ghorbanpour, M. (2024). Nanotechnology approaches for the remediation of agricultural polluted soils. *ACS Omega, 9*(12), 13522–13533. https://doi.org/10.1021/acsomega.3c09776

Díez, A. G., Rincón-Iglesias, M., Lanceros-Méndez, S., Reguera, J., & Lizundia, E. (2022). Multicomponent magnetic nanoparticle engineering: The role of structure-property relationship in advanced applications. *Materials Today Chemistry, 26*, 101220.

Duque Londono, C., Cones, S. F., Deng, J., et al. (2024). Bioadhesive interface for marine sensors on diverse soft fragile species. *Nature Communications, 15*, 2958. https://doi.org/10.1038/s41467-024-46833-4

Felgueiras, H. P. (2021). An insight into biomolecules for the treatment of skin infectious diseases. *Pharmaceutics, 13*(7), 1012. https://doi.org/10.3390/pharmaceutics13071012

Fuwad, A., Ryu, H., Malmstadt, N., Kim, S. M., & Jeon, T. J. (2019). Biomimetic membranes as potential tools for water purification: Preceding and future avenues. *Desalination, 458*, 97–115. https://doi.org/10.1016/j.desal.2019.02.003

García-Mayagoitia, S., Pérez-Hernández, H., Medina-Pérez, G., Campos-Montiel, R. G., & Fernández-Luqueño, F. (2020). Bio-nanomaterials in the air pollution treatment. In *Nanomaterials for air remediation* (pp. 227–248). Elsevier. https://doi.org/10.1016/B978-0-12-818821-7.00012-9

Garhwal, R., Shady, S. F., Ellis, E. J., Ellis, J. Y., Leahy, C. C., McCarthy, S. P., Crawford, K. S., & Gaines, P. (2012). Sustained ocular delivery of ciprofloxacin using nanospheres and conventional contact lens materials. *Investigative Ophthalmology & Visual Science, 53*, 1341–1352. https://doi.org/10.1167/iovs.11-8215

Ghazi, A. N., Mohsin, K., Muhammad, A. I., Nouman, N., Muhammad, A., Sabah, H., Tayyab, A., Kareem, M., Syed S. H. (2023). Reviewing black phosphorus for biomedical and optoelectronic applications. In: Inorganic Chemistry Communications,*160*, 111912. https://doi.org/10.1016/j.inoche.2023.11191

Gözde, K. A., Dulta, K., Chauhan, P., & Chauhan, P. K. (2022). Biomimetic material-based biosensor for environmental monitoring. In *Nanobiosensors for environmental monitoring: Fundamentals and application* (pp. 191–202). Springer International Publishing. https://link.springer.com/chapter/10.1007/978-3-031-16106-3_10

Gupta, R., & Xie, H. (2018). Nanoparticles in daily life: Applications, toxicity and regulations. *Journal of Environmental Pathology, Toxicology and Oncology, 37*(3), 209–230. https://doi.org/10.1615/JEnvironPatholToxicolOncol.2018026009

Harun-Ur-Rashid, M., Jahan, I., Foyez, T., & Imran, A. B. (2023). Bio-inspired nanomaterials for micro/nanodevices: A new era in biomedical applications. *Micromachines, 14*(9), 1786. https://doi.org/10.3390/mi14091786. https://link.springer.com/chapter/10.1007/978-3-031-16106-3_7

Hu, R., Tang, R., Xu, J., & Lu, F. (2018). Chemical nanosensors based on molecularly-imprinted polymers doped with silver nanoparticles for the rapid detection of caffeine in wastewater. *Analytica Chimica Acta, 1034*, 176–183. https://doi.org/10.1016/j.aca.2018.06.012

Idil, N., Bakhshpour, M., Perçin, I., & Mattiasson, B. (2021). Whole cell recognition of Staphylococcus aureus using biomimetic SPR sensors. *Biosensors, 11*(5), 140. https://doi.org/10.3390/bios11050140

Ispas, C. R., Crivat, G., & Andreescu, S. (2012). Recent developments in enzyme-based biosensors for biomedical analysis. *Analytical Letters, 45*(2–3), 168–186.

Izuegbunam, C. L., Wijewantha, N., Wone, B., Ariyarathne, M. A., Sereda, G., & Wone, B. W. M. (2021). A nano-biomimetic transformation system enables *in planta* expression of a reporter gene in mature plants and seeds. *Nanoscale Advances, 3*(11), 3240–3250. https://doi.org/10.1039/d1na00107h

Jiang, M., Qi, Y., Liu, H., & Chen, Y. (2018). The role of nanomaterials and nanotechnologies in wastewater treatment: A bibliometric analysis. *Nanoscale Research Letters, 13*(1), 1–13.

Kapil, S., Bhattu, M., Vinayak, A., Pal, N., & Sharma, V. (2022). Nanobiosensors' potentialities for environmental monitoring. In *Nanobiosensors for environmental monitoring: Fundamentals and application* (pp. 41–74). Springer International Publishing. https://link.springer.com/chapter/10.1007/978-3-031-16106-3_3

Kavita, V. (2017). DNA biosensors-a review. *Journal of Bioengineering and Biomedical Science, 7*(2), 222.

Kumar, R., Rauwel, P., & Rauwel, E. (2021). Nanoadsorbants for the removal of heavy metals from contaminated water: Current scenario and future directions. *Processes, 9*(8), 1379. https://doi.org/10.3390/pr9081379

Li, M., Wang, Q., Shi, X., Hornak, L. A., & Wu, N. (2011). Detection of mercury(II) by quantum dot/DNA/gold nanoparticle ensemble based nanosensor via nanometal surface energy transfer. *Analytical Chemistry, 83*(18), 7061–7065. https://doi.org/10.1021/ac2019014

Li, M., Zhou, X., Guo, S., & Wu, N. (2013). Detection of lead (II) with a "turn-on" fluorescent biosensor based on energy transfer from CdSe/ZnS quantum dots to graphene oxide. *Biosensors & Bioelectronics, 43*, 69–74. https://doi.org/10.1016/j.bios.2012.11.039

Li, H., Zheng, G., & Zhu, S. (2018). Construction of an organelle-like nanodevice via supramolecular self-assembly for robust biocatalysts. *Microbial Cell Factories, 17*, 26. https://doi.org/10.1186/s12934-018-0873-3

Manjceevan, A., & Velauthamurty, K. (2022). Role and significance of nanobiosensors for environmental remediation. In *Nanobiosensors for environmental monitoring: Fundamentals and application* (pp. 131–148). Springer International Publishing.

Mohamed, E. F., & Awad, G. (2022). Development of nano-sensor and biosensor as an air pollution detection technique for the foreseeable future. In *Comprehensive analytical chemistry* (Vol. 99, pp. 163–188). Elsevier. https://doi.org/10.1016/bs.coac.2021.11.003

Oluwaseun Adetunji, C., Mathew, J. T., Inobeme, A., Olaniyan, O. T., RB Singh, K., Abimbola, O. F., et al. (2022). Microbial and plant cell biosensors for environmental monitoring. In *Nanobiosensors for environmental monitoring: Fundamentals and application* (pp. 175–190). Springer International Publishing. https://link.springer.com/chapter/10.1007/978-3-031-16106-3_9

Pathak, V. M., Verma, V. K., Rawat, B. S., Kaur, B., Babu, N., Sharma, A., Dewali, S., Yadav, M., Kumari, R., Singh, S., Mohapatra, A., Pandey, V., Rana, N., & Cunill, J. M. (2022). Current status of pesticide effects on environment, human health and it's eco-friendly management as bioremediation: A comprehensive review. *Frontiers in Microbiology, 13*, 962619. https://doi.org/10.3389/fmicb.2022.962619

Qin, H., Teng, Y., Dai, R., Wang, A., & Liu, J. (2024). Glycan-based scaffolds and nanoparticles as drug delivery system in cancer therapy. *Frontiers in Immunology, 15*, 1395187.

Raj, S., & Shankaran, D. R. (2016). Curcumin based biocompatible nanofibers for lead ion detection. *Sensors and Actuators B: Chemical, 226*, 318–325. https://doi.org/10.1016/j.snb.2015.12.006

Rigo, A. A., Cezaro, A. M. D., Muenchen, D. K., Martinazzo, J., Manzoli, A., Steffens, J., & Steffens, C. (2020). Heavy metals detection in river water with cantilever nanobiosensor. *Journal of Environmental Science and Health, Part B, 55*(3), 239–249. https://doi.org/10.1080/03601234.2019.1685318

Riquelme, M. V., Leng, W., Carzolio, M., Pruden, A., & Vikesland, P. (2017). Stable oligonucleotide-functionalized gold nanosensors for environmental biocontaminant monitoring. *Journal of Environmental Sciences, 62*, 49–59. https://doi.org/10.1016/j.jes.2017.08.005

Roy, A., Sharma, A., Yadav, S., Jule, L. T., & Krishnaraj, R. (2021). Nanomaterials for remediation of environmental pollutants. *Bioinorganic Chemistry and Applications, 2021*, 1764647. https://doi.org/10.1155/2021/1764647

Salouti, M., & Khadivi Derakhshan, F. (2020). Biosensors and nanobiosensors in environmental appli-cations. In M. Ghorbanpour, P. Bhargava, A. Varma, & D. Choudhary (Eds.), *Biogenic nano-particles and their use in agro-ecosystems*. Springer. https://doi.org/10.1007/978-981-15-2985-6_26

Scharfenberg, S. (2023). Biomimetic air quality sensor device. Technical Disclosure Commons.https://www.tdcommons.org/dpubs_series/5906

Singh, P., Sankhla, M. S., Vanisree, C. R., Parihar, K., Jadhav, E. B., & Verma, S. K. (2022). Nanobiosensors and industrial wastewater treatments. In *Nanobiosensors for environmental monitoring: Fundamentals and application* (pp. 339–361). Springer International Publishing. https://doi.org/10.1007/978-3-031-16106-3_17

Slocik, J. M., Zabinski, J. J. S., Phillips, D. M., & Naik, R. R. (2008). Colorimetric response of peptidefunctionalized gold nanoparticles to metal ions. *Small, 4*(5), 548–551. https://doi.org/10.1002/smll.200700920

Sreekanth, S. P., Alodhayb, A., Assaifan, A. K., Alzahrani, K. E., Muthuramamoorthy, M., Alkhammash, H. I., Pandiaraj, S., Alswieleh, A. M., Van Le, Q., Mangaiyarkarasi, R., Grace, A. N., & Raghavan, V. (2021). Multiwalled carbon nanotube-based nanobiosensor for the detection of cadmium in water. *Environmental Research, 197*, 111148. https://doi.org/10.1016/j.envres.2021.111148

Torres-Pérez, S. A., Torres-Pérez, C. E., Pedraza-Escalona, M., Pérez-Tapia, S. M., & Ramón-Gallegos, E. (2020). Glycosylated nanoparticles for cancer-targeted drug delivery. *Frontiers in Oncology, 10*, 605037. https://doi.org/10.3389/fonc.2020.605037

Turdean, G. L. (2011). Design and development of biosensors for the detection of heavy metal toxicity. *International Journal of Electrochemistry, 2011*(1), 343125.

Williams, P. C. M., Beardsley, J., Isaacs, D., Preisz, A., & Marais, B. J. (2023). The impact of climate change and biodiversity loss on the health of children: An ethical perspective. *Frontiers in Public Health, 10*, 1048317. https://doi.org/10.3389/fpubh.2022.1048317

Xiang, Y., & Lu, Y. (2011). Using personal glucose meters and functional DNA sensors to quantify a variety of analytical targets. *Nature Chemistry, 3*, 697–703. https://doi.org/10.1038/nchem.1092

Xue, X., Wang, F., & Liu, X. (2008). One-step, room temperature, colorimetric detection of mercury (Hg2+) using DNA/nanoparticle conjugates. *Journal of the American Chemical Society, 130*(11), 3244–3245. https://doi.org/10.1021/ja076716c

Yadav, R., Kushwah, V., Gaur, M. S., Bhadauria, S., Berlina, A. N., Zherdev, A. V., & Dzantiev, B. B. (2020). Electrochemical aptamer biosensor for As3+ based on apta deep trapped Ag-Au alloy nanoparticlesimpregnated glassy carbon electrode. *International Journal of Environmental Analytical Chemistry, 100*(6), 623–634. https://doi.org/10.1080/03067319.2019.1638371

Yang, L., Sun, H., Wang, X., Yao, W., Zhang, W., & Jiang, L. (2019). An aptamer based aggregation assay for the neonicotinoid insecticide acetamiprid using fluorescent upconversion nanoparticles and DNA functionalized gold nanoparticles. *Mikrochimica Acta, 186*(5), 308. https://doi.org/10.1007/s00604-019-3422-9

Yu, I., Mori, T., Ando, T., Harada, R., Jung, J., Sugita, Y., & Feig, M. (2016). Biomolecular interactions modulate macromolecular structure and dynamics in atomistic model of a bacterial cytoplasm. *eLife, 5*, e19274. https://doi.org/10.7554/eLife.19274

Zhang, X., Liu, J., Kelly, S. J., Huang, X., & Liu, J. (2014). Biomimetic snowflake-shaped magnetic micro–/nanostructures for highly efficient adsorption of heavy metal ions and organic pollutants from aqueous solution. *Journal of Materials Chemistry A, 2*(30), 11759–11767. https://doi.org/10.1039/C4TA02058H

Zhang, T., Yu, H., Li, J., Song, H., Wang, S., Zhang, Z., & Chen, S. (2020). Green light–triggered antimicrobial cotton fabric for wastewater disinfection. *Materials Today Physics, 15*, 100254. https://doi.org/10.1016/j.mtphys.2020.100254

Zhu, Y., Zeng, G.-m., Zhang, Y., Tang, L., Chen, J., Cheng, M., Zhang, L.-h., He, L., Guo, Y., He, X.-x., Lai, M.-Y., & He, Y.-B. (2014). Highly sensitive electrochemical sensor using a MWCNTs/GNPs-modified electrode for lead (II) detection based on Pb2+-induced G-rich DNA conformation. *Analyst, 139*(19), 5014–5020. https://doi.org/10.1039/C4AN00874J

Nanotechnology for Carbon Capture and Climate Change Mitigation in Nano-Ecosystems

Anas Mahmud, Rad Mosharrof Mim, Abu Bin Imran, and Al-Nakib Chowdhury

Abstract

Climate change is a prioritized global issue in this era, necessitating innovative and efficient approaches to mitigate its impact. Addressing climate change requires creative solutions to reduce greenhouse gas emissions, especially carbon dioxide (CO_2). Nanotechnology has emerged as a promising field in the battle against climate change, and at the forefront of this field are nano-ecosystems. These nano-ecosystems utilize nanomaterials and biomimicry to replicate the effectiveness of natural ecosystems in mitigating the effects of climate change. This chapter will explore the underlying principles, materials, technologies, and applications of nanotechnology in the context of nano-ecosystems for climate change mitigation. The scientific foundations of nano-ecosystems, the nanomaterials employed, and their roles in capturing and mitigating greenhouse gases will also be discussed. Furthermore, the chapter will address the challenges, ethical considerations, and future prospects within this crucial area of research.

Keywords

Climate change · Greenhouse gas · Carbon capture · Renewable energy · Membrane separation

A. Mahmud · R. M. Mim · A. B. Imran · A.-N. Chowdhury (✉)
Department of Chemistry, Bangladesh University of Engineering and Technology, Dhaka, Bangladesh
e-mail: nakib@chem.buet.ac.bd

D. K. Gupta et al. (eds.), *Nano-Ecosystems*,
https://doi.org/10.1007/978-981-95-5033-3_6

1 Introduction

Anthropogenic CO_2, CH_4, and N_2O emissions are responsible for climate change, a global threat that has stressed numerous areas since the pre-industrial period (Intergovernmental Panel on Climate Change (IPCC), 2023a). The excess use of fossil fuels in industries is greatly responsible for the rising concentration of emitted gases in the atmosphere. The enhanced emission of the gases raises the temperature, which causes global warming. For example, the global mean surface temperature has increased by 1.4 °C in the urban industrialized area, which is estimated by the Intergovernmental Panel on Climate Change (IPCC) in its Sixth Assessment Report (AR6) (CEDA Archive Web Browser, 2023; Intergovernmental Panel on Climate Change (IPCC), 2023b). Moreover, the sea level has been rising nearly 17 mm/year over the last 100 years, which is another strong evidence of global warming (Williams, 2013). Different extreme natural events, such as droughts and storms, are also indications of ecological deprivations. Another frequent and long-lasting issue of global warming is heat waves, which negatively impact the mass of people, including maternal and newborn health (Chersich et al., 2023). However, the primary gas responsible for global warming, CO_2, originates in the atmosphere through the combustion of fossil fuels that contain a lot of carbon content. Therefore, to deal with climate change and expand the required actions and investment, the United Nations Framework Convention on Climate Change (UNFCCC) reached an agreement in 2015 at a conference in Paris. The Paris Agreement, considered a crucial treaty, introduces unprecedented measures to address environmental harm from human activities. Its primary goal is to reaffirm the global commitment to tackling climate change and ensure that the temperature increase stays below 2 °C above pre-industrial levels (Chien et al., 2021; Sharma et al., 2020). In addition, the agreement commits to lowering chaotic greenhouse gas (GHG) emissions and emphasizes the imperative of achieving net-zero CO_2 emissions by the year 2050.

As a result, recycling or capturing CO_2 has been prioritized globally due to its adverse environmental impact. Despite this urgency, the world still heavily relies on oil-based commodities such as coal (41%), combustible gas (21%), and crude oil (5%). Consequently, scientists are actively exploring feasible sources, which implies the conversion of greenhouse CO_2 gas into valuable hydrocarbons to ensure long-term sustainability. As a result, carbon capture and storage appears to be a recent and efficient innovation (Bui et al., 2018). The main idea behind carbon capture and storage involves capturing high-density CO_2 emitted from industries to prevent the release of CO_2 into the atmosphere. Further, the captured gas is then compressed and transported to a store to recycle for future practical applications (Samanta et al., 2012; Keith, 2009). This chapter mainly focuses on insights into mitigating anthropogenic CO_2 emissions. In addition, it also contains various approaches of storage and material-based capture techniques in the context of nano-ecosystem. Furthermore, it provides practical solutions to address the complex issue of CO_2 emissions and their environmental consequences.

2 Effects on Climate Change

2.1 Environmental Impacts

The rise in atmospheric CO_2 levels, attributed to industrialization and civilization, has become a concerning issue for all living species on Earth (Khan, 2018). This ongoing emission of CO_2 poses a significant threat to climate and the environment, leading to severe consequences such as the disruption of atmospheric gas balance. The resulting phenomena affect ecosystems and the natural processes crucial for supporting life on our planet (Reitze Jr, 2010; Hopwood & Cohen, 2008; Dong et al., 2021). Furthermore, this environmental challenge directly threatens human well-being, affecting food security, health, and economic stability (Mitchell, 1989).

Many human activities like deforestation cause CO_2 emission, the inflammation of different industries like smoke (Lammel & Graßl, 1995), smog (Schneider, 1989) or other particles, cement production, and the burning of fossil fuels like coal, oil and natural gas. The high concentration of CO_2 in the environment turned into global warming, which truly changes the pattern of the environment (McKinstry Jr, 2004). Global environmental challenges encompass rising sea levels attributed to melting polar ice caps, species extinction, and shifting water resource availability. These issues are prevalent worldwide and demand urgent attention (Olesen & Bindi, 2002). CO_2 emissions primarily stem from various sectors, including transportation, industry, commercial and residential activities, land use, and forestry (Ciscar et al. 2011). The transportation sector contributes significantly to greenhouse gas emissions, with the burning of fossil fuels in cars, trucks, ships, trains, and planes being a major source. Over 94% of transportation fuel is derived from petroleum, predominantly gasoline, and diesel (Hardy, 2003). In the commercial and residential sectors, emissions arise from the combustion of fossil fuels for heating purposes and the use of gases in refrigeration and cooling systems within buildings. Nonbuilding-specific emissions, such as waste handling, also contribute to this sector's greenhouse gas footprint. Moreover, when accounting for electricity end-use, emissions from commercial and residential buildings further escalate, given the substantial share of electricity consumption (Hönisch et al., 2012; Hardy, 2003). These issues are closely linked to the broader context of climate change (World Health Organization, 1994). The impact of CO_2 on the environment manifests in various ways, including elevated temperatures, intensified storms, prolonged droughts, rising sea levels, biodiversity loss, ozone layer depletion, and environmental displacement. As concentrations of greenhouse gases, particularly CO_2, increase, the Earth's surface experiences a gradual temperature rise (Johnson et al., 2020; Howard et al., 2013; de Gruijl & Leun, 2000; Mitchell, 1989). This global warming leads to widespread hot weather and heatwaves, increasing the risks of heat-related illnesses.

Additionally, the increased temperature contributes to the rapid spread of wildfires. Rising temperatures also influence the water cycle, causing more moisture to evaporate and exacerbating extreme rainfall patterns, leading to destructive storms (Bograd et al., 2003). These storms, in turn, threaten homes, communities, and livelihoods, resulting in substantial human and economic losses. Climate change further impacts water availability, contributing to shortages and posing ecological vulnerabilities to ecosystems (Abbass et al., 2022; Kumar & Majid, 2020). The warming of the ocean, attributed to its absorption of heat, leads to an expansion in volume and a subsequent rise in sea levels. Melting ice sheets further contribute to this phenomenon, directly threatening coastal and island communities. Meanwhile, the ocean serves as a sink for CO_2, absorbing it from the atmosphere (Reganold et al., 1990). The increasing levels of CO_2 also lead to ocean acidification, endangering marine life. This complex interplay of environmental factors underscores the multifaceted challenges posed by climate change. Global warming contributes to the depletion of the ozone layer, a crucial protective barrier often likened to the Earth's umbrella (Kogo et al., 2021). This layer shields us from harmful radiation originating from outer space. However, the rise in CO_2 levels disrupts this delicate balance, leading to the deterioration of the ozone layer. Major pollutants such as COx, NOx, and SOx play a significant role in the depletion of the ozone layer (Saunois et al., 2016). The consequences of climate change extend to the survival of species on land and in the ocean, with risks intensifying as temperatures increase. Climate change has accelerated the rate of species loss, surpassing historical records by a factor of 1000 (Duyck et al., 2018). Various threats associated with climate change, including forest fires, extreme weather events, and the proliferation of invasive pests and diseases, contribute to this alarming trend. While some species may adapt and relocate successfully, others face a more precarious fate (Malhi et al., 2021). The acidification of the ocean due to climate change poses a significant risk to marine resources, which are vital for sustaining billions of people. Snow and ice cover changes in Arctic regions disrupt traditional food supplies from herding, hunting, and fishing. Heat stress exacerbates water and grassland availability challenges, leading to declining crop yields and affecting livestock (Smith, 2019; Pan et al., 2018; Mendelsohn, 2014). The interconnected impacts of climate change highlight the urgency of addressing these complex environmental challenges. These environmental changes significantly threaten the world's species/lives, health and economic stability (Ziska, 2022). To eradicate these effects, it is necessary to develop an aim to control and reduce the concentration of CO_2 emission in all major or minor sectors of pollution. Sustainable energy sources measure the efficiency of controllable carbon capture and storage techniques that help to uptake the leading position of solving the environmental effects of CO_2 emission (Hatfield & Prueger, 2015).

2.2 Agricultural Impacts

The agricultural sector is a fundamental pillar of nations, playing a crucial role in sustaining the world's food supply (Hatfield & Prueger, 2011). Agriculture is a universal presence, connecting every country across the globe. However, the sector faces significant impacts from global environmental changes, with environmental surprises closely intertwined with agricultural activities (Gruza et al., 1999). The changing climate, marked by elevated levels of greenhouse CO_2 gas, has resulted in numerous consequences worldwide. Agriculture and fisheries are intricately linked to climate conditions. While some regions may experience increased crop yields due to rising temperatures and elevated CO_2 levels, realizing these benefits requires favorable conditions such as nutrient levels, soil moisture, and water availability (Nieder & Benbi, 2022). The changing climate, characterized by alterations in the frequency and severity of droughts and floods, poses challenges for farmers and ranchers, threatening food safety. Simultaneously, warming water temperatures are anticipated to shift the habitat ranges of various fish and shellfish species, potentially disrupting ecosystems (Motha, 2011). Climate change may necessitate reconsidering traditional approaches to growing crops, raising animals, and catching fish. The potential challenges posed by climate change need to be understood in conjunction with evolving factors, including changes in farming practices and technology. Addressing the multifaceted impacts on agriculture requires a comprehensive and adaptive approach to ensure sustainable food production in a changing climate (Harun-Ur-Rashid & Imran, 2023; Leduc et al., 2013).

Climate change significantly influences crop yields, with various factors at play in crop production (Laffoley & Baxter, 2019). Key elements such as CO_2 levels, temperature, light, water, and chlorophyll play pivotal roles in enhancing plant growth, but their impact varies under different conditions. Elevated levels of CO_2, higher temperatures, and increased light exposure generally contribute to improved plant growth. However, it is crucial to note that these benefits are contingent upon optimal conditions. Temperature, in particular, plays a critical role in crop production. Deviations from the optimal temperature range can hamper production, especially when temperatures do not reach the desired levels. While elevated CO_2 levels have been associated with increased plant growth, there are potential drawbacks. For instance, studies indicate that elevated CO_2 can reduce protein and nitrogen content in plants such as alfalfa and soybeans, reducing overall quality (Hoegh-Guldberg, 1999). The complex interplay of these factors underscores the importance of understanding the nuanced relationships between climate variables and crop production. As climate change continues to alter environmental conditions, adapting agricultural practices to optimize these factors becomes essential for ensuring sustainable and resilient crop yields (Pandey, 2020). The optimal temperature range for plant growth typically falls between 35 and 45 °C, and deviations from this range, along with extreme temperatures and precipitation variations, can hinder crop development (Ghorani-Azam et al., 2016). The increasing levels of CO_2 contribute to intensified global warming, leading to the melting of polar ice caps and subsequent unpredictable occurrences of droughts and precipitation (Manisalidis et al., 2020). Changes in the traditional patterns of seasons, such as delayed summers and unbearable temperatures, result in dry soils that impede proper crop growth (Kabir et al., 2014).

Ozone layer depletion adds another layer of complexity, exposing the environment to elevated radiation levels, including microwaves and ultraviolet rays. This heightened radiation can adversely impact environmental conditions, leading to increased disease vulnerability and reduced fertility (A Scientific Assessment, 2016). Some regions may experience prolonged and intense droughts due to elevated summer temperatures and reduced precipitation, while heat waves can negatively affect marine livestock (Permentier et al., 2017). Changes in temperature and seasons also influence the timing of reproduction and migration in various species. Moreover, water sources are gradually becoming more acidic due to increased atmospheric CO_2 (Stolp, 2022). The rising acidity threatens fish and shellfish by weakening their shells, a process exacerbated by removing calcium from seawater. This acidification also endangers sensitive ecosystems that particular fish and shellfish rely on. The increased presence of CO_2 in water sources contributes to the acidification process (Duarte et al., 2020). The absorption of CO_2 into seawater results in the formation of carbonic acid (H_2CO_3), a weak acid that dissociates into hydrogen ions (H^+) and bicarbonate ions (HCO_3^-). This ocean acidification negatively impacts corals' ability to recover from bleaching events as it reduces the availability of calcium carbonate needed for their growth and recovery (Zappulla, 2013).

The intricate interplay of these factors underscores the far-reaching consequences of environmental changes on terrestrial and marine ecosystems. The sustainability of the agricultural era can be alarming because of the increasing concentration of CO_2. Marine and rural areas are affected by production challenges and food safety issues (McGrath et al., 2022). To alleviate these challenges, the agricultural and aquaculture sectors must implement effective measures as these two domains largely contribute to global food issues worldwide.

2.3 Health Impacts

Aspiring to lead a healthy life is a universal desire, with the hope of residing in a society characterized by robust well-being. Pursuing health often involves adopting practices such as maintaining a balanced diet and exercising regularly. However, it is essential to recognize that true health is intricately connected to the environment one inhabits. Environmental pollution, encompassing air, soil, and water contamination, poses significant obstacles to daily life and the aspiration for a healthy existence. Among these, air pollution is particularly detrimental as the air quality directly impacts our health. Numerous sources contribute to air pollution, with substances like SO_x, NO_x, and CO_x serving as primary culprits. The escalating levels of anthropogenic CO_2 have profound implications for human health, leading to various complications (Table 1). These may include headaches, dizziness, restlessness, difficulty breathing, increased heart rate, elevated blood pressure, and more severe consequences such as coma and convulsions (Ebi et al., 2021; Hyeon et al., 2015). The rise in CO_2 levels also correlates with increased temperature, resulting in heat-related illnesses or fatalities. Additionally, precipitation patterns change contributes to water-related diseases, including dengue fever, West Nile fever, Japanese encephalitis, Chikungunya fever, malaria, Lyme borreliosis, and hantavirus infection (Chao, 2012). Climate impacts extend to waterborne pathogens and parasites, with CO_2 affecting their reproduction and toxin production by harmful algal and cyanobacterial blooms. Altered water temperatures further lead to variations in the presence of waterborne Vibrio bacteria and harmful algal toxins, posing threats in different seasons or locations.

Table 1 Adverse effects of CO_2 levels on human health

10,000 ppm (1.0%)	Typically, no effects, possible drowsiness
15,000 ppm (1.5%)	Mild respiratory stimulation for some people
30,000 ppm (3.0%)	Moderate respiratory stimulation and increased heart rate and blood
40,000 ppm (1.5%)	Immediately dangerous to life or health
50,000 ppm (1.5%)	Strong respiratory stimulation, dizziness, confusion, and headache
80,000 ppm (1.5%)	Dimmed sight, sweating, tremor, unconsciousness, and possible death

While CO_2 is minimally toxic when inhaled, its primary health impact lies in its role as a simple asphyxiant, displacing normal oxygen in the air. Mild exposure to CO_2 may result in symptoms such as headaches and drowsiness. However, as concentrations increase, the health consequences become more severe, encompassing a range of symptoms (Bell et al., 2018). Recognizing and addressing the relationship between environmental factors, especially CO_2 levels, and human health is crucial for fostering a healthy and sustainable living environment.

Persistent exposure to elevated levels of atmospheric CO_2 can result in diverse health effects, encompassing inflammation, bone and kidney composition alterations, respiratory acidosis, behavioral and physiological changes, and oxidative stress. Chronic exposure may lead to deoxygenation of hemoglobin in circulating blood, causing mild hypoxemia and methemoglobin formation, potentially resulting in oxidative damage within red blood cells (Bobb et al., 2014). This damage triggers the activation of macrophages, releasing pro-inflammatory cytokines strongly associated with comorbidities such as cardiovascular diseases. Analogously, chronic, low-grade metabolic acidosis induced by an increased acid load from the diet can impact blood pH within the normal range but toward the lower end (Savage et al., 2021; Springer, 2020; Campbell-Lendrum & Prüss-Ustün, 2019). Exposure to CO_2 levels exceeding 600 ppm has been linked to the "sick-building syndrome," characterized by symptoms like irritation, fatigue, anxiety, headaches, and compromised cognitive performance (Griffin, 2003). Elevated CO_2 levels have also been associated with sleep apnea, emotional responses such as fear and panic, verified in humans and rats. Increased CO_2 concentration, a consequence of climate change, contributes to health impacts through heatwaves and extreme weather events. Vulnerable populations face elevated heat stress during heatwaves, increasing the risk of heat-related mortality, heat stroke, hyperthermia, low-quality sleep, acute kidney injury, and complications with pregnancy (Raganati et al., 2021). Additionally, it can exacerbate preexisting cardiovascular and respiratory diseases. CO_2 adversely affects cognitive function, with levels around 1000 ppm leading to reduced concentration and focus, causing discomfort from breathing stuffy air. Higher levels, approximately 2500 ppm, significantly impact cognitive thought processes, particularly for tasks requiring deeper thinking (Suomalainen et al., 2013).

Symptoms vary widely even among healthy individuals, with the seriousness dependent on CO_2 concentration and exposure duration. Due to its odorless nature and lack of irritation, CO_2 has poor warning properties (Wang et al., 2011). Thankfully, conditions resulting from low to moderate exposures are generally reversible upon removal from the high CO_2 environment. Beyond the direct health effects, the increasing concentration of CO_2 due to climate change has broader repercussions, leading to death and illness from more frequent extreme weather events, disruptions in food systems, increases in zoonoses and food-, water-, and vector-borne diseases, and mental health issues (Buhre et al., 2005). Moreover, climate change undermines social determinants of good health, affecting livelihoods, equality, access to healthcare, and social support structures. Vulnerable populations, including women, children, ethnic minorities, poor communities, migrants, older individuals, and those with underlying health conditions, disproportionately bear the burden of these climate-sensitive health risks (Gabrielli et al., 2020; Mac Dowell et al., 2017; Shah et al., 2013).

3 Nanotechnology-Enabled Carbon Capture Techniques

The twentieth century witnessed a significant surge in CO_2 pollution, which accelerated at an alarming rate in the twenty-first century due to rapid industrialization and urbanization. This drastic increase has resulted in catastrophic impacts on global economics and ecosystems, contributing to a rise in the global average temperature beyond the normal range, reaching 0.6–1 °C. Approximately 60% of CO_2 emissions worldwide can be attributed to large stationary sources, further exacerbating the challenges posed by climate change. The consequences, including rising sea levels, floods, and droughts, present extreme global dangers to all living organisms (Roy et al., 2023). In response, carbon capture becomes increasingly crucial to create a suitable living environment for organisms worldwide.

3.1 Carbon Capture Technology

The fundamental principle of carbon capture technology involves the capture of dense CO_2 emissions from polluted sectors, either pre- or post-combustion, preventing their release into the atmosphere. Subsequently, the captured CO_2 undergoes compression and is safely sequestered in underground sediment layers. With continuous monitoring, this sequestration method aims to isolate the CO_2 from the atmosphere permanently. Additionally, there is potential for recycling the stored CO_2 for future utilization. CO_2 capture technology is broadly categorized into pre-combustion, post-combustion, and oxy-fuel combustion technologies (McLaughlin et al., 2023).

3.1.1 Pre-combustion Technology

This method involves extracting CO_2 from flue gas after the combustion process. The primary hurdle for post-combustion CO_2 capture lies in its substantial parasitic load. The CO_2 concentration in combustion flue gas is typically low, so the capture unit incurs significant energy penalties and associated costs to attain the required CO_2 concentration for transportation. The US national energy technology

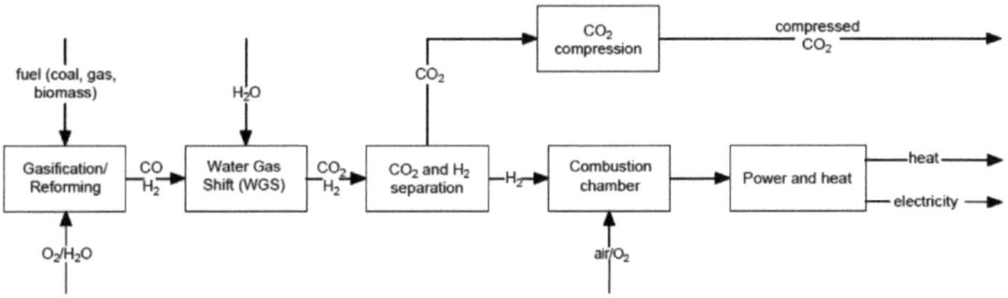

Fig. 1 Schematic illustration of a pre-combustion technology. (Reproduced with permission from Qiao et al. (2015))

laboratory has approximated that implementing post-combustion CO_2 capture could raise the cost of electricity production by as much as 70% (Akimoto et al., 2021; Harun-Ur-Rashid et al., 2021). A schematic illustration of a pre-combustion process is depicted in Fig. 1.

3.1.2 Post-combustion Technology

In this method, the fuel, typically coal or natural gas, undergoes pretreatment before combustion. For coal, this pretreatment involves gasification in a low-oxygen environment within a gasifier, producing syngas mainly composed of CO and H_2 while being relatively free from other pollutant gases. The syngas then undergoes the water gas shift reaction with steam, generating more H_2 and converting CO to CO_2. Subsequently, the H_2 is burned in the presence of air, resulting in mainly N_2 and water vapor (Karimi et al., 2023). The pre-combustion capture technique applies to integrated gasification combined cycle power plants utilizing coal as a fuel source. In case of natural gas, primarily composed of CH_4, it can be reformed into syngas containing H_2 and CO. The content of H_2 can be increased through the water gas shift reaction and the subsequent process mirrors that described earlier for coal.

A performance and cost analysis was conducted on advanced combined cycle gas turbine plants utilizing natural gas, incorporating a pre-combustion CO_2 capture system. The obtained CO_2 capture efficiency reached 80%, with the cost of avoided CO_2 reaching \$29 per ton for this advanced design concept. Reactions occurring in post-combustion (Nakao et al., 2019) are given below:

$$\text{Coal} \rightarrow \text{CO} + \text{H}_2 \left(\text{Gasification}\right)$$

$$\text{CO} + \text{H}_2\text{O} \rightarrow \text{H}_2 + \text{CO}_2 \left(\text{water} - \text{gas}\,\text{shift}\right)$$

$$\text{CH}_4 + \text{H}_2\text{O} \rightarrow \text{CO} + \text{H}_2 \left(\text{Reform}\right)$$

3.1.3 Oxy-fuel Combustion

Oxy-fuel combustion involves using oxygen instead of air for the combustion process. This substitution reduces the presence of nitrogen in the exhaust gas, positively influencing subsequent separation processes. A notable advantage of this

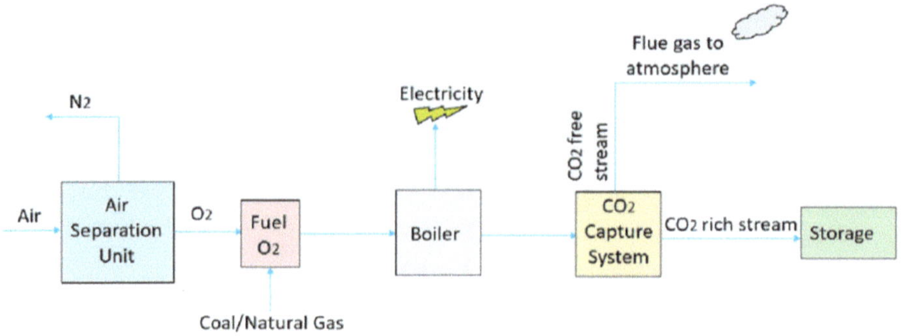

Fig. 2 A schematic illustration of an oxy-fuel technology. (Reproduced with permission from Elhenawy et al. (2020))

approach is a substantial reduction in thermal NO*x* (Garcia et al., 2022). With the use of pure oxygen for combustion, the primary components of the flue gases are CO_2, water, particulates, and SO_2. Traditional methods, such as electrostatic precipitators and flue gas desulphurization, can effectively remove particulates and SO_2. The remaining gases, characterized by a high concentration of CO_2, can be compressed, transported, and stored (Wagaarachchige et al., 2023). Figure 2 illustrates an oxy-fuel combustion process.

3.2 Criteria for a Suitable Method

Carbon capture and storage technology plays a crucial role in mitigating CO_2 emissions from the environment, but it is essential to consider the limitations of various methods to determine their feasibility. The pre-combustion method focuses on coal gasification plants to enhance sorption efficiency for high CO_2 concentrations from industrial production (Tamidi et al., 2023). In contrast, post-combustion and oxy-fuel combustion technologies primarily address coal-fired and gas-fired plants. Oxy-fuel combustion technology, being more mature and easily retrofitted into existing plants, is efficient for high CO_2 concentrations, enhancing absorption efficiency (Nandy et al., 2016). However, it has drawbacks, including a high-efficiency drop, energy penalty, costly production of cryogenic O_2, and potential equipment corrosion issues. Post-combustion technology, while mature and easily retrofitted, faces challenges due to its low efficiency in CO_2 capture (Jomekian & Behbahani, 2021). Despite being a mature process, its feasibility is hindered by the low efficiency of CO_2 capture. Oxy-fuel technology also presents disadvantages, including a significant efficiency drop, energy penalty, costly production of cryogenic O_2, and potential equipment corrosion issues. Among carbon capture and storage technologies, pre-combustion technology emerges as the most feasible due to its versatility (Waheed et al., 2016). While it has its drawbacks, the high efficiency in capturing CO_2 makes it a tremendous process. Comparatively, post-combustion and oxy-fuel technologies have similar costs. However, for gas-fired plants, the cost per ton of CO_2 avoided for post-combustion capture is almost 50% lower than the other two capture technologies (Yu et al., 2012). Moreover, post-combustion CO_2 capture is generally the least efficient option, incurring an energy penalty of about 8% and 6% for coal-fired and gas-fired plants, respectively. Understanding the advantages and limitations of each method is crucial for making informed decisions on the most suitable carbon capture and storage technology for specific applications.

3.3 CO_2 Segregation Technology

The separation technology of CO_2 holds significant importance for transportation and conversion to supply alternative energy storage across various sectors. Advanced technologies have been developed for the transport of CO_2, including wet scrubbers, dry regenerable sorbents, membranes, and cryogenics. Segregation becomes a crucial process in utilizing valuable fuels derived from CO_2. Several segregation methods have been developed, and some of these are described next.

3.3.1 Adsorption

Adsorption processes based on a liquid absorbent and a solid sorbent bind CO_2 on their surfaces. Several conditions are crucial for the effectiveness of this adsorption process, with the sorbents requiring larger specific surface areas and high regeneration abilities to facilitate efficient CO_2 adsorption. Various sorbents are utilized in these processes, including molecular sieves, activated carbon, zeolites, calcium oxides, hydrotalcites, and lithium zirconate (Pröll et al. 2016). The adsorbed CO_2 can be recovered by manipulating the pressure swing adsorption or temperature swing adsorption of the system containing the CO_2-saturated sorbent. Pressure swing adsorption can recover up to 85% CO_2, where CO_2 is adsorbed in the solid sorbent under specific pressure conditions. Higher pressure proves to be efficient for the adsorption process in CO_2 capture. Additionally, the gas is released in a hot air system or through stream injections by adjusting the temperature. Regeneration time is critical because a longer contact time allows more CO_2 to be absorbed in the sorbent (Fernandez et al., 2022).

3.3.2 Absorption

Absorption is another technique employed to segregate CO_2 from flue gas, using a liquid sorbent for the separation process. The sorbent can be regenerated through van der Waals attraction or electrolytic systems. This method stands out as the most mature approach for CO_2 separation, with typical organic amine-based sorbents proving particularly feasible for transportation (Webley et al., 2017). Various inorganic aqueous mediums, such as sodium bicarbonate, potassium bicarbonate, and sodium carbonate, are commonly used as capture media for CO_2. Confirming the capture of CO_2 involves injecting CO_2 into a basic medium and monitoring the pH change. An effective organic liquid sorbent for CO_2 capture is alkanolamines, with a capture efficiency of up to 90%. Alkanolamines like monoethanolamide (MEA) and diethanolamine (DEA) are highly efficient for capturing CO_2. Other sorbents include piperazine and anion-functionalized ionic liquid (Pashaei & Ghaemi, 2022; Yeo et al., 2012). Due to its higher volatility, piperazine has been observed to react faster than MEA. However, it is important to note that amine emissions may degrade into nitrosamines and nitramines, posing potential risks to human health. The chilled ammonia process utilizes aqueous ammonium salt to capture CO_2, and the captured CO_2 can be regenerated at elevated temperatures and pressures, reducing downstream compression requirements (Drage et al., 2009).

3.3.3 Chemical Looping Combustion

Looping combustion is an intriguing method that involves a metal oxide-based component instead of direct oxygen for combustion in an oxy-fuel cell system. In this system, a metal oxide is an oxygen carrier instead of pure oxygen. The metal oxide undergoes reduction and oxidation processes to produce CO_2. The metal is then cycled through another stage and recycled within the storage system. Water produced as a by-product is also obtained without impurities, resulting in pure CO_2. A wide variety of metal oxides are feasible and active for this looping combustion process, including Fe_2O_3, NiO, CuO, and Mn_2O_3. The effectiveness of different metal oxides is assessed in terms of their selectivity in product formation, transformation, and conversion (Zeng et al., 2022). A schematic illustration of chemical looping combustion is presented in Fig. 3.

3.3.4 Membrane Separation

The membrane separation technique is efficient in isolating pure CO_2 by allowing membranes to selectively pass CO_2 while preventing other components from passing through. The membrane, typically made of polymer, consists of a thin selective layer bonded to a thicker, nonselective layer that provides mechanical support. This configuration is utilized to separate CO_2 from other natural gases like O_2 and N_2. Compared to ordinary methods, this technique increases separation efficiency from 82% to 88% (Rahman, 2016). Various materials, including ceramics, metallic membranes, and polymeric membranes, are employed in the membrane diffusion process, contributing to significantly improved CO_2 separation compared to liquid absorption processes. The membrane system is particularly effective for liquids, eliminating the need for specific temperature and pressure conditions. The flow pressure across the solution and membrane determines the efficiency. However, this system is vulnerable to the flow of flue gas or other impurities, posing a significant challenge in its application (Peredo-Mancilla et al., 2019). A schematic model for membrane separation is shown in Fig. 4.

Among others, adsorption-based carbon capture and storage technology is discussed next.

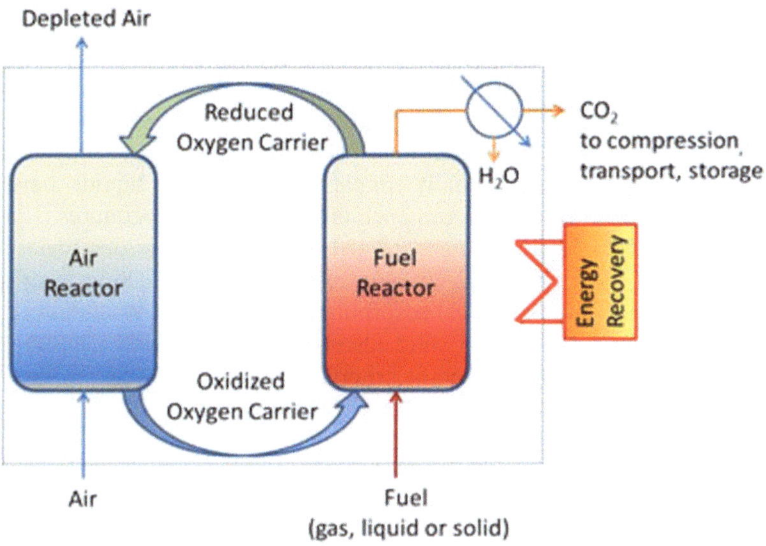

Fig. 3 A schematic representation of chemical looping combustion. (Reproduced with permission from IFPEN (2019))

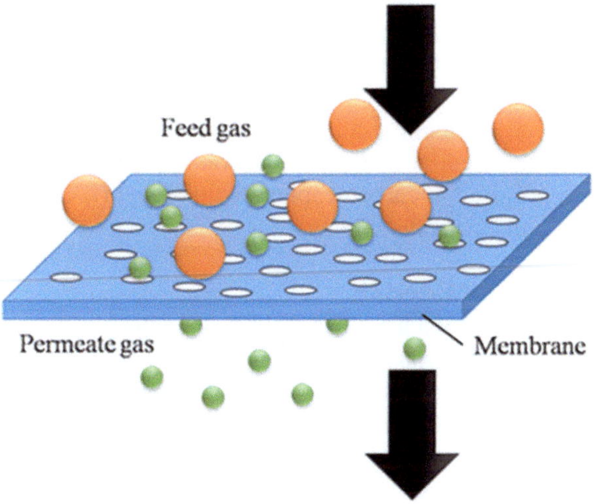

Fig. 4 A schematic representation of the membrane separation technique. (Reproduced with permission from Ji and Zhao (2017))

3.4 Adsorption

In adsorption, gases come into contact with porous particles capable of adsorbing the gas due to surface affinity, separating it from the gas mixture. Adsorption techniques are considered complex separation methods. Compared to other methods, adsorption processes are operationally simple, do not produce liquids, have lower energy costs for regeneration, and can operate at different temperatures (Lin et al., 2017). However, there are challenges related to corrosion. Various materials are used to implement carbon capture technology using adsorption. The sorbent adsorption mechanism is based on physical adsorption, where van der Waals forces play a role in adsorbing on the surface of solid adsorbents. To effectively capture CO_2 on the solid surface of the sorbent, specific criteria should be followed:

(a) High adsorption capacity
(b) High selectivity for CO_2
(c) Adequate adsorption/desorption kinetics
(d) Stability during repeated adsorption/desorption cycling
(e) Low cost

Adsorption isotherms are used to evaluate adsorption capacity, and knowing the adsorption equilibrium capacity helps determine the amount of adsorbent required. Adsorption selectivity, represented by the ratio of CO_2 capacity to other bulk gas components, is a crucial property (Boer et al., 2023). Selectivity directly impacts the purity of the captured CO_2 during the separation process. A good adsorbent with high selectivity shows fast adsorption kinetics. Adsorption kinetics helps determine how stable the material is over multiple cycles. The cost of an absorbent is a critical factor in its evaluation (Aouaini et al., 2022). Low cost, good stability, selectivity, and high capacity are essential factors for carbon capture technologies. There are two types of materials used in carbon capture technology: (i) carbonaceous materials and (ii) non-carbonaceous materials.

3.4.1 Carbonaceous Materials

Carbonaceous materials play a leading role in CO_2 capture, with a diverse range of carbon-based adsorbent materials, including activated carbons, graphenes, and carbon nanotubes. Carbon-based materials are advantageous for carbon capture due to their low cost and abundance. Activated carbon, carbon nanomaterials, and doped carbon materials are all efficient choices for capturing carbon, benefiting from the abundance of carbon materials.

Activated Carbon

Activated carbons are widely used as adsorbents in various industrial applications such as gas purification, water treatment, and air pollution monitoring. These carbons are derived from carbon-containing biological materials like wood, coconut shells, sawdust, and olive stones. One of the key advantages of activated carbon is its low cost as it is derived from abundant natural materials (Acevedo et al., 2020).

Activated carbon, whether in bulk or nano states, effectively captures carbon. The production of activated carbon involves two main steps: carbonization and activation. Carbonization is achieved through thermal treatment (500–1200 °C) in an inert atmosphere, forming carbonaceous material called char, which initially has poor surface properties. The activation process involves mixing the char with an activating agent and heating it to a desired temperature (Thommes et al. 2015). Activation can be achieved through physical or chemical methods, creating high porosity in the materials and increasing adsorption capacity. Due to pore size or morphological limitations, Virgin carbonaceous materials often face challenges in carbon capture.

However, functionalized carbon materials exhibit high carbon capture capacity, especially those doped with heteroatoms like nitrogen and sulfur. Nitrogen doping enhances material conductivity, ultimately improving CO_2 adsorption. While also contributing to conductivity, sulfur requires thorough washing of the precursor or doped materials since it reacts with CO_2 (Ioannidou & Zabaniotou, 2007). Doping with both nitrogen and sulfur proves to be a promising technique for higher CO_2 capture efficiency. Additionally, phosphorous doping can be efficient, but the dopant amount and temperature need to be carefully controlled. Phosphorous-doped materials can be tuned to provide more active sites and increased selectivity (Tsai et al., 1997).

3.4.2 Non-carbonaceous Materials

Non-carbonaceous materials refer to materials that are devoid of carbon. These materials encompass metal-based clusters, metal oxides, and metal–organic frameworks (MOFs), offering alternative options for exploring various shapes, sizes, and conformational changes to enhance the CO_2 capture capacity in carbon capture technologies.

Zeolite

Zeolites are highly microporous crystalline framework materials known for their unique properties, including pore size, molecular sieving ability, and varied chemical composition influencing their adsorption performance. Silicate frameworks in zeolites result in a negative charge, while aluminum imparts a positive charge with exchangeable cations within the pore structure (Yu et al., 2017). Zeolites utilize alkali cations to create strong electrostatic interactions with acidic molecules like CO_2. Thus, modifying the Si/Al ratio and extra-framework cations can significantly impact CO_2 adsorptive properties (Mason et al., 2011). However, the adsorption capacity of zeolites is affected by the presence of moisture in a gas stream, which can have a detrimental effect on CO_2 adsorption. Moisture on zeolite surfaces can block access for CO_2 adsorption. Studies have shown that the CO_2 adsorption capacity of CaX zeolite drastically decreases with increasing water vapor concentration. Zeolites with a low Si/Al ratio are considered promising adsorbents for CO_2 adsorption and separation applications. Nevertheless, CO_2 adsorption on zeolites has limitations, including sensitivity to temperature, pressure, and water vapor. The capacity of zeolites for CO_2 adsorption decreases with increasing temperature and increases with the gas-phase partial pressure of CO_2. Additionally, zeolites are highly sensitive to water content in flue gas, requiring extensive drying before CO_2 capture (Sumida et al., 2012).

Metal–Organic Framework (MOF)
MOFs are structures composed of metal clusters coordinated with organic ligands, forming various structures in terms of dimensions, size, and shape. MOFs are renowned for their extensive porous structures incorporating metal ions and organic subunits (Maia et al., 2021). The repetition of organic subunits or linkers leads to diverse structural dimensions, introducing disorder and increasing capture capacity. MOFs exhibit robustness, high surface area, thermal and chemical stabilities, high void volume, and low densities. These frameworks find applications in gas storage, ion exchange, molecular separation, and catalysis (Saha et al., 2012). MOFs consist of metal-based nodes forming a hybrid network that links different organic groups. Depending on the nature of organic linkers, the well-defined crystal structure, strong coordination bonds, and tunability distinguish MOFs from other materials like zeolites. The synthesis of MOFs is relatively easy and feasible, employing various methods such as solvothermal, microwave, sonication, mechanochemical, and solid-start synthesis. Challenges in MOF preparation include optimizing reaction conditions for high yield and crystallinity, influenced by parameters like temperature, solvent, time, reagent concentration, and solvent pH (Martín-Calvo et al., 2012). Structurally, MOFs offer various interfaces, including rigid frameworks, flexible/dynamic frameworks, surface-functionalized frameworks, and open metal sites.

Permanent porosity enhances stability and reactivity, allowing tunable porosity for selective adsorption or desorption. Flexible MOFs, although complex due to selective adsorption, can restore adsorption of gas molecules after solvent-induced transformations, benefiting CO_2 capture (Demir et al., 2022; Lian et al., 2019). The choice of metal in MOFs significantly influences CO_2 capture technology. Tuning and optimizing the metal behavior impact morphological disorganization or orientation of selectivity. Researchers often evaluate the metal center's role in the capacity and selectivity of CO_2 adsorption. Surface-functionalized frameworks, achieved through ligand modification or coordination, enhance MOFs' capacity and selectivity for CO_2 adsorption by increasing the affinity for CO_2. Functional groups on the surface enhance the selective interaction between CO_2 and the functionalized molecule (Maji et al., 2005).

4 Challenges and Future Directions

CO_2 capture is crucial to addressing global environmental challenges but comes with several significant challenges. One primary challenge is the substantial amount of energy required for CO_2 consumption. Methodological issues further complicate capture technology, including the high energy and material costs of moving large volumes of air through absorbent structures, leading to elevated capture expenses. The segregation of CO_2 from various sources, whether from stationary points or distributed sources, requires thorough investigation and overcoming obstacles. CO_2 capture methods can generally be categorized into chemical and physical approaches, each with its own challenges. High energy costs for oxygen production and susceptibility to air leakage into the system are notable issues. Additionally, special materials are needed to withstand high flame temperatures, and while recycled CO_2 can help moderate these temperatures, it may not be a sufficient solution. The efficiency of solvothermal methods can decrease due to the low CO_2 loading capacity of the solvent. Challenges extend to the adsorbent, reactor, and system scales. Processes like adsorption and reduction demand high energy for adsorbent regeneration and CO_2 activation, respectively. Amine degradation in the presence of nitrogen oxides (NOx), sulfur oxides (SOx), O_2, particulate matter, and corrosion in amine equipment pose further challenges. The solvent regeneration process is energy-intensive, and the waste stream generated can be hazardous. Increasing selectivity often necessitates a larger surface area for the medium, as higher purity requirements lead to a faster decrease in driving force. Purity remains a significant concern in CO_2 capture technology. Addressing these challenges requires innovative solutions and ongoing research efforts.

5 Conclusion

The generation of energy from nonrenewable sources harms the environment, impacting both ecosystems and public health. Technologies that minimize the emission of harmful chemicals and particulates are crucial to address this issue. In evaluating the technical and economic feasibility of CO_2 capture processes, adsorption techniques are considered challenging separation methods. Despite their technical limitations and moderate efficiency in gas separation, advancements are being made to reduce costs and enhance the overall viability of these processes. Absorption processes involving the transfer of gases by solubilizing them in the liquid phase of a mixture are among the most promising and mature technologies for large-scale application. The ability of materials to absorb gases plays a key role in post-capture processes, making absorption an efficient and economically viable technology. Membrane separation relies on the selectivity of materials constituting the membranes, with parameters such as drum permeability, particle size, selectivity, and pore volume influencing membrane efficiency. This presents a challenge in membrane separation processes. Chemical capture, another technology under development, involves chemical reactions with CO_2 to produce final products, predominantly carbonates, contributing to the generation of carbonates for sustainability.

Both carbon-based and non-carbon-based materials exhibit efficiency in capturing CO_2. The porous structure of carbon materials allows easy penetration of CO_2 into their surface, while the functionality of non-carbon materials facilitates the adsorption of CO_2 on their surface. Commonly used materials for capture include metal–organic frameworks and zeolites, which, due to their inorganic and organic composition, can convert CO_2 through phase-changing reactions. Monitoring the development of new technologies and optimizing existing ones is crucial for future projections in carbon capture, enabling accurate estimations of time, cost, and environmental impacts. This analysis aids in quantifying existing technologies and informing decision-making processes, ultimately contributing to reducing anthropogenic CO_2 emissions.

Acknowledgments A.N. Chowdhury and A.B. Imran express sincere gratitude to the Committee for Advanced Studies and Research (CASR) at BUET for their support.

References

A Scientific Assessment. (2016). *The impacts of climate change on human health in the United States*. U.S Global Change Research Program.

Abbass, K., Qasim, M. Z., Song, H., Murshed, M., Mahmood, H., & Younis, I. (2022). A review of the global climate change impacts, adaptation, and sustainable mitigation measures. *Environmental Science and Pollution Research, 29*(28), 42539–42559.

Acevedo, S., Giraldo, L., & Moreno-Piraján, J. C. (2020). Adsorption of CO2 on activated carbons prepared by chemical activation with cupric nitrate. *ACS Omega, 5*(18), 10423–10432.

Akimoto, K., Sano, F., Oda, J., Kanaboshi, H., & Nakano, Y. (2021). Climate change mitigation measures for global net-zero emissions and the roles of CO_2 capture and utilization and direct air capture. *Energy and Climate Change, 2*, 100057.

Aouaini, F., Bouaziz, N., Alfwzan, W., Khemiri, N., Elqahtani, Z., & Lamine, A. B. (2022). Adsorption of CO_2 on ZSM-5 zeolite: Analytical investigation via a multilayer statistical physics model. *Applied Sciences, 12*(3), 1558.

Bell, J. E., Brown, C. L., Conlon, K., Herring, S., Kunkel, K. E., Lawrimore, J., et al. (2018). Changes in extreme events and the potential impacts on human health. *Journal of the Air & Waste Management Association, 68*(4), 265–287.

Bobb, J. F., Obermeyer, Z., Wang, Y., & Dominici, F. (2014). Cause-specific risk of hospital admission related to extreme heat in older adults. *JAMA, 312*(24), 2659–2667.

Boer, D. G., Langerak, J., & Pescarmona, P. P. (2023). Zeolites as selective adsorbents for CO_2 separation. *ACS Applied Energy Materials, 6*(5), 2634–2656.

Bograd, S. J., Checkley, D. A., Jr., & Wooster, W. S. (2003). CalCOFI: A half century of physical, chemical, and biological research in the California Current System. *Deep Sea Research Part II: Topical Studies in Oceanography, 50*, 2349–2353.

Buhre, B. J., Elliott, L. K., Sheng, C. D., Gupta, R. P., & Wall, T. F. (2005). Oxy-fuel combustion technology for coal-fired power generation. *Progress in Energy and Combustion Science, 31*(4), 283–307.

Bui, M., Adjiman, C. S., Bardow, A., Anthony, E. J., Boston, A., Brown, S., Fennell, P. S., Fuss, S., Galindo, A., & Hackett, L. A. (2018). Carbon capture and storage (CCS): The way forward. *Energy & Environmental Science, 11*(5), 1062–1176.

Campbell-Lendrum, D., & Prüss-Ustün, A. (2019). Climate change, air pollution and noncommunicable diseases. *Bulletin of the World Health Organization, 97*(2), 160.

CEDA Archive Web Browser. (2023). Available at https://data.ceda.ac.uk/badc/ar6_wg1/data/spm/spm_02/v20210809. Accessed December 15, 2023.

Chao, J. (2012). *Elevated indoor carbon dioxide impairs decision-making performance*. Lawrence Berkeley National Laboratory.

Chersich, M. F., Scorgie, F., Filippi, V., Luchters, S., & Climate Change and Heat-Health Study Group. (2023). Increasing global temperatures threaten gains in maternal and newborn health in Africa: A review of impacts and an adaptation framework. *International Journal of Gynecology & Obstetrics, 160*(2), 421–429. https://doi.org/10.1002/ijgo.14381

Chien, F., Anwar, A., Hsu, C.-C., Sharif, A., Razzaq, A., & Sinha, A. (2021). The role of Information and Communication Technology in encountering environmental degradation: Proposing an SDG framework for the BRICS countries. *Technology in Society, 65*, 101587.

Ciscar, J. C., Iglesias, A., Feyen, L., Szabó, L., Van Regemorter, D., Amelung, B., et al. (2011). Physical and economic consequences of climate change in Europe. *Proceedings of the National Academy of Sciences, 108*(7), 2678–2683.

de Gruijl, F., & Leun, J. (2000). Environment and health: 3. Ozone depletion and ultraviolet radiation. *CMAJ, 163*(7), 851–855.

Demir, H., Aksu, G. O., Gulbalkan, H. C., & Keskin, S. (2022). MOF membranes for CO_2 capture: Past, present and future. *Carbon Capture Science & Technology, 2*, 100026.

Dong, Y., Coleman, M., & Miller, S. A. (2021). Greenhouse Gas Emissions from Air Conditioning and Refrigeration Service Expansion in Developing Countries. *Annual Review of Environment and Resources, 46*, 59–83. https://doi.org/10.1146/annurev-environ-012220-034103

Drage, T. C., Kozynchenko, O., Pevida, C., Plaza, M. G., Rubiera, F., Pis, J. J., Snape, C. E., & Tennison, S. (2009). Developing activated carbon adsorbents for pre-combustion CO_2 capture. *Energy Procedia, 1*(1), 599–605.

Duarte, C. M., Jaremko, Ł., & Jaremko, M. (2020). Hypothesis: Potentially systemic impacts of elevated CO_2 on the human proteome and health. *Frontiers in Public Health, 8*, 543322.

Duyck, S., Lennon, E., Obergassel, W., & Savaresi, A. (2018). Human rights and the Paris Agreement's implementation guidelines: Opportunities to develop a rights-based approach. *Carbon & Climate Law Review, 12*(3), 191–202.

Ebi, K. L., Vanos, J., Baldwin, J. W., Bell, J. E., Hondula, D. M., Errett, N. A., Hayes, K., Reid, C. E., Saha, S., Spector, J., & Berry, P. (2021). Extreme weather and climate change: Population health and health system implications. *Annual Review of Public Health, 42*(1), 293–315. https://doi.org/10.1146/annurev-publhealth-012420-105026

Elhenawy, S. E. M., Khraisheh, M., AlMomani, F., & Walker, G. (2020). Metal-organic frameworks as a platform for CO2 capture and chemical processes: Adsorption, membrane separation, catalytic-conversion, and electrochemical reduction of CO_2. *Catalysts, 10*(11), 1293.

Fernandez, D. C., Morales, D. S., Jiménez, J. R., & Fernández-Rodriguez, J. M. (2022). CO_2 adsorption by organohydrotalcites at low temperatures and high pressure. *Chemical Engineering Journal, 431*, 134324.

Gabrielli, P., Gazzani, M., & Mazzotti, M. (2020). The role of carbon capture and utilization, carbon capture and storage, and biomass to enable a net-zero-CO_2 emissions chemical industry. *Industrial and Engineering Chemistry Research, 59*(15), 7033–7045. https://doi.org/10.1021/acs.iecr.9b06579

Garcia, J. A., Villen-Guzman, M., Rodriguez-Maroto, J. M., & Paz-Garcia, J. M. (2022). Technical analysis of CO_2 capture pathways and technologies. *Journal of Environmental Chemical Engineering, 10*, 108470.

Ghorani-Azam, A., Riahi-Zanjani, B., & Balali-Mood, M. (2016). Effects of air pollution on human health and practical measures for prevention in Iran. *Journal of Research in Medical Sciences, 21*, 65.

Griffin, J. M. (Ed.). (2003). *Global climate change: The science, economics and politics* (Vol. 4). Edward Elgar Publishing.

Howard, J. et al. (2013). *Oceans and marine resources in a changing climate.* Oceanography and Marine Biology: An Annual Review, 51, 71-192

Gruza, G., Rankova, E., Razuvaev, V., & Bulygina, O. (1999). Indicators of climate change for The Russian Federation. *Climatic Change, 42*, 219–242.

Hardy, J. T. (2003). *Climate change: Causes, effects, and solutions.* Wiley.

Harun-Ur-Rashid, M., & Imran, A. B. (2023). Engineered nanomaterials for energy conversion cells. In *Applications of emerging nanomaterials and nanotechnology* (Vol. 148). Materials Research Forum LLC.

Harun-Ur-Rashid, M., Foyez, T., & Imran, A. B. (2021). Fabrication of stretchable composite thin film for superconductor applications. In *Sensors for stretchable electronics in nanotechnology* (pp. 63–78). CRC Press.

Hatfield, J. L., & Prueger, J. H. (2011). Agroecology: Implications for plant response to climate change. In *Crop adaptation to climate change* (pp. 27–43). Wiley.

Hatfield, J. L., & Prueger, J. H. (2015). Temperature extremes: Effect on plant growth and development. *Weather and Climate Extremes, 10*, 4–10.

Hoegh-Guldberg, O. (1999). Climate change, coral bleaching and the future of the world's coral reefs. *Marine and Freshwater Research, 50*, 839–866.

Hönisch, B., Ridgwell, A., Schmidt, D. N., Thomas, E., Gibbs, S. J., Sluijs, A., Zeebe, R., Kump, L., Martindale, R. C., Greene, S. E., Kiessling, W., Ries, J., Zachos, J. C., Royer, D. L., Barker, S., Marchitto, T. M., Moyer, R., Pelejero, C., Ziveri, P., Foster, G. L., & Williams, B. (2012). The geological record of ocean acidification. *Science, 335*(6072), 1058–1063. https://doi.org/10.1126/science.1208277

Hopwood, N., & Cohen, J. (2008). *Greenhouse gases and society. Retrieved April 8, 2008.* https://oa.mg/work/2554944231.

Hyeon, J. E., Kim, S. W., Park, C., & Han, S. O. (2015). Efficient biological conversion of carbon monoxide (CO) to carbon dioxide (CO2) and for utilization in bioplastic production by *Ralstonia eutropha* through the display of an enzyme complex on the cell surface. *Chemical Communications, 51*(50), 10202–10205.

IFPEN. (2019). *CO_2 capture: Clarifying chemical looping combustion (CLC).* IFPEN. Available at https://www.ifpenergiesnouvelles.com/article/co2-capture-clarifying-chemical-looping-combustion-clc. Accessed July 10, 2024.

Intergovernmental Panel on Climate Change (IPCC). (2023a). The Earth's energy budget, climate feedbacks and climate sensitivity. In *Climate change 2021—The physical science basis: Working Group I contribution to the sixth assessment report of the Intergovernmental Panel on Climate Change* (pp. 923–1054). Cambridge University Press.

Intergovernmental Panel on Climate Change (IPCC). (2023b). Changing state of the climate system. In *Climate change 2021—The physical science basis: Working Group I contribution to the sixth assessment report of the Intergovernmental Panel on Climate Change* (pp. 287–422). Cambridge University Press.

Ioannidou, O., & Zabaniotou, A. (2007). Agricultural residues as precursors for activated carbon production—A review. *Renewable and Sustainable Energy Reviews, 11*(9), 1966–2005.

Ji, G., & Zhao, M. (2017). Membrane separation technology in carbon capture. In *Recent advances in carbon capture and storage* (pp. 59–90). IntechOpen.

Johnson, J. E., Allain, V., Basel, B., Bell, J. D., Chin, A., Dutra, L. X. C., Hooper, E., Loubser, D., Lough, J., Moore, B. R., & Nicol, S. (2020). Impacts of climate change on marine resources in the Pacific Island region. In L. Kumar (Ed.), *Climate change and impacts in the Pacific, Springer climate* (pp. 359–402). Springer International Publishing.

Jomekian, A., & Behbahani, R. M. (2021). Experimental, modeling and AspenPlus simulation of different configurations of membrane separation systems for highly loaded CO2 selective Pebax 1657-ZIF-8 membrane. *Journal of Membrane Science and Research, 7*(3), 209–223.

Kabir, R., Khan, H. T. A., Ball, E., & Caldwell, K. (2014). Climate change and public health situations in the coastal areas of Bangladesh. *International Journal of Social Science Studies, 2*, 109.

Karimi, M., Shirzad, M., Silva, J. A., & Rodrigues, A. E. (2023). Carbon dioxide separation and capture by adsorption: A review. *Environmental Chemistry Letters, 21*, 2041–2084.

Keith, D. W. (2009). Why capture CO_2 from the atmosphere? *Science, 325*(5948), 1654–1655. https://doi.org/10.1126/science.1175680

Khan, M. I. (2018). Comparative Well-to-Tank energy use and greenhouse gas assessment of natural gas as a transportation fuel in Pakistan. *Energy for Sustainable Development, 43*, 38–59.

Kogo, B. K., Kumar, L., & Koech, R. (2021). Climate change and variability in Kenya: A review of impacts on agriculture and food security. *Environment, Development and Sustainability, 23*, 23–43.

Kumar, J., C.R., & Majid, M. A. (2020). Renewable energy for sustainable development in India: Current status, future prospects, challenges, employment, and investment opportunities. *Energy, Sustainability and Society, 10*(1), 2. https://doi.org/10.1186/s13705-019-0232-1

Laffoley, D., & Baxter, J. M. (2019). *Ocean deoxygenation: Everyone's problem: Causes, impacts, consequences and solutions: Summary for policy makers*. International Union for Conservation of Nature (IUCN).

Lammel, G., & Graßl, H. (1995). Greenhouse effect of NOX. *Environmental Science and Pollution Research, 2*, 40–45.

Leduc, A. O. H. C., Munday, P. L., Brown, G. E., & Ferrari, M. C. O. (2013). Effects of acidification on olfactory-mediated behaviour in freshwater and marine ecosystems: A synthesis. *Philosophical Transactions of the Royal Society B: Biological Sciences, 368*(1627), 20120447. https://doi.org/10.1098/rstb.2012.0447

Lian, X., Xu, L., Chen, M., Wu, C., Li, W., Huang, B., & Cui, Y. (2019). Carbon dioxide captured by metal organic frameworks and its subsequent resource utilization strategy: A review and prospect. *Journal of Nanoscience and Nanotechnology, 19*(6), 3059–3078.

Lin, Z., Wei, J., Geng, L., Mei, D., & Liao, L. (2017). Adsorption of carbon dioxide by a novel amine impregnated ZSM-5/KIT-6 composite. *RSC Advances, 7*(86), 54422–54430.

Mac Dowell, N., Fennell, P. S., Shah, N., & Maitland, G. C. (2017). The role of CO_2 capture and utilization in mitigating climate change. *Nature Climate Change, 7*(4), 243–249.

Maia, R. A., Louis, B., Gao, W., & Wang, Q. (2021). CO_2 adsorption mechanisms on MOFs: A case study of open metal sites, ultra-microporosity and flexible framework. *Reaction Chemistry & Engineering, 6*(7), 1118–1133.

Maji, T. K., Mostafa, G., Matsuda, R., & Kitagawa, S. (2005). Guest-induced asymmetry in a metal−organic porous solid with reversible single-crystal-to-single-crystal structural transformation. *Journal of the American Chemical Society, 127*(49), 17152–17153. https://doi.org/10.1021/ja0561439

Malhi, G. S., Kaur, M., & Kaushik, P. (2021). Impact of climate change on agriculture and its mitigat ion strategies: A review. *Sustainability, 13*(3), 1318.

Manisalidis, I., Stavropoulou, E., Stavropoulos, A., & Bezirtzoglou, E. (2020). Environmental and health impacts of air pollution: A review. *Frontiers in Public Health, 8*, 14.

Martín-Calvo, A., Lahoz-Martín, F. D., & Calero, S. (2012). Understanding carbon monoxide capture using metal–organic frameworks. *The Journal of Physical Chemistry C, 116*(11), 6655–6663.

Mason, J. A., Sumida, K., Herm, Z. R., Krishna, R., & Long, J. R. (2011). Evaluating metal–organic frameworks for post-combustion carbon dioxide capture via temperature swing adsorption. *Energy & Environmental Science, 4*(8), 3030–3040.

McGrath, E. R., Frings-Meuthen, P., Sibonga, J., Heer, M., Clement, G. R., Mulder, E., Smith, S. M., & Zwart, S. R. (2022). Bone metabolism during strict head-down tilt bed rest and exposure to elevated levels of ambient CO_2. *npj Microgravity, 8*(1), 57.

McKinstry, R. B., Jr. (2004). Laboratories for local solutions for global problems: State, local and private leadership in developing strategies to mitigate the causes and effects of climate change. *Penn State Environmental Law Review, 12*, 15.

McLaughlin, H., Littlefield, A. A., Menefee, M., Kinzer, A., Hull, T., Sovacool, B. K., et al. (2023). Carbon capture utilization and storage in review: Sociotechnical implications for a carbon reliant world. *Renewable and Sustainable Energy Reviews, 177*, 113215.

Mendelsohn, R. (2014). The impact of climate change on agriculture in Asia. *Journal of Integrative Agriculture, 13*(4), 660–665.

Mitchell, J. F. (1989). The "greenhouse" effect and climate change. *Reviews of Geophysics, 27*(1), 115–139.

Motha, R. P. (2011). The impact of extreme weather events on agriculture in the United States. In *Challenges and opportunities in agrometeorology* (pp. 397–407). Springer.

Nakao, S., Yogo, K., Goto, K., Kai, T., & Yamada, H. (2019). *Advanced CO_2 capture technologies: Absorption, adsorption, and membrane separation methods*. Springer.

Nandy, A., Loha, C., Gu, S., Sarkar, P., Karmakar, M. K., & Chatterjee, P. K. (2016). Present status and overview of chemical looping combustion technology. *Renewable and Sustainable Energy Reviews, 59*, 597–619.

Nieder, R., & Benbi, D. K. (2022). Reactive nitrogen compounds and their influence on human health: An overview. *Reviews on Environmental Health, 37*(2), 229–246. https://doi.org/10.1515/reveh-2021-0021
Olesen, J. E., & Bindi, M. (2002). Consequences of climate change for European agricultural productivity, land use and policy. *European Journal of Agronomy, 16*(4), 239–262.
Pan, C., Ahammed, G. J., Li, X., & Shi, K. (2018). Elevated CO_2 improves photosynthesis under high temperature by attenuating the functional limitations to energy fluxes, electron transport and redox homeostasis in tomato leaves. *Frontiers in Plant Science, 9*, 1739.
Pandey, D. (2020). Agricultural sustainability and climate change nexus. In *Contemporary environmental issues and challenges in era of climate change* (pp. 77–97).
Pashaei, H., & Ghaemi, A. (2022). Review of CO_2 capture using absorption and adsorption technologies. *Iranian Journal of Chemistry and Chemical Engineering, 41*(8), 2771–2789.
Peredo-Mancilla, D., Ghimbeu, C. M., Ho, B. N., Jeguirim, M., Hort, C., & Bessieres, D. (2019). Comparative study of the CH_4/CO_2 adsorption selectivity of activated carbons for biogas upgrading. *Journal of Environmental Chemical Engineering, 7*(5), 103368.
Permentier, K., Vercammen, S., Soetaert, S., & Schellemans, C. (2017). Carbon dioxide poisoning: A literature review of an often-forgotten cause of intoxication in the emergency department. *International Journal of Emergency Medicine, 10*(1), 14.
Pröll, T., Schöny, G., Sprachmann, G., & Hofbauer, H. (2016). Introduction and evaluation of a double loop staged fluidized bed system for post-combustion CO_2 capture using solid sorbents in a continuous temperature swing adsorption process. *Chemical Engineering Science, 141*, 166–174.
Qiao, X., Liao, S., You, C., & Chen, R. (2015). Phosphorus and nitrogen dual doped and simultaneously reduced graphene oxide with high surface area as efficient metal-free electrocatalyst for oxygen reduction. *Catalysts, 5*(2), 981–991.
Raganati, F., Miccio, F., & Ammendola, P. (2021). Adsorption of carbon dioxide for post-combustion capture: A review. *Energy & Fuels, 35*(16), 12845–12868.
Rahman, M. A. (2016). *Development of a pressure swing adsorption (PSA) process for CO_2 capture from flue gas* [Doctoral dissertation, University of South Carolina].
Reganold, J. P., Papendick, R. I., & Parr, J. F. (1990). Sustainable agriculture. *Scientific American, 262*(6), 112–121.
Reitze, A. W., Jr. (2010). Federal control of greenhouse gas emissions. *Environmental Law, 40*, 1261.
Roy, P., Mohanty, A. K., & Misra, M. (2023). Prospects of carbon capture, utilization and storage for mitigating climate change. *Environmental Science: Advances, 2*(3), 409–423.
Saha, S., Chandra, S., Garai, B., & Banerjee, R. (2012). Carbon dioxide capture by metal organic frameworks. *Indian Journal of Chemistry-Part A Inorganic Physical Theoretical and Analytical, 51*(9), 1223.
Samanta, A., Zhao, A., Shimizu, G. K., Sarkar, P., & Gupta, R. (2012). Post-combustion CO_2 capture using solid sorbents: A review. *Industrial & Engineering Chemistry Research, 51*(4), 1438–1463. Available at https://pubs.acs.org/doi/10.1021/ie200686q. Accessed December 15, 2023.
Saunois, M., Jackson, R. B., Bousquet, P., Poulter, B., & Canadell, J. G. (2016). The growing role of methane in anthropogenic climate change. *Environmental Research Letters, 11*(12), 120207. https://doi.org/10.1088/1748-9326/11/12/120207
Savage, A., Bambrick, H., McIver, L., & Gallegos, D. (2021). Climate change and socioeconomic determinants are structural constraints to agency in diet-related non-communicable disease prevention in Vanuatu: A qualitative study. *BMC Public Health, 21*(1), 1231. https://doi.org/10.1186/s12889-021-11245-2
Schneider, S. H. (1989). The greenhouse effect: Science and policy. *Science, 243*(4892), 771–781.
Shah, K., Moghtaderi, B., & Wall, T. (2013). Effect of flue gas impurities on the performance of a chemical looping based air separation process for oxy-fuel combustion. *Fuel, 103*, 932–942.
Sharma, R., Sinha, A., & Kautish, P. (2020). Examining the impacts of economic and demographic aspects on the ecological footprint in south and southeast Asian countries. *Environmental Science and Pollution Research, 27*, 36970–36982.

Smith, T. J. (2019). Corn, cows, and climate change: How federal agriculture subsidies enable factory farming and exacerbate US greenhouse gas emissions. *Washington Journal of Environmental Law & Policy, 9*, 26.

Springer, R. (2020). *Exploring the connections between wellbeing, non-communicable diseases and climate change: A case study of Barbados*. University of Waterloo.

Stolp, H. (2022). Developing in a bad atmosphere: A link between long-term exposure to elevated atmospheric CO_2 and hyperactivity. *The Journal of Physiology, 600(6), 1275–1276.*

Sumida, K., Rogow, D. L., Mason, J. A., McDonald, T. M., Bloch, E. D., Herm, Z. R., et al. (2012). Carbon dioxide capture in metal–organic frameworks. *Chemical Reviews, 112*(2), 724–781.

Suomalainen, M. S., Arasto, A., Teir, S., & Siitonen, S. (2013). Improving a pre-combustion CCS concept in gas turbine combined cycle for CHP production. *Energy Procedia, 37*, 2327–2340.

Tamidi, A. M., Lau, K. K., Yusof, S. M. M., Azmi, N., Zakariya, S., & Patthi, U. (2023). Enhancement of CO_2 absorption process using high-frequency ultrasonic waves. *Sustainability, 15*(14), 11064. https://doi.org/10.3390/su151411064

Thommes, M., Kaneko, K., Neimark, A. V., Olivier, J. P., Rodriguez-Reinoso, F., Rouquerol, J., & Sing, K. S. (2015). Physisorption of gases, with special reference to the evaluation of surface area and pore size distribution (IUPAC Technical Report). *Pure and Applied Chemistry, 87*(9-10), 1051–1069.

Tsai, W. T., Chang, C. Y., & Lee, S. L. (1997). Preparation and characterization of activated carbons from corn cob. *Carbon, 35*, 1198–1200.

Wagaarachchige, J. D., Idris, Z., Khatibzadeh, A., Drageset, A., Jens, K. J., & Halstensen, M. (2023). Demonstration of CO2 capture process monitoring and solvent degradation detection by chemometrics at the Technology Centre Mongstad CO_2 capture plant. *Industrial & Engineering Chemistry Research, 62*, 9747–9754.

Waheed, N., Mushtaq, A., Tabassum, S., Gilani, M. A., Ilyas, A., Ashraf, F., et al. (2016). Mixed matrix membranes based on polysulfone and rice husk extracted silica for CO_2 separation. *Separation and Purification Technology, 170*, 122–129.

Wang, M., Lawal, A., Stephenson, P., Sidders, J., & Ramshaw, C. (2011). Post-combustion CO2 capture with chemical absorption: A state-of-the-art review. *Chemical Engineering Research and Design, 89*(9), 1609–1624.

Webley, P., Singh, R., & Xiao, P. (2017). Adsorption processes for CO_2 capture: An overview. In *CO2 summit III: Pathways to carbon capture, utilization, and storage deployment*. ECI.

Williams, S. J. (2013). Sea-level rise implications for coastal regions. *Journal of Coastal Research, 63*(63), 184–196.

World Health Organization. (1994). *Ultraviolet radiation: An authoritative scientific review of environmental and health effects of UV, with reference to global ozone layer depletion*. World Health Organization.

Yeo, T. H. C., Tan, I. A. W., & Abdullah, M. O. (2012). Development of adsorption air-conditioning technology using modified activated carbon—A review. *Renewable and Sustainable Energy Reviews, 16*(5), 3355–3363.

Yu, C. H., Huang, C. H., & Tan, C. S. (2012). A review of CO_2 capture by absorption and adsorption. *Aerosol and Air Quality Research, 12*(5), 745–769.

Yu, J., Xie, L.-H., Li, J.-R., Ma, Y., Seminario, J. M., & Balbuena, P. B. (2017). CO_2 capture and separations using MOFs: Computational and experimental studies. *Chemical Reviews, 117*(14), 9674–9754. https://doi.org/10.1021/acs.chemrev.6b00626

Zappulla, D. (2013). The stress of global warming on human health: pH homeostasis, the linkage between breathing & feeding via CO_2 economy. In *Air pollution: Sources, prevention and health effects*. Nova Science Publishers.

Zeng, H., Qu, X., Xu, D., & Luo, Y. (2022). Porous adsorption materials for carbon dioxide capture in industrial flue gas. *Frontiers in Chemistry, 10*, 939701.

Ziska, L. H. (2022). Rising carbon dioxide and global nutrition: Evidence and action needed. *Plants, 11*(7), 1000.

Nanotechnology Novel Trends for Soil Revitalization

Sukanya Ghosh and M. Jaison

Abstract

Soil contamination and its loss in productivity via degradation is a relevant and persistent problem in current timescale and has become a global challenge. The severity increases with climate crisis, loss of habitat, and increased consumption through anthropogenic activities, which further makes reclamation a major hurdle. Contaminated soils resulting mostly from human-induced activities make land less suitable for farming, which threatens global food security. As population continues to increase worldwide, more food production demands exploitation of a dwindling nonrenewable resource, that is, soil. More food production leads to mismanagement of fertile soils thereby, introducing xenobiotes on the likes of heavy metals, persistent organic pollutants, pesticides, and contagious microorganisms. Such additions chemically alter soil properties and can pose serious health concern when it enters the food chain via bioaccumulation of such toxins in living tissue. Moreover, the expected food grain production is also lowered due to such degradation. Nanotechnology-mediated soil remediation can bring about sustainable rejuvenation of soil pollution. These technologies are simple in application, economic, and highly effective in small doses and scalable. These simple-to-manufacture nanoproducts, in addition to reclaim soil, also aid in plant–soil interactions positively by enhancing metabolism, production of protein-like substances, and increased nutrient use efficiency. This chapter addresses the exploration of nanotech-mediated soil remediation through both biological and nonbiological nanoproducts.

S. Ghosh (✉)
GBPUAT, Pantnagar, Uttarakhand, India

M. Jaison
BCKV, Mohanpur, West Bengal, India

D. K. Gupta et al. (eds.), *Nano-Ecosystems*,
https://doi.org/10.1007/978-981-95-5033-3_7

Keywords

Nanotechnology · Nanoproducts · Soil remediation · Heavy metals

1 Introduction

1.1 Present Outline of Soil Spoilation

Soil is a nonrenewable resource that is made naturally furnished slowly over the years by breakdown of rocks through the influence of many factors acting upon it. It is the most crucial reserve that sustains ecosystems both in terrestrial and aquatic habitats. An essential media for biomass growth, productivity, and food supply, soil administers widespread activities that establish life itself on our planet (Abhilash et al., 2012; Ganga et al., 2023). These vital processes encompass hydrology cycle, biogeochemical cycles, climate change, removal of toxicants from soil systems, life-supporting gaseous balance, maintenance of biodiversity, and ecosystem rehabilitation (Gil-Díaz et al., 2014; Gavrilescu, 2021). The terms "soil" and "land" are often interchangeably used as both native and man-made afflictions on it nearly affects it likewise. Soil health management is key to securing food abundance, agriculture sustenance, and maintaining equilibrium within the biosphere (Lal, 2015; Maharjan et al., 2020).

In the present circumstances, soil deterioration and subsequent contamination constitute a significant threat to agricultural potential and nutritional security (Lal, 2009; Shaheb et al., 2021). The primary catalysts that contribute to soil deterioration are erosive processes, over-exploitation of grasslands by unaccounted grazing, logging, desertification, and urban expansion (León & Osorio, 2014; Wu et al., 2022a, b). Apart from these, chemical pollution discharged into soils through release of pesticide use, polyaromatic hydrocarbons, organic contaminants, and heavy metals directly assist soil fertility depletion. Inordinate usage of plant protectants and plant nutrients prompts their buildup and subsequent loss from soil system to air through release of greenhouse gas emission or into water bodies as run-off thereby, entering newer lands. Emergence of heavy metals from manufacturing plants not only spoils the locale area but also remote areas (Ghormade et al., 2011; Stephenson et al., 2001; Raffa et al., 2021). Natural hazards related to environ are concomitant to polluted soils can be a considerable menace to the total ecological system. The abovementioned explanations and subsequent anthropogenic impacts led to an approximately 30% land spoilage (whether devalued or polluted) with successive expansion of stages of decline (Huettermann et al., 2009; Brevik et al., 2020). The extensive harm to soil ecosystem not just voices concern but parallelly gives a pressing need for restoration of these defiled assets. With the passage of time, man-made xenobiotes will adulterate soil to a degree where crop cultivation becomes infeasible on such a resource (Tripathi et al., 2015; Bayabil et al., 2022). A polluted land not only makes it unfitting for plant growth and production of food but promotes serious danger to corrupt the food chain through bioaccumulation and consequential toxin amplification within living tissues of all organisms reliant on the food chain

(Lu et al., 2015; Haller & Jonsson, 2020). Ingestion of agricultural produce on contaminated soil may lead to serious implications for human population (Sharma & Nagpal, 2020). To avoid such circumstances, most polluted soils in vast areas are left deserted, creating extensive detriment to the agricultural industry. With the passage of year after year, such degradation brings about a reduction of fertile, productive soils that contest increased food production from such diminishing land for a flourishing population along with avoidance of further land destruction and promoting recuperation of existing damaged lands. A need for the moment is an urgent restitution of such tarnished soils that can be in service to meet the food and energy demands of the present times (Nehra et al., 2021).

1.2 Utilization of Nanotech for Soil Restoration

Nanotechnology is a fast-progressing breakthrough technology propagating in all spectrums of life (Fig. 1). Nanotech products hold distinctive properties from the material they are derived from (Parveen et al., 2016; Rao et al., 2017; Khan et al., 2017; Arshad et al., 2022; Joudeh & Linke, 2022). These engineered particles, which are nano in size, find a wide array of applications such as food processing, manufacturing goods, medicine, energy, and contamination removal from soil and water and agricultural operations (De et al., 2008; Kuiken, 2010; Shi et al., 2010; Gogos et al., 2012; De Oliveira et al., 2014; Mathur et al., 2022; Yin et al., 2023; Vijayaram et al., 2024).

The widespread utilization of nanoproducts for environ pollution abatement is probably derived from the compelling need of a speedy, efficient, low-cost, and readily accessible technology that can promise quick cleanup without major hassles (Cai et al., 2019; Anani et al., 2023; Das et al., 2024). The commonly used nanomaterials for soil restorative purposes are (a) nanoparticles that convert toxic forms of heavy metals to their nontoxic molecule forms; (b) nanoparticles that capture and degrade persistent organic molecules; (c) sensors of nanosize to detect residual

Fig. 1 Nanoparticles can tackle most persistent xenobiotes in soil

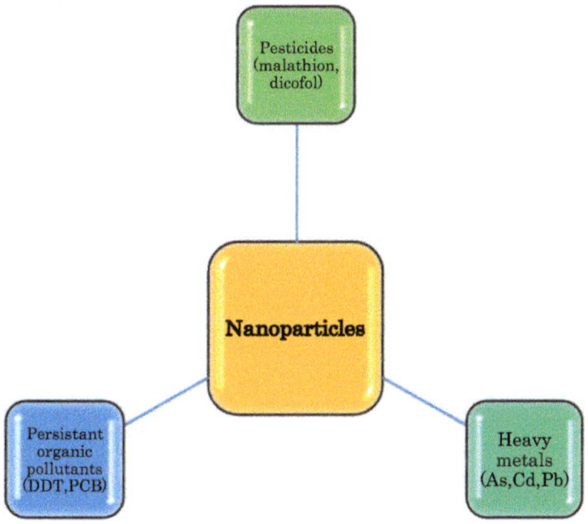

pesticide in soils; and (d) nanotech-aided soil bioremediation (Khan, 2020; Vu & Mulligan, 2023). Frequently used nanoparticles associated with soil decontamination are titanium dioxide, carbon nanotubes, zerovalent cations (Fe), and other nanosized carbon allotropes (fullerenes) (Pan & Xing, 2012; Cai et al., 2019; Rizwan et al., 2021; Arsenov et al., 2023). Nano-assisted rectification requires reactive species when involved in soil can either detoxify or transform the toxicant (Yan et al., 2013; Zheng et al., 2023). Because of their ultra small size, nanoparticles possess high surface area, good sorption rate, quick diffusion rates, and smaller pore diameter, creating a unique catalyst with a good absorbent capacity (Tang et al., 2014; Gong et al., 2018; Szczyglewska et al., 2023). Although soil restoration can employ both onsite and offsight remedial measure, nanoremediation techniques are mostly in situ practice, which makes it more fetching and economical method for broader employment (Jiang et al., 2018; Majumder & George, 2023; Premarathne et al., 2023).

2 Nanotechnology Contribution to the Deletion of Soil Contaminated with Heavy Metals

From the viewpoint of noxiousness, some common occurrences of heavy metals available in soil are arsenic, mercury, chromium, lead, cadmium, vanadium, molybdenum, and tin. They are all naturally occurring mineral forms in earth's crust, which through native process of rock weathering are found in trace amounts in soils (Violante et al., 2010; Pujari & Kapoor, 2021) (Fig. 2). They pose a risk when the levels in soils increase radically through anthropogenic processes and usage of materials known for their higher content of heavy metals such as pesticides and fertilizer usage in agriculture, mining, dyes, paints, wrongful disposal of waste, fossil fuel usage, and aerosol deposition through burning (Fayiga & Saha, 2016; Men et al., 2018; Pacyna et al., 2009).

Carbon-derived nanotubes and metal or metal-based oxides are the most common nanoparticles to alleviate heavy metal problem from soils. Metal and metal derivatives are classified as inorganic while carbon derivatives like nanotubes, multiwalled, graphene, or fullerenes are carbon-based nanoparticles (Fig. 3). Hybrid composites are also a novel manufacturing way to enhance the effectiveness of nanoparticles.

The high adsorbent and immobilization characteristics of nanomaterials make it ideal to remove heavy metal contamination from soils and return to their former glory for crop production (Gil-Díaz et al., 2014; Naderi Peikam & Jalali, 2019; Rajput et al., 2022; Golia, 2023). The most broadly studied nanoparticle for the elimination of inorganic and organic contaminant from soil is nano zerovalent iron (Fu et al., 2017; Li et al., 2017; Galdames et al., 2020; Mohanapriya et al., 2023). Usage of nano zerovalent iron to reduce chromium hexavalent to chromium trivalent form with an efficiency of 14.5%–86% in tannery wastewater was reported (Singh et al., 2012). Gil-Díaz et al. (2014) reported to immobilize and reduce lead concentrate in leachate through use of nano zerovalent iron. Similarly, Su et al.

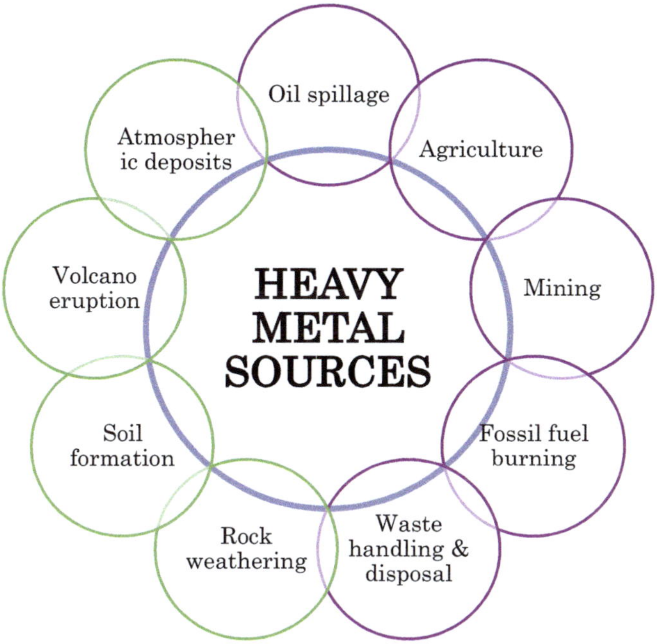

Fig. 2 Anthropogenic (green) and natural (purple) sources of heavy metals

(2016a, b) found significant reduction in chromium content in soil through employing biochar-mediated nano zerovalent iron. Gil-Díaz et al.'s 2016 application of nanoscale zerovalent iron has shown promising results in reducing the availability of heavy metals in contaminated soil such as arsenic that was reduced by more than 82%, lead availability was reduced by over 82%, reduction in zinc availability ranged between 31% and 75%, and lastly, lowest reduction was observed for cadmium, ranging between 13% and 42%. Gil-Díaz et al. (2017a, b) showed successful application of nanoscale zerovalent iron in mitigating arsenic contamination in highly polluted brownfield soil. When 10% nZVI was applied, it significantly reduced the uptake of arsenic by plants. This reduction occurred by decreasing the content of available arsenic fractions while increasing the amount of arsenic in the residual fraction. The immobilization of arsenic was attributed to the formation of adsorption complexes. Specifically, ferric arsenates formed between the corrosion products of Fe0 (iron) from the nZVI's shell. These complexes effectively trapped and immobilized the arsenic, preventing it from being taken up by plants. Barley plants cultivated in these As-polluted soils treated with nZVI exhibited several benefits. The plants showed improved growth and development, there were no negative effects on nutrient content, and the plants took up less arsenic, which is crucial for environmental and human health. Importantly, no adverse impacts were observed on the soil physicochemical and biological properties. This indicates that the nZVI treatment was effective without compromising the overall health of the soil ecosystem. Researchers (Madhavi et al., 2014) developed a composite material by

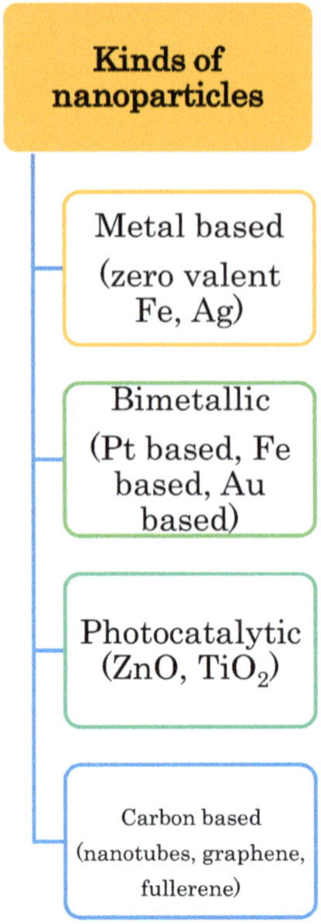

Fig. 3 Various kinds of nanoparticles commonly used for heavy metal remediation

combining iron nanoparticles (nZVI) with acetylene black and carboxymethyl cellulose (CMC) that effectively removed 98.6% of Cr(VI) within just 2 hours at a concentration of 50 mg/L. The maximum adsorption capacity was 116.30 mg/g. It outperformed other traditional and newly reported materials in terms of Cr(VI) adsorption activity and stands as an effective treatment option for Cr(VI)-contaminated soils.

The addition of farmyard manure (FYM) further facilitated Cr(VI) reduction. Hydroquinones, present in FYM, acted as the major source of electron donors for the reduction of Cr(VI) to Cr(III) in the soil. The use of carboxymethyl cellulose (CMC)-stabilized Fe–Mn binary oxide nanoparticles has shown remarkable effectiveness in treating arsenic-contaminated soil. The CMC coating ensures stability and enhances their performance in soil remediation. Fe–Mn binary oxide nanoparticles (NPs) significantly reduced the water-leachable arsenic by 91%–96%. The toxicity characteristic leaching procedure (TCLP) leachability of As(III)-laden soil was reduced by an impressive 94%–98%. Even the As(III) remaining in the soil bed experienced a substantial reduction of 78% (An & Zhao, 2012). The utilization of starch-stabilized iron-based nanoparticles, which included zerovalent iron (ZVI), magnetite (Fe_3O_4), and iron sulfide (FeS), has shown promise in the immobilization of arsenic (As) in contaminated soils. As the Fe/As molar ratio increased, the bioaccessibility of arsenic decreased significantly and leachability of arsenic in soil was also lowered. These nanoparticles effectively immobilized arsenic, making it less available for uptake by plants or leaching into groundwater (Zhang et al., 2010). Liu and Zhao (2007) reported the use of Fe(II) phosphate nanoparticles has shown promise in reducing the leachability and bioaccessibility of soil-bound copper (Cu(II)) and lead (Pb(II)). Again, in Liu and Zhao (2013), Liu and Zhao focused on the synthesis of a novel class of calcium phosphate nanoparticles. These nanoparticles demonstrated remarkable effectiveness in reducing the TCLP-leachable fraction of lead (Pb) in contaminated soil. Specifically, the fraction decreased from an initial value of 66% down to just 10% after treatment. The studies conducted by the authors highlight that the small particle size of these NPs contributes to their enhanced mobility in soil, which make them effective for in situ soil remediation. By immobilizing or reducing the mobility of these contaminants (Cd, Cu, Zn, and radioactive nuclides like U), NPs contribute to environmental protection. NPs can serve as nanofertilizers to fulfill the phosphorus (P) requirements of crops. Importantly, it helps mitigate eutrophication, a common issue associated with excessive fertilizer use.

Zhu et al. (2016) remediated chromium-contaminated soils using zerovalent nano iron–copper particles and found that the activity of the nanoremediation is related to soil pH and temperature. Removal rates were higher at high-temperature gradients. Use of iron sulphide nanoparticles synthesized with caboxymethyl cellulose remediated chromium hexavalent contamination in groundwater through adsorption (Wang et al., 2019). Moharem et al. (2019) reported reduced chromium and mercury mobility in calcareous soils by water treatment residual nanoparticles. Mallampati et al. (2013) achieved complete immobilization of arsenic and lead in Japanese soils through the use of nanosized mix of iron–calcium–calcium oxide. Zhu et al. (2016) reported that bimetallic zerovalent iron–copper nanoparticle could remove chromium VI at the rate of 99% in 10 min time from contaminated soils. Ghrair et al. (2010) reported cadmium (Cd) sorption by Na-zeolitic nanotuff is highly influenced by pH. The Na-zeolitic nanotuff increased Cd sorption by up to three times compared to the raw zeolitic tuff when applied to soils. Naderi Peikam and Jalali (2019) investigated three nanoparticles on their immobilization capacity of heavy metals in soil. SiO_2 NPs were found to be most effective for immobilizing the three metals (Zn, Cd, and Ni) in calcareous soils. The high surface area of SiO2 likely contributes to its effectiveness. Al_2O_3 NPs showed maximum mobility reduction for Cd and Zn in noncalcareous soils. The high site density of Al_2O_3 likely plays a key role in immobilizing these metals. Bradl (2005) reported that soil with high carbonate calcium content has a high metal immobilization capacity. Liming of soil (adding calcium carbonate) increases its adsorption potential, thereby reducing metal bioavailability. A list of nanoparticles used for heavy metal remediation by various researchers for their study is provided in Table 1.

Table 1 Heavy metal removal by different nanoparticles

Nanomaterial used	Heavy metal target	Highlight	Reference
nZVI	Cd(II)	NPs had a maximal adsorption capacity of 769.2 mg g^{-1} at 297 K	Boparai et al. (2011)
Oxalic acid nZVI	Cd(II), Pb(II), and Zn(II)	High removal efficiency for Cd, Pb, and Zn (46.66%, 48.88%, and 47.01%, respectively) in farm soil	Cao et al. (2018)
TiO_2 NPs	Cd(II)	In barley, PGPR +500 mg kg^{-1} TiO_2 NPs immobilized 1235 μg pot^{-1} Cd(II)	Zand et al. (2020)
RNS-FeS	Pb, As, Cd	It captured As, Pb, and Cd in soil by 80.1%, 97.1%, and 85%, respectively	Cao et al. (2020)
Fulvic acid-aided hydroxyapatite nanofluid (25:1)	Cd	Removal efficiency of Cd by >94%	Li et al. (2019)
CFGo	As, Cd, Cu, Pb, and Zn	Reduced metalloid bioavailability	Janeiro-Tato et al. (2022)
SA-FeSSi beads	Cd, Pb, Ni, Cr	Adsorption sites for metals enhanced by these beads	Wu et al. (2020)
DDTC-LDH	Cr (VI)	Cr(VI) levels reduced to 4.01 reduced from 134.26	Zhang et al. (2022)
S-nZVI	Cd	Exchangeable Cd was reduced >97.6% over 30-day incubation	Guo et al. (2021)
Tea polyphenols rGO-ZnO	Pb(II)	Efficient adsorption of Pb(II) by 98.92%	Zheng et al. (2018)
Magnetic nanocomposite	As(III)	Improved immobilization of As(II) in soils	Cui et al. (2019)
Fe_3O_4/ZIF-8	Cu(II), As(III)	High adsorption of Cu(II) by 33.48 mg g^{-1} and As(III) by 21.12 mg g^{-1}	Wu et al. (2021)
MoS_2/Fe_3O_4	Pb(II)	Effectively alleviated Pb(II) by 57.15%	Wang et al. (2020)
Magnetite NPs	As(V)	As(V) reduced content by 42.5%	Baragaño et al. (2020)
BC-nZVI	Cd(II), Pd(II)	Immobilized metal by 90%	Qian et al. (2022)
CTBC/GO-nZVI	Cu	After 14 days Cu content in soil reduced by 14%	Mandal et al. (2020)
CMC-FeS@biochar	Cr(VI)	Immobilized Cr(VI) by 94.7%	Lyu et al. (2018)

3 Organic Pollutant and Pesticide Remediation from Soils Using Nanomaterials

An absolute prohibition has been imposed by the Stockholm Convention on Persistent Organic Pollutants since 2004 due to regressive effects of pesticides and organic pollutants following their introduction in soils by humans. A considerable effort has thereafter been undertaken to entirely eradicate these pollutants from soil worldwide through nano-remediated measures. A total of 12 such toxicants have been listed, 8 of which are identified as pesticides (Allinson et al., 2016; Prabhu & Lakshmipraba, 2022). Due to the bioaccumulating nature across trophic levels and deposition in fatty tissues of living beings, these pollutants bring in irreversible damage to biota and disrupt the entire food chain (Gupta & Gupta, 2020). While many pesticides have been banned long since, their residual persistence poses dangers in the environment due to longer degradation time within the soil system. They also spread across newer soils due to global circulation via atmospheric deposition. The method by which these pollutants are rectified by nanoparticles is through photocatalysis. In the presence of light, nanoparticles catalyze the toxic forms of the contaminants rendering them into benign forms of carbon dioxide, dinitrogen, and water. Some good nanoparticles that act as efficient photocatalyst are zinc oxide and titanium oxide (Fujishima et al., 2008; Šebesta et al., 2021; Mony et al., 2022).

Yu et al. (2007) investigated the photocatalytic degradation of three organochlorinated pesticides with nano titanium dioxide (TiO_2). The nanoparticles of TiO_2 achieved the results by surface deterioration of these molecules by attracting radicles of peroxide and transfer of electrons. Carbamate and organophosphorus pesticide degradation was achieved by photocatalytic effect of nano TiO_2 doped rhenium in soil at the rate of up to 30% (Rui et al., 2010). Also, nano TiO_2 decreased the half-life of many pesticides without sacrificing its efficacy (Gogos et al., 2012). Utilization of multiwalled carbon nanotubes caused catalytic breakdown of herbicide metolachlor, along with its complete mineralization, thereby neutralizing its effects in soil (Restivo et al., 2012). Soils spiked with elevated levels of DDT were efficiently degraded with nano zerovalent iron at a level of 50% and 24%, respectively, in legacy and aged soils (El-Temsah et al., 2016). Tilston et al. (2013) reported the increased remediation of organochlorine in soil with nano zerovalent iron as well as a reduction in the mineralizing activity of chloroaromatic microbes. Gu et al. (2012) reported a reduction in the half-life of phenanthrene by photocatalysis of nano-TiO_2 from 46 to 31 h. Chen et al. (2014) showed that bimetallic lead–iron nanoparticles effectively removed polychlorinated biphenyls compared to nano zerovalent iron in in situ soils. Dien et al. (2013) achieved breakdown of organochlorine pesticide pentachlorophenol in sandy soils with an efficiency of 90% using bimetallic iron biocomposite. Table 2 shows a list of nanoparticles employed for the removal of organic pollutants.

3.1 Biological Sensing of Residual Pesticides Using Nanotechnology

Nano-based sensors have been envisaged to have promising prospects for resurrecting agricultural soils. Nano biobased sensors can have varied applications of pesticide detection, toxicant-level monitoring, pathogen occurrence in fields, as well as in storage units (Baruah & Dutta, 2009; Zia et al., 2022). These nanosensors can make substantial participation for soil productivity and soil quality, one of them being the detection of residual quantities of pesticide. Thereby, they can relatively

Table 2 Use of some nanoparticles in pesticide and persistent organic pollutant (POP) removal

Nanoparticle	Pesticide/POP	Effect in soil	Reference
MWCNTs	Metolachlor	Improved mineralization and decreased toxicity of metachlor on ozonation	Restivo et al. (2012)
nZVI	Organochlorine Aroclor	Inhibited the microorganism activity of chloraromatic nature	Tilston et al. (2013)
nZVI	Organochlorine	Neutralization of solvent organochlorine	Pan and Xing (2012)
Mesoporous silica	Pesticide	Catalytic degradation of pesticide	Kong et al. (2021)
CuO and NiO	Lambda-cyhalothrin	Degradation of pesticide by 99% for CuO and 89% for NiO	Iqbal et al. (2021)
TiO_2-rGO	Methomyl, isoproturon, pyrimethanil, and alachlor	In <25 min, complete breakdown of pesticide at low doses was achieved	Luna-Sanguino et al. (2020)
Fe–ZnO	Chlorpyrifos	About 93% breakdown achieved in 60 min at 10 ppm concentration of pesticide	Khan et al. (2018)
nZVI-alfalfa	PCB28	PCB degradation in soil by 72.4%–93.5%	Wu et al. (2022a, b)
Zerovalent MG-graphite	HCB	Total degradation of aromatic compound at room temperature under 30 min	Garbou et al. (2019)
Graphene oxide Ag NP	Chlordane	Complete degradation removal from solution at room temperature	Sarno et al. (2017)

lower toxin penetration in food chain and reduce foodborne poisoning. Nanosensors can provide quality detection of pesticide sensitive at low concentrations in a plethora of environment (Sharma et al., 2021). Therefore, they are more popular as monitoring devices than conventional techniques of chromatography and spectroscopic methods. The advantage is also perceived through economics of undertaking such soil detection tests where nanosensors are more selective, nonlaborious, quick, and field suitable (Liu et al., 2008; Dar et al., 2020). Nanosensors can be distinguished into three types based on their sensory mechanism: (a) organophosphorus hydrolase, (b) cholinesterase inhibitors, and (c) immunoassays (Liu et al., 2008). Most nanosensors are geared toward the detection of acetylcholinesterase-based pesticides like carbamate and organophosphate.

Simonian et al. (2005) used a clever combination of enzyme conjugation, gold nanoparticles, and fluorescent decoys for detecting organophosphate paraoxon. The enzyme organophosphate hydrolase hydrolyzes paraoxon into simpler molecules. Gold nanoparticles enhance signals and interact with biomolecules. The fluorescent decoy mimics paraoxon and competes with it. When paraoxon binds to OPH, the decoy fluoresces. The presence of paraoxon reduces the fluorescence signal. The sensor exhibited high sensitivity even at very low concentrations of paraoxon. Su et al. (2008) reported that silicon-based gold nanoparticles that they form stable structure with high specific surface area and adjustable pore structures. With high electrical conductivity, the nanocomplex was used to detect dichlorvos, an organophosphate pesticide, even at an incredibly low concentration of 8 ng/L. Wang et al. (2010) effectively combined silver nanoparticles (AgNPs) with silicon nanowire arrays (SiNWs) to create an efficient surface-enhanced Raman scattering (SERS) substrate. The SiNWs provide a high surface area, while the AgNPs enhance the Raman signal. The AgNPs are coated onto the SiNWs, creating a structured surface that promotes strong Raman enhancement. The developed SERS sensor demonstrated remarkable sensitivity even at incredibly low concentrations of 10^{-7} mol/L (0.02 ppm) for carbaryl, a commonly applied pesticide in agricultural soils. Cesarino et al. (2012) created a biosensor tailored for detecting carbamate pesticides. They immobilized ChE on a core–shell-modified glassy electrode. The core–shell structure likely consists of multiwalled carbon nanotubes (MWCNTs) and polyaniline (PANI). The biosensor exhibited impressive detection limits of 1.4 mol/L for carbaryl and 0.95 mol/L for methomyl, another carbamate pesticide. These low detection limits mean that the biosensor can identify these pesticides even at very low concentrations. Liu et al. (2011) created a novel nanosensor relying on immobilization of acetylcholinesterase on graphene oxide–gold composite for identifying the presence of carbamate. Appreciable sensitivity was observed on the usage of the sensor but also a good reproducibility of the results was also obtained. Wu et al. (2013) innovated a biosensor of disposable nature based on graphene oxide nanocomposite. It effectively detected dichlorvos with high precision on a real-time basis and with good reproducibility of results. Dong et al. (2013) created a biosensor made of multiwalled carbon nanotube with chitosan and detected methyl parathion based on its cholinesterase inhibitory reaction in soil and water samples. Table 3 depicts some nanosensors successfully developed and utilized for pesticide detection in soil environment.

Although the range of application for nano biosensors is immense, its effective implementation is reliant on accessibility of the nanomaterials that are sensitive enough to detect pesticide residues in soil, simplicity in the construction of a

Table 3 Nanosensors for residual pesticide detection in soil

Nanosensors employed	Target pesticide	Remark	Reference
Chitosan–iron oxide nanocomposite	Malathion	80%–90% recovery of pesticide from soil	Prabhakar et al. (2016)
AP-algae with flower-shaped ZnO NP	Chlorpyrifos	Reliable and reproducible results in the detection of pesticide in soil	Pabbi et al. (2018)
FT@CNT	Parathion	Good stability, sensitive detection at low dose of pesticide	Tümay et al. (2021)
CdTe QDs	Methidathion	Precise monitoring of pesticide in soil	Yang et al. (2021)
Carbonaceous nanoaggregate probe	Paraoxon-ethyl	Good recovery, fast monitoring in situ	Cioffi et al. (2021)
AuNPs-SERS	Chlorpyrifos	Ultrasensitive to low dosage of pesticide, recovery rate >95%	He et al. (2019)

reliable sensor, desirable repeatability and surety of results, economics of employment, and, lastly, exposure risk of novel nanomaterials into the soil environment (Khot et al., 2012; El-Chaghaby & Rashad, 2023). Creation of a smart nanosensor that incorporates release of fertilizer and plant protectants efficiently detects moisture status of soil and signals the soil nutritional deficiency might help eliminate the usage of sensors of pesticide residue detection in soil altogether. Also, these sensors can act as an early warning system for plant growers in terms of determining fertilizer rates, frequency of pesticide application, etc. (Yadav et al., 2023).

4 Enhancement of Bioremediation Through Nanotechnology

Bioremediation is the clean-up process of soil–water–air ecosystems with the aid biological organisms (flora or fauna). The organisms can be of macro or micro in size. Usually, plant biomass is involved in such large-scale clean-up operations, which is termed phytoremediation (Pilon-Smits, 2005; Mallikarjunaiah et al., 2020). Many promising plant species have been identified and successfully used such as tomato, cabbage, sunflower, poplar, and alfalfa, which could degrade toxicants and capture them from soil system (Mitton et al., 2014; Pande et al., 2020). The methods deployed by these phytoremediators for soil sanitization are phytoextraction, phytovolatilization, phytostabilization, and phytodegradation (Song et al., 2019). Phytoextraction involves using plants to extract and accumulate pollutants in their tissues (Chaney et al., 2020). Phytovolatilization is a process where plants absorb and transform pollutants into volatile forms. These volatile compounds are then transported to the leaves and released through transpiration (Zayed et al., 2020). Phytostabilization occurs in situ, where plant roots immobilize pollutants within the rhizosphere (the soil zone influenced by the roots) (Galal et al., 2020). Unlike within the plant itself, phytostabilization aims to reduce pollutant mobility by processes such as root adsorption, complex formation, or precipitation. Ultimately, this helps minimize the risk of pollutants entering groundwater or the food chain. Phytodegradation refers to the direct breakdown of organic pollutants by enzymatic activities within plants (Kafle et al., 2022).

Soil environment conservation initiatives using phytoremediation have received positive attention from the public and government due to co-benefits generated from deploying this measure. They can be reduced by carbon footprint, increased soil organic carbon stock, soil stabilization under erosive conditions, beneficial microbial activity, and bioenergy providence (Su et al., 2022). The drawback of this technique is that such clean-up measures are time consuming, laborious, and effects can be subdued due to climatic conditions, medium over which they grow and plant sensitivity to already present toxicant in contaminated soil (Liu et al., 2020). Amalgamation of nanotechnology with bioremediation using microbe and plants has been more pronounced in recent years for its quick wins, higher degradability, and viability than embracing a technology singly (Gul et al., 2022). Onsite applications of nanoparticles have demonstrated influence on the contaminant uptake and its degradation behavior by the organism involved in remediation.

Predominantly, in nano-mediated phytoremediation, carbon nanotubes and metallic oxides are used. Due to their large specific surface area, carbon nanotubes exhibit excellent adsorption capacity for various pollutants, especially hydrophobic organic pollutants. Organic compounds infused with carbon nanotubes have the potential to serve as effective carriers for contaminants, particularly organic pollutants (Chung et al., 2022). Their presence can significantly alter the fate and transport of these pollutants in the environment. De La Torre-Roche et al., 2012 reported that full exposure of fullerene C_{60} to DDT for plants of tomato, soybean, and zucchini increased its uptake in roots and leaves between 30%–65%. In another study conducted by the same investigator (De La Torre-Roche et al., 2013), using fullerenes and multiwalled carbon nanotubes proved that DDT uptake in plants ranged from 21% to 80%, but it varies with from plant-to-plant species and carbon nanoparticle nature. Hamdi et al. (2015) employed multiple types of carbon nanotubes to detect phytotoxicity of chlordane in lettuce and reported that different concentrations and types of the nanotubes influenced toxicant level in the plant. Nonfunctionalized nanotube decreased content by 88%, whereas functionalized nanotubes were more modest in uptake by 57%. Wu et al. (2016) studied the effects of bimetallic nickel–iron nanoparticles on the uptake of polybrominated diphenyl ethers by Chinese cabbage from contaminated soils. Results showed that fresh nanoparticles had higher toxicity than aged particles. Hussain et al. (2019) revealed that nanoparticles of iron oxide can alleviate cadmium toxicity in wheat. After 120 days of growth, biomass harvested showed reduced cadmium content in plant tissues grown under doped pot culture soil and biofortification of iron in plant with increased yield. López-Luna et al. (2016) reviewed the uptake of citrate-coated magnetite nanoparticles in wheat and alleviation of cadmium and chromium accumulation levels. The nanoparticles were successful in reducing cadmium–chromium content in plant tissue. Cai et al. (2017) evaluated four types of titanium dioxide nanoparticles on lead bioaccumulation in rice seedlings. Though the remediation was achieved in reducing lead levels in rice, it also contributed to the build-up of the nanoparticles itself in the roots that can create possible hazards to health safety. Table 4 is a summarized list of nanoparticles used in bioremediation of soil.

5 An Anticipated Optimism and Doubts for the Future

While nanotechnology provides a range of innovative solutions for environmental clean-up sustainably with special reference to amend contaminated soils, its associated unpredictability and concerns related to risks should be reviewed thoroughly (Coutris et al., 2020). One of the major concerns that is a constant debate is the introduction of novel man-made material in surroundings and its diffusion into newer ecosystems (Malakar & Snow, 2020; Bathi et al., 2021). The likelihood and potential of these nanoparticles to enter food chain by virtue of its size and motility through soil make it a hesitant choice of food producers and consumers for mass use

Table 4 Bioremediation of pollutants with nanoparticles

Nanoparticles used	Target pollutant	Effect	Reference
CNTs	Cd	Reduced Cd toxicity in **Ramie seedlings**, enhanced growth	Gong et al. (2019)
ZnO	Cd and Pb	Good growth of ***Leucaena* seedlings,** reduced Cd and Pb content in dry matter	Venkatachalam et al. (2017)
Si NP	As(V)	Reduced As(V) content in **maize**, improved heavy metal tolerance stress	Tripathi et al. (2016)
TiO_2	Cd	Increased photosynthetic activity in **soybean,** reduced Cd stress in plants	Singh and Lee (2016)
nZVI	As, Cd, Pb	Reduced heavy metal uptake in **rye** and **sunflower**	Vítková et al. (2018)
Modified carbon black	Cd, petroleum	Reduced bioavailability of heavy metal in soil, improved growth of ***Suaeda salsa***	Cheng et al. (2019)
Multiwall carbon nanotube	Cd and Pb	Root and shoot growth increment in **sunflower** and **Brassica,** reduced bioaccumulation in plants	Oloumi et al. (2018)
nZVI	Endosulfan	An 82% removal efficiency in 7 days achieved, reduced accumulation in **lemongrass, tulsi**	Pillai and Kottekottil (2016)

of the nanotechnology to reclaim soils (Dash & Kundu, 2020). The eventual outcomes of such occurrences are not easily deducible, need long-term exploration, and may not be deleterious in the future. Currently, there exist no international laws or policies specifically addressing nanotechnology, manufactured nanomaterials, and their environmental release. This regulatory gap poses challenges in ensuring responsible and safe utilization of nanotechnology across borders (Cummings et al., 2021). Hence, a comprehensive understanding of life cycle analysis of the nanoparticles and their conduct on mixing with soil and water media in native state is vital for safe use of this technology (Singh & Gurjar, 2022).

Assuming that soil decontamination or remediation by use of nanotechnology gives a brighter prospect in relatively faster timescale, with scaling up potential over large areas without the added baggage of logistics, in-depth strategic assessment on a long-term basis is needed for a successful deployment and to mitigate adversaries.

References

Abhilash, P. C., Powell, J. R., Singh, H. B., & Singh, B. K. (2012). Plant–microbe interactions: Novel applications for exploitation in multipurpose remediation technologies. *Trends in Biotechnology, 30*(8), 416–420.

Allinson, G., Allinson, M., Bui, A., Zhang, P., Croatto, G., Wightwick, A., Rose, G., & Walters, R. (2016). Pesticide and trace metals in surface waters and sediments of rivers entering the Corner Inlet Marine National Park, Victoria, Australia. *Environmental Science and Pollution Research, 23*, 5881–5891.

An, B., & Zhao, D. (2012). Immobilization of As (III) in soil and groundwater using a new class of polysaccharide stabilized Fe–Mn oxide nanoparticles. *Journal of Hazardous Materials, 211*, 332–341.

Anani, O. A., Inobeme, A., Osarenotor, O., Olisaka, F. N., Aidonojie, P. A., Olatunji, E. O., & Habib, A. I. (2023). Application of microorganisms as biofactories to produce biogenic nanoparticles for environmental cleanup: Currents advances and challenges. *Current Nanoscience, 19*(6), 770–782.

Arsenov, D., Beljin, J., Jović, D., Maletić, S., Borišev, M., & Borišev, I. (2023). Nanomaterials as endorsed environmental remediation tools for the next generation: Eco-safety and sustainability. *Journal of Geochemical Exploration, 253*, 107283.

Arshad, I., Noor, A., Rashid, H., Hussan, M. U., Waqas, M., Fatima, N., Anwer, A., Ijaz, V., Qayyum, A., Arshad, F., & Choudhry, S. (2022). A comprehensive review on role of nanotechnology in the soil pollutants remediation. *Agricultural Sciences Journal, 4*(1), 17–38.

Baragaño, D., Alonso, J., Gallego, J. R., Lobo, M. C., & Gil-Díaz, M. (2020). Magnetite nanoparticles for the remediation of soils co-contaminated with As and PAHs. *Chemical Engineering Journal, 399*, 125809.

Baruah, S., & Dutta, J. (2009). Nanotechnology applications in pollution sensing and degradation in agriculture: A review. *Environmental Chemistry Letters, 7*, 191–204.

Bathi, J. R., Moazeni, F., Upadhyayula, V. K., Chowdhury, I., Palchoudhury, S., Potts, G. E., & Gadhamshetty, V. (2021). Behavior of engineered nanoparticles in aquatic environmental samples: Current status and challenges. *Science of the Total Environment, 793*, 148560.

Bayabil, H. K., Teshome, F. T., & Li, Y. C. (2022). Emerging contaminants in soil and water. *Frontiers in Environmental Science, 10*, 873499.

Boparai, H. K., Meera Joseph, M., & O'Carroll, D. M. (2011). Kinetics and thermodynamics of cadmium ion removal by adsorption onto nano zerovalent iron particles. *Journal of Hazardous Materials, 186*, 458–465.

Bradl, H. B. (2005). Sources and origins of heavy metals. In *Interface science and technology* (Vol. 6, pp. 1–27). Elsevier.

Brevik, E. C., Slaughter, L., Singh, B. R., Steffan, J. J., Collier, D., Barnhart, P., & Pereira, P. (2020). Soil and human health: Current status and future needs. *Air, Soil and Water Research, 13*, 1178622120934441.

Cai, F., Wu, X., Zhang, H., Shen, X., Zhang, M., Chen, W., Gao, Q., White, J. C., Tao, S., & Wang, X. (2017). Impact of TiO2 nanoparticles on lead uptake and bioaccumulation in rice (*Oryza sativa* L.). *NanoImpact, 5*, 101–108.

Cai, C., Zhao, M., Yu, Z., Rong, H., & Zhang, C. (2019). Utilization of nanomaterials for in-situ remediation of heavy metal (loid) contaminated sediments: A review. *Science of the Total Environment, 662*, 205–217.

Cao, Y., Zhang, S., Zhong, Q., Wang, G., Xu, X., Li, T., Wang, L., Jia, Y., & Li, Y. (2018). Feasibility of nanoscale zero-valent iron to enhance the removal efficiencies of heavy metals from polluted soils by organic acids. *Ecotoxicology and Environmental Safety, 162*, 464–473.

Cao, P., Qiu, K., Zou, X., Lian, M., Liu, P., Niu, L., Yu, L., Li, X., & Zhang, Z. (2020). Mercapto propyltrimethoxysilane-and ferrous sulfate-modified nano-silica for immobilization of lead and cadmium as well as arsenic in heavy metal-contaminated soil. *Environmental Pollution, 266*, 115152.

Cesarino, I., Moraes, F. C., Lanza, M. R., & Machado, S. A. (2012). Electrochemical detection of carbamate pesticides in fruit and vegetables with a biosensor based on acetylcholinesterase immobilised on a composite of polyaniline–carbon nanotubes. *Food Chemistry, 135*(3), 873–879.

Chaney, R. L., Li, Y. M., Brown, S. L., Homer, F. A., Malik, M., Angle, J. S., Baker, A. J., Reeves, R. D., & Chin, M. (2020). Improving metal hyperaccumulator wild plants to develop commercial phytoextraction systems: Approaches and progress. In *Phytoremediation of contaminated soil and water* (pp. 129–158). CRC Press.

Chen, X., Yao, X., Yu, C., Su, X., Shen, C., Chen, C., Huang, R., & Xu, X. (2014). Hydrodechlorination of polychlorinated biphenyls in contaminated soil from an e-waste recycling area, using nanoscale zerovalent iron and Pd/Fe bimetallic nanoparticles. *Environmental Science and Pollution Research, 21*, 5201–5210.

Cheng, J., Sun, Z., Yu, Y., Li, X., & Li, T. (2019). Effects of modified carbon black nanoparticles on plant-microbe remediation of petroleum and heavy metal co-contaminated soils. *International Journal of Phytoremediation, 21*(7), 634–642.

Chung, J. H., Hasyimah, N., & Hussein, N. (2022). Application of carbon nanotubes (CNTs) for remediation of emerging pollutants—A review. *Tropical Aquatic and Soil Pollution, 2*(1), 13–26.

Cioffi, A., Mancini, M., Gioia, V., & Cinti, S. (2021). Office paper-based electrochemical strips for organophosphorus pesticide monitoring in agricultural soil. *Environmental Science & Technology, 55*(13), 8859–8865.

Coutris, C., Ševců, A., & Joner, E. J. (2020). Ecotoxicity of nanomaterials used for remediation. In *Advanced nano-bio technologies for water and soil treatment* (pp. 573–585). Springer.

Cui, J., Jin, Q., Li, Y., & Li, F. (2019). Oxidation and removal of As (III) from soil using novel magnetic nanocomposite derived from biomass waste. *Environmental Science: Nano, 6*(2), 478–488.

Cummings, C. L., Kuzma, J., Kokotovich, A., Glas, D., & Grieger, K. (2021). Barriers to responsible innovation of nanotechnology applications in food and agriculture: A study of US experts and developers. *NanoImpact, 23*, 100326.

Dar, F. A., Qazi, G., & Pirzadah, T. B. (2020). Nano-biosensors: NextGen diagnostic tools in agriculture. In *Nanobiotechnology in agriculture: An approach towards sustainability* (pp. 129–144). Springer.

Das, K., Dey, D., & Mondal, P. (2024). The headway of nanotechnologymediated approaches toward environmental clean-up. In *Environmental contaminants* (pp. 277–313). Apple Academic Press.

Dash, S. R., & Kundu, C. N. (2020). Promising opportunities and potential risk of nanoparticle on the society. *IET Nanobiotechnology, 14*(4), 253–260.

De La Torre-Roche, R., Hawthorne, J., Deng, Y., Xing, B., Cai, W., Newman, L. A., Wang, C., Ma, X., & White, J. C. (2012). Fullerene-enhanced accumulation of p, p′-DDE in agricultural crop species. *Environmental Science & Technology, 46*(17), 9315–9323.

De La Torre-Roche, R., Hawthorne, J., Deng, Y., Xing, B., Cai, W., Newman, L. A., Wang, Q., Ma, X., Hamdi, H., & White, J. C. (2013). Multiwalled carbon nanotubes and C60 fullerenes differentially impact the accumulation of weathered pesticides in four agricultural plants. *Environmental Science & Technology, 47*(21), 12539–12547.

De Oliveira, J. L., Campos, E. V. R., Bakshi, M., Abhilash, P. C., & Fraceto, L. F. (2014). Application of nanotechnology for the encapsulation of botanical insecticides for sustainable agriculture: Prospects and promises. *Biotechnology Advances, 32*(8), 1550–1561.

De, M., Ghosh, P. S., & Rotello, V. M. (2008). Applications of nanoparticles in biology. *Advanced Materials, 20*(22), 4225–4241.

Dien, N. T., De Windt, W., Buekens, A., & Chang, M. B. (2013). Application of bimetallic iron (BioCAT slurry) for pentachlorophenol removal from sandy soil. *Journal of Hazardous Materials, 252*, 83–90.

Dong, J., Fan, X., Qiao, F., Ai, S., & Xin, H. (2013). A novel protocol for ultra-trace detection of pesticides: Combined electrochemical reduction of Ellman's reagent with acetylcholinesterase inhibition. *Analytica Chimica Acta, 761*, 78–83.

El-Chaghaby, G. A., & Rashad, S. (2023). Nanosensors in agriculture: Applications, prospects, and challenges. In *Handbook of nanosensors: Materials and technological applications* (pp. 1–29). Springer.

El-Temsah, Y. S., Sevcu, A., Bobcikova, K., Cernik, M., & Joner, E. J. (2016). DDT degradation efficiency and ecotoxicological effects of two types of nano-sized zero-valent iron (nZVI) in water and soil. *Chemosphere, 144*, 2221–2228.

Fayiga, A. O., & Saha, U. K. (2016). Arsenic hyperaccumulating fern: Implications for remediation of arsenic contaminated soils. *Geoderma, 284*, 132–143.

Fu, R., Zhang, X., Xu, Z., Guo, X., Bi, D., & Zhang, W. (2017). Fast and highly efficient removal of chromium (VI) using humus-supported nanoscale zero-valent iron: Influencing factors, kinetics and mechanism. *Separation and Purification Technology, 174*, 362–371.

Fujishima, A., Zhang, X., & Tryk, D. A. (2008). TiO2 photocatalysis and related surface phenomena. *Surface Science Reports, 63*(12), 515–582.

Galal, T. M., Al-Sodany, Y. M., & Al-Yasi, H. M. (2020). Phytostabilization as a phytoremediation strategy for mitigating water pollutants by the floating macrophyte *Ludwigia stolonifera* (Guill. & Perr.) PH Raven. *International Journal of Phytoremediation, 22*(4), 373–382.

Galdames, A., Ruiz-Rubio, L., Orueta, M., Sánchez-Arzalluz, M., & Vilas-Vilela, J. L. (2020). Zero-valent iron nanoparticles for soil and groundwater remediation. *International Journal of Environmental Research and Public Health, 17*(16), 5817.

Ganga, A., Roder, L. R., Paganini, E. A., Jani, A. D., Abreu-Junior, C. H., Nogueira, T. A. R., & Capra, G. F. (2023). Filming a hidden resource: The soil in the seventh art narrative. *Geoderma, 440*, 116710.

Garbou, A. M., Liu, M., Zou, S., & Yestrebsky, C. L. (2019). Degradation kinetics of hexachlorobenzene over zero-valent magnesium/graphite in protic solvent system and modeling of degradation pathways using density functional theory. *Chemosphere, 222*, 195–204.

Gavrilescu, M. (2021). Water, soil, and plants interactions in a threatened environment. *Water, 13*(19), 2746.

Ghormade, V., Deshpande, M. V., & Paknikar, K. M. (2011). Perspectives for nano-biotechnology enabled protection and nutrition of plants. *Biotechnology Advances, 29*(6), 792–803.

Ghrair, A. M., Ingwersen, J., & Streck, T. (2010). Immobilization of heavy metals in soils amended by nanoparticulate zeolitic tuff: Sorption-desorption of cadmium. *Journal of Plant Nutrition and Soil Science, 173*(6), 852–860.

Gil-Díaz, M., Ortiz, L. T., Costa, G., Alonso, J., Rodríguez-Membibre, M. L., Sánchez-Fortún, S., Pérez-Sanz, A., Martín, M., & Lobo, M. C. (2014). Immobilization and leaching of Pb and Zn in an acidic soil treated with zerovalent iron nanoparticles (nZVI): Physicochemical and toxicological analysis of leachates. *Water, Air, & Soil Pollution, 225*, 1–13.

Gil-Díaz, M., Alonso, J., Rodríguez-Valdés, E., Gallego, J. R., & Lobo, M. C. (2017a). Comparing different commercial zero valent iron nanoparticles to immobilize As and Hg in brownfield soil. *Science of the Total Environment, 584*, 1324–1332.

Gil-Díaz, M., Pinilla, P., Alonso, J., & Lobo, M. C. (2017b). Viability of a nanoremediation process in single or multi-metal (loid) contaminated soils. *Journal of Hazardous Materials, 321*, 812–819.

Gogos, A., Knauer, K., & Bucheli, T. D. (2012). Nanomaterials in plant protection and fertilization: Current state, foreseen applications, and research priorities. *Journal of Agricultural and Food Chemistry, 60*(39), 9781–9792.

Golia, E. E. (2023). The impact of heavy metal contamination on soil quality and plant nutrition. Sustainable management of moderate contaminated agricultural and urban soils, using low cost materials and promoting circular economy. *Sustainable Chemistry and Pharmacy, 33*, 101046.

Gong, X., Huang, D., Liu, Y., Peng, Z., Zeng, G., Xu, P., et al. (2018). Remediation of contaminated soils by biotechnology with nanomaterials: Bio-behavior, applications, and perspectives. *Critical Reviews in Biotechnology, 38*(3), 455–468.

Gong, X., Huang, D., Liu, Y., Zeng, G., Wang, R., Xu, P., Zhang, C., Cheng, M., Xue, W., & Chen, S. (2019). Roles of multiwall carbon nanotubes in phytoremediation: Cadmium uptake and oxidative burst in *Boehmeria nivea* (L.) Gaudich. *Environmental Science: Nano, 6*(3), 851–862.

Gu, J., Dong, D., Kong, L., Zheng, Y., & Li, X. (2012). Photocatalytic degradation of phenanthrene on soil surfaces in the presence of nanometer anatase TiO_2 under UV-light. *Journal of Environmental Sciences, 24*(12), 2122–2126.

Gul, M. Z., Rupula, K., & Beedu, S. R. (2022). Nano-phytoremediation for soil contamination: An emerging approach for revitalizing the tarnished resource. In *Phytoremediation* (pp. 115–138). Academic Press.

Guo, Y., Li, X., Liang, L., Lin, Z., Su, X., & Zhang, W. (2021). Immobilization of cadmium in contaminated soils using sulfidated nanoscale zero-valent iron: Effectiveness and remediation mechanism. *Journal of Hazardous Materials, 420*, 126605.

Gupta, S., & Gupta, K. (2020). Bioaccumulation of pesticides and its impact on biological systems. In *Pesticides in crop production: Physiological and biochemical action* (pp. 55–67). Wiley.

Haller, H., & Jonsson, A. (2020). Growing food in polluted soils: A review of risks and opportunities associated with combined phytoremediation and food production (CPFP). *Chemosphere, 254*, 126826.

Hamdi, H., De La Torre-Roche, R., Hawthorne, J., & White, J. C. (2015). Impact of non-functionalized and amino-functionalized multiwall carbon nanotubes on pesticide uptake by lettuce (*Lactuca sativa* L.). *Nanotoxicology, 9*(2), 172–180.

He, Y., Xiao, S., Dong, T., & Nie, P. (2019). Gold nanoparticles with different particle sizes for the quantitative determination of chlorpyrifos residues in soil by SERS. *International Journal of Molecular Sciences, 20*(11), 2817.

Huettermann, A., Orikiriza, L. J., & Agaba, H. (2009). Application of superabsorbent polymers for improving the ecological chemistry of degraded or polluted lands. *CLEAN–Soil, Air, Water, 37*(7), 517–526.

Hussain, A., Ali, S., Rizwan, M., ur Rehman, M. Z., Qayyum, M. F., Wang, H., & Rinklebe, J. (2019). Responses of wheat (*Triticum aestivum*) plants grown in a Cd contaminated soil to the application of iron oxide nanoparticles. *Ecotoxicology and Environmental Safety, 173*, 156–164.

Iqbal, A., Haq, A. U., Cerrón-Calle, G. A., Naqvi, S. A. R., Westerhoff, P., & Garcia-Segura, S. (2021). Green synthesis of flower-shaped copper oxide and nickel oxide nanoparticles via *Capparis decidua* leaf extract for synergic adsorption-photocatalytic degradation of pesticides. *Catalysts, 11*(7), 806.

Janeiro-Tato, I., Baragaño, D., Lopez-Anton, M. A., Rodríguez, E., Peláez, A. I., García, R., & Gallego, J. R. (2022). Goethite-based carbon foam nanocomposites for concurrently immobilizing arsenic and metals in polluted soils. *Chemosphere, 301*, 134645.

Jiang, D., Zeng, G., Huang, D., Chen, M., Zhang, C., Huang, C., & Wan, J. (2018). Remediation of contaminated soils by enhanced nanoscale zero valent iron. *Environmental Research, 163*, 217–227.

Joudeh, N., & Linke, D. (2022). Nanoparticle classification, physicochemical properties, characterization, and applications: A comprehensive review for biologists. *Journal of Nanbiotechnology, 20*, 262.

Kafle, A., Timilsina, A., Gautam, A., Adhikari, K., Bhattarai, A., & Aryal, N. (2022). Phytoremediation: Mechanisms, plant selection and enhancement by natural and synthetic agents. *Environmental Advances, 8*, 100203.

Khan, F. A. (2020). Nanomaterials: Types, classifications, and sources. In *Applications of nanomaterials in human health* (pp. 1–13). Springer.

Khan, M. N., Mobin, M., Abbas, Z. K., AlMutairi, K. A., & Siddiqui, Z. H. (2017). Role of nanomaterials in plants under challenging environments. *Plant Physiology and Biochemistry, 110*, 194–209.

Khan, S. H., Pathak, B., & Fulekar, M. H. (2018). Synthesis, characterization and photocatalytic degradation of chlorpyrifos by novel Fe: ZnO nanocomposite material. *Nanotechnology For Environmental Engineering, 3*, 1–14.

Khot, L. R., Sankaran, S., Maja, J. M., Ehsani, R., & Schuster, E. W. (2012). Applications of nanomaterials in agricultural production and crop protection: A review. *Crop Protection, 35*, 64–70.

Kong, X. P., Zhang, B. H., & Wang, J. (2021). Multiple roles of mesoporous silica in safe pesticide application by nanotechnology: A review. *Journal of Agricultural and Food Chemistry, 69*(24), 6735–6754.

Kuiken, T. (2010). Cleaning up contaminated waste sites: Is nanotechnology the answer? *Nano Today, 1*(5), 6–8.

Lal, R. (2009). Soil degradation as a reason for inadequate human nutrition. *Food Security, 1*, 45–57.

Lal, R. (2015). Restoring soil quality to mitigate soil degradation. *Sustainability, 7*(5), 5875–5895.

León, J. D., & Osorio, N. W. (2014). Role of litter turnover in soil quality in tropical degraded lands of Colombia. *The Scientific World Journal, 2014*, 1.

Li, S., Wang, W., Liang, F., & Zhang, W. X. (2017). Heavy metal removal using nanoscale zero-valent iron (nZVI): Theory and application. *Journal of Hazardous Materials, 322*, 163–171.

Li, Q., Chen, X., Chen, X., Jin, Y., & Zhuang, J. (2019). Cadmium removal from soil by fulvic acid-aided hydroxyapatite nanofluid. *Chemosphere, 215*, 227–233.

Liu, R., & Zhao, D. (2007). In situ immobilization of Cu (II) in soils using a new class of iron phosphate nanoparticles. *Chemosphere, 68*(10), 1867–1876.

Liu, R., & Zhao, D. (2013). Synthesis and characterization of a new class of stabilized apatite nanoparticles and applying the particles to in situ Pb immobilization in a fire-range soil. *Chemosphere, 91*(5), 594–601.

Liu, S., Yuan, L., Yue, X., Zheng, Z., & Tang, Z. (2008). Recent advances in nanosensors for organophosphate pesticide detection. *Advanced Powder Technology, 19*(5), 419–441.

Liu, T., Su, H., Qu, X., Ju, P., Cui, L., & Ai, S. (2011). Acetylcholinesterase biosensor based on 3-carboxyphenylboronic acid/reduced graphene oxide–gold nanocomposites modified electrode for amperometric detection of organophosphorus and carbamate pesticides. *Sensors and Actuators B: Chemical, 160*(1), 1255–1261.

Liu, S., Yang, B., Liang, Y., Xiao, Y., & Fang, J. (2020). Prospect of phytoremediation combined with other approaches for remediation of heavy metal-polluted soils. *Environmental Science and Pollution Research, 27*, 16069–16085.

López-Luna, J., Silva-Silva, M. J., Martinez-Vargas, S., Mijangos-Ricardez, O. F., González-Chávez, M. C., Solís-Domínguez, F. A., & Cuevas-Díaz, M. C. (2016). Magnetite nanoparticle (NP) uptake by wheat plants and its effect on cadmium and chromium toxicological behavior. *Science of the Total Environment, 565*, 941–950.

Lu, Y., Song, S., Wang, R., Liu, Z., Meng, J., Sweetman, A. J., Jenkins, A., Ferrier, R. C., Li, H., Luo, W., & Wang, T. (2015). Impacts of soil and water pollution on food safety and health risks in China. *Environment International, 77*, 5–15.

Luna-Sanguino, G., Ruíz-Delgado, A., Tolosana-Moranchel, A., Pascual, L., Malato, S., Bahamonde, A., & Faraldos, M. (2020). Solar photocatalytic degradation of pesticides over TiO_2-rGO nanocomposites at pilot plant scale. *Science of the Total Environment, 737*, 140286.

Lyu, H., Zhao, H., Tang, J., Gong, Y., Huang, Y., Wu, Q., & Gao, B. (2018). Immobilization of hexavalent chromium in contaminated soils using biochar supported nanoscale iron sulfide composite. *Chemosphere, 194*, 360–369.

Madhavi, V., Prasad, T. N. V. K. V., Reddy, B. R., Reddy, A. V. B., & Gajulapalle, M. (2014). Conjunctive effect of CMC–zero-valent iron nanoparticles and FYM in the remediation of chromium-contaminated soils. *Applied Nanoscience, 4*, 477–484.

Maharjan, B., Das, S., & Acharya, B. S. (2020). Soil health gap: A concept to establish a benchmark for soil health management. *Global Ecology and Conservation, 23*, e01116.

Majumder, S., & George, S. (2023). Multi functional properties of halloysite nano-clays in food safety and security. In *Materials science and engineering in food product development* (pp. 261–277). Wiley.

Malakar, A., & Snow, D. D. (2020). Nanoparticles as sources of inorganic water pollutants. In *Inorganic pollutants in water* (pp. 337–370). Elsevier.

Mallampati, S. R., Mitoma, Y., Okuda, T., Sakita, S., & Kakeda, M. (2013). Total immobilization of soil heavy metals with nano-Fe/Ca/CaO dispersion mixtures. *Environmental Chemistry Letters, 11*, 119–125.

Mallikarjunaiah, S., Pattabhiramaiah, M., & Metikurki, B. (2020). Application of nanotechnology in the bioremediation of heavy metals and wastewater management. In *Nanotechnology for food, agriculture, and environment* (pp. 297–321). Springer.

Mandal, S., Pu, S., He, L., Ma, H., & Hou, D. (2020). Biochar induced modification of graphene oxide & nZVI and its impact on immobilization of toxic copper in soil. *Environmental Pollution, 259*, 113851.

Mathur, S., Singh, D., & Ranjan, R. (2022). Remediation of heavy metal (loid) contaminated soil through green nanotechnology. *Frontiers in Sustainable Food Systems, 6*, 932424.

Men, C., Liu, R., Xu, F., Wang, Q., Guo, L., & Shen, Z. (2018). Pollution characteristics, risk assessment, and source apportionment of heavy metals in road dust in Beijing, China. *Science of the Total Environment, 612*, 138–147.

Mitton, F. M., Miglioranza, K. S., Gonzalez, M., Shimabukuro, V. M., & Monserrat, J. M. (2014). Assessment of tolerance and efficiency of crop species in the phytoremediation of DDT polluted soils. *Ecological Engineering, 71*, 501–508.

Mohanapriya, V., Sakthivel, R., Pham, N. D. K., Cheng, C. K., Le, H. S., & Dong, T. M. H. (2023). Nanotechnology-A ray of hope for heavy metals removal. *Chemosphere, 311*, 136989.

Moharem, M., Elkhatib, E., & Mesalem, M. (2019). Remediation of chromium and mercury polluted calcareous soils using nanoparticles: Sorption–desorption kinetics, speciation and fractionation. *Environmental Research, 170*, 366–373.

Mony, C., Kaur, P., Rookes, J. E., Callahan, D. L., Eswaran, S. V., Yang, W., & Manna, P. K. (2022). Nanomaterials for enhancing photosynthesis: Interaction with plant photosystems and scope of nanobionics in agriculture. *Environmental Science: Nano, 9*(10), 3659–3683.

Naderi Peikam, E., & Jalali, M. (2019). Application of three nanoparticles (Al_2O_3, SiO_2 and TiO_2) for metal-contaminated soil remediation (measuring and modeling). *International journal of Environmental Science and Technology, 16*, 7207–7220.

Nehra, M., Dilbaghi, N., Marrazza, G., Kaushik, A., Sonne, C., Kim, K. H., & Kumar, S. (2021). Emerging nanobiotechnology in agriculture for the management of pesticide residues. *Journal of Hazardous Materials, 401*, 123369.

Oloumi, H., Mousavi, E. A., & Nejad, R. M. (2018). Multi-wall carbon nanotubes effects on plant seedlings growth and cadmium/lead uptake in vitro. *Russian Journal of Plant Physiology, 65*, 260–268.

Pabbi, M., Kaur, A., Mittal, S. K., & Jindal, R. (2018). A surface expressed alkaline phosphatase biosensor modified with flower shaped ZnO for the detection of chlorpyrifos. *Sensors and Actuators B: Chemical, 258*, 215–227.

Pacyna, J. M., Pacyna, E. G., & Aas, W. (2009). Changes of emissions and atmospheric deposition of mercury, lead, and cadmium. *Atmospheric Environment, 43*(1), 117–127.

Pan, B., & Xing, B. (2012). Applications and implications of manufactured nanoparticles in soils: A review. *European Journal of Soil Science, 63*(4), 437–456.

Pande, V., Pandey, S. C., Sati, D., Pande, V., & Samant, M. (2020). Bioremediation: An emerging effective approach towards environment restoration. *Environmental Sustainability, 3*, 91–103.

Parveen, K., Banse, V., & Ledwani, L. (2016). Green synthesis of nanoparticles: Their advantages and disadvantages. In *AIP conference proceedings* (Vol. 1724, No. 1). AIP Publishing.

Pillai, H. P., & Kottekottil, J. (2016). Nano-phytotechnological remediation of endosulfan using zero valent iron nanoparticles. *Journal of Environmental Protection, 7*(05), 734.

Pilon-Smits, E. (2005). Phytoremediation. *Annual Review of Plant Biology, 56*, 15–39.

Prabhakar, N., Thakur, H., Bharti, A., & Kaur, N. (2016). Chitosan-iron oxide nanocomposite based electrochemical aptasensor for determination of malathion. *Analytica Chimica Acta, 939*, 108–116.

Prabhu, R. N., & Lakshmipraba, J. (2022). Persistent organic pollutants (part I): The "dirty dozen"–sources and adverse effects. In *Organic pollutants: Toxicity and solutions* (pp. 1–27). Springer.

Premarathne, D. M. M. M., Priyadarshana, G., Karunarathne, K. P. S. W., & Jeewantha, D. M. C. (2023). A review of nanotechnology on agriculture and the food industry. *In Proceedings of the 4th International Conference on Advancements in Computing and Technology (ICACT).*

Pujari, M., & Kapoor, D. (2021). Heavy metals in the ecosystem: Sources and their effects. In *Heavy metals in the environment* (pp. 1–7). Elsevier.

Qian, W., Liang, J. Y., Zhang, W. X., Huang, S. T., & Diao, Z. H. (2022). A porous biochar supported nanoscale zero-valent iron material highly efficient for the simultaneous remediation of cadmium and lead contaminated soil. *Journal of Environmental Sciences, 113*, 231–241.

Raffa, C. M., Chiampo, F., & Shanthakumar, S. (2021). Remediation of metal/metalloid-polluted soils: A short review. *Applied Sciences, 11*(9), 4134.

Rajput, V. D., Minkina, T., Upadhyay, S. K., Kumari, A., Ranjan, A., Mandzhieva, S., Sushkova, S., Singh, R. K., & Verma, K. K. (2022). Nanotechnology in the restoration of polluted soil. *Nanomaterials, 12*(5), 769.

Rao, B. G., Mukherjee, D., & Reddy, B. M. (2017). Novel approaches for preparation of nanoparticles. In *Nanostructures for novel therapy* (pp. 1–36). Elsevier.

Restivo, J., Órfão, J. J. M., Armenise, S., Garcia-Bordejé, E., & Pereira, M. F. R. (2012). Catalytic ozonation of metolachlor under continuous operation using nanocarbon materials grown on a ceramic monolith. *Journal of Hazardous Materials, 239*, 249–256.

Rizwan, M., Shoukat, A., Ayub, A., Razzaq, B., & Tahir, M. B. (2021). Types and classification of nanomaterials. In *Nanomaterials: Synthesis, characterization, hazards and safety* (pp. 31–54). Elsevier.

Rui, Z. E. N. G., Jingguo, W. A. N. G., Jianyu, C. U. I., Lin, H. U., & Kangguo, M. U. (2010). Photocatalytic degradation of pesticide residues with RE^{3+}-doped nano-TiO_2. *Journal of Rare Earths, 28*, 353–356.

Sarno, M., Casa, M., Cirillo, C., & Ciambelli, P. (2017). Complete removal of persistent pesticide using reduced graphene oxide–silver nanocomposite. *Chemical Engineering Transactions, 60*, 151–156.

Šebesta, M., Kolenčík, M., Sunil, B. R., Illa, R., Mosnáček, J., Ingle, A. P., & Urík, M. (2021). Field application of ZnO and TiO_2 nanoparticles on agricultural plants. *Agronomy, 11*(11), 2281.

Shaheb, M. R., Venkatesh, R., & Shearer, S. A. (2021). A review on the effect of soil compaction and its management for sustainable crop production. *Journal of Biosystems Engineering, 46*, 1–23.

Sharma, A., & Nagpal, A. K. (2020). Contamination of vegetables with heavy metals across the globe: Hampering food security goal. *Journal of Food Science and Technology, 57*, 391–403.

Sharma, P., Pandey, V., Sharma, M. M. M., Patra, A., Singh, B., Mehta, S., & Husen, A. (2021). A review on biosensors and nanosensors application in agroecosystems. *Nanoscale Research Letters, 16*, 1–24.

Shi, J., Votruba, A. R., Farokhzad, O. C., & Langer, R. (2010). Nanotechnology in drug delivery and tissue engineering: From discovery to applications. *Nano Letters, 10*(9), 3223–3230.

Simonian, A. L., Good, T. A., Wang, S. S., & Wild, J. R. (2005). Nanoparticle-based optical biosensors for the direct detection of organophosphate chemical warfare agents and pesticides. *Analytica Chimica Acta, 534*(1), 69–77.

Singh, D., & Gurjar, B. R. (2022). Nanotechnology for agricultural applications: Facts, issues, knowledge gaps, and challenges in environmental risk assessment. *Journal of Environmental Management, 322*, 116033.

Singh, J., & Lee, B. K. (2016). Influence of nano-TiO2 particles on the bioaccumulation of Cd in soybean plants (*Glycine max*): A possible mechanism for the removal of Cd from the contaminated soil. *Journal of Environmental Management, 170*, 88–96.

Singh, R., Misra, V., & Singh, R. P. (2012). Removal of Cr (VI) by nanoscale zero-valent iron (nZVI) from soil contaminated with tannery wastes. *Bulletin of Environmental Contamination and Toxicology, 88*, 210–214.

Song, B., Xu, P., Chen, M., Tang, W., Zeng, G., Gong, J., Zhang, P., & Ye, S. (2019). Using nanomaterials to facilitate the phytoremediation of contaminated soil. *Critical Reviews in Environmental Science and Technology, 49*(9), 791–824.

Stephenson, G. R., Coats, J. R., & Yamamoto, H. (2001). Pesticide use and world food production: Risks and benefits. In *Proceedings of the 2000 national meeting of the expert committee on weeds Comité d'experts en malherbologie* (pp. 9–15). Citeseer.

Su, S., He, Y., Zhang, M., Yang, K., Song, S., Zhang, X., Fan, C., & Lee, S. T. (2008). High-sensitivity pesticide detection via silicon nanowires-supported acetylcholinesterase-based electrochemical sensors. *Applied Physics Letters, 93*(2), 023113.

Su, H., Fang, Z., Tsang, P. E., Fang, J., & Zhao, D. (2016a). Stabilisation of nanoscale zero-valent iron with biochar for enhanced transport and in-situ remediation of hexavalent chromium in soil. *Environmental Pollution, 214*, 94–100.

Su, H., Fang, Z., Tsang, P. E., Zheng, L., Cheng, W., Fang, J., & Zhao, D. (2016b). Remediation of hexavalent chromium contaminated soil by biochar-supported zero-valent iron nanoparticles. *Journal of Hazardous Materials, 318*, 533–540.

Su, R., Wang, Y., Huang, S., Chen, R., & Wang, J. (2022). Application for ecological restoration of contaminated soil: Phytoremediation. *International Journal of Environmental Research and Public Health, 19*(20), 13124.

Szczyglewska, P., Feliczak-Guzik, A., & Nowak, I. (2023). Nanotechnology–general aspects: A chemical reduction approach to the synthesis of nanoparticles. *Molecules, 28*(13), 4932.

Tang, W. W., Zeng, G. M., Gong, J. L., Liang, J., Xu, P., Zhang, C., & Huang, B. B. (2014). Impact of humic/fulvic acid on the removal of heavy metals from aqueous solutions using nanomaterials: A review. *Science of the Total Environment, 468*, 1014–1027.

Tilston, E. L., Collins, C. D., Mitchell, G. R., Princivalle, J., & Shaw, L. J. (2013). Nanoscale zerovalent iron alters soil bacterial community structure and inhibits chloroaromatic biodegradation potential in Aroclor 1242-contaminated soil. *Environmental Pollution, 173*, 38–46.

Tripathi, V., Fraceto, L. F., & Abhilash, P. C. (2015). Sustainable clean-up technologies for soils contaminated with multiple pollutants: Plant-microbe-pollutant and climate nexus. *Ecological Engineering, 82*, 330–335.

Tripathi, D. K., Singh, S., Singh, V. P., Prasad, S. M., Chauhan, D. K., & Dubey, N. K. (2016). Silicon nanoparticles more efficiently alleviate arsenate toxicity than silicon in maize cultiver and hybrid differing in arsenate tolerance. *Frontiers in Environmental Science, 4*, 46.

Tümay, S. O., Şenocak, A., Sarı, E., Şanko, V., Durmuş, M., & Demirbas, E. (2021). A new perspective for electrochemical determination of parathion and chlorantraniliprole pesticides via carbon nanotube-based thiophene-ferrocene appended hybrid nanosensor. *Sensors and Actuators B: Chemical, 345*, 130344.

Venkatachalam, P., Jayaraj, M., Manikandan, R., Geetha, N., Rene, E. R., Sharma, N. C., & Sahi, S. V. (2017). Zinc oxide nanoparticles (ZnONPs) alleviate heavy metal-induced toxicity in Leucaena leucocephala seedlings: A physiochemical analysis. *Plant Physiology and Biochemistry, 110*, 59–69.

Vijayaram, S., Razafindralambo, H., Sun, Y. Z., Vasantharaj, S., Ghafarifarsani, H., Hoseinifar, S. H., & Raeeszadeh, M. (2024). Applications of green synthesized metal nanoparticles—A review. *Biological Trace Element Research, 202*(1), 360–386.

Violante, A. U. D. N., Cozzolino, V. U. D. N., Perelomov, L. P. S. U., Caporale, A. G., & Pigna, M. U. D. N. (2010). Mobility and bioavailability of heavy metals and metalloids in soil environments. *Journal of Soil Science and Plant Nutrition, 10*(3), 268–292.

Vítková, M., Puschenreiter, M., & Komárek, M. (2018). Effect of nano zero-valent iron application on As, Cd, Pb, and Zn availability in the rhizosphere of metal (loid) contaminated soils. *Chemosphere, 200*, 217–226.

Vu, K. A., & Mulligan, C. N. (2023). Remediation of oil-contaminated soil using Fe/Cu nanoparticles and biosurfactants. *Environmental Technology, 44*(22), 3446–3458.

Wang, X. T., Shi, W. S., She, G. W., Mu, L. X., & Lee, S. T. (2010). High-performance surfaceenhanced Raman scattering sensors based on Ag nanoparticles-coated Si nanowire arrays for quantitative detection of pesticides. *Applied Physics Letters, 96, 053104*.

Wang, T., Liu, Y., Wang, J., Wang, X., Liu, B., & Wang, Y. (2019). In-situ remediation of hexavalent chromium contaminated groundwater and saturated soil using stabilized iron sulfide nanoparticles. *Journal of Environmental Management, 231*, 679–686.

Wang, Z., Zhang, J., Wen, T., Liu, X., Wang, Y., Yang, H., Sun, J., Feng, J., Dong, S., & Sun, J. (2020). Highly effective remediation of Pb (II) and Hg (II) contaminated wastewater and soil by flower-like magnetic MoS_2 nanohybrid. *Science of the Total Environment, 699*, 134341.

Wu, S., Huang, F., Lan, X., Wang, X., Wang, J., & Meng, C. (2013). Electrochemically reduced graphene oxide and Nafion nanocomposite for ultralow potential detection of organophosphate pesticide. *Sensors and Actuators B: Chemical, 177*, 724–729.

Wu, J., Xie, Y., Fang, Z., Cheng, W., & Tsang, P. E. (2016). Effects of Ni/Fe bimetallic nanoparticles on phytotoxicity and translocation of polybrominated diphenyl ethers in contaminated soil. *Chemosphere, 162*, 235–242.

Wu, B., Wang, Z., Peng, D., Wang, Y., He, T., Tang, H., & Xu, H. (2020). Removal and recovery of heavy metals from soil with sodium alginate coated FeSSi nanocomposites in a leaching process. *Journal of Hazardous Materials, 398*, 122732.

Wu, Q., Wang, D., Chen, C., Peng, C., Cai, D., & Wu, Z. (2021). Fabrication of Fe_3O_4/ZIF-8 nanocomposite for simultaneous removal of copper and arsenic from water/soil/swine urine. *Journal of Environmental Management, 290*, 112626.

Wu, Y., Li, X., Yu, L., Wang, T., Wang, J., & Liu, T. (2022a). Review of soil heavy metal pollution in China: Spatial distribution, primary sources, and remediation alternatives. *Resources, Conservation and Recycling, 181*, 106261.

Wu, T., Liao, X., Zou, Y., Liu, Y., Yang, K., White, J. C., & Lin, D. (2022b). Fe-based nanomaterial transformation to amorphous Fe: Enhanced alfalfa rhizoremediation of PCBs-contaminated soil. *Journal of Hazardous Materials, 425*, 127973.

Yadav, S., Sawarni, N., Dahiya, T., Rana, J. S., Sharma, M., & Batra, B. (2023). Nanoagriculture: Advantages and drawbacks. In *Agricultural and environmental nanotechnology: Novel technologies and their ecological impact* (pp. 3–42). Springer Nature Singapore.

Yan, W., Lien, H. L., Koel, B. E., & Zhang, W. X. (2013). Iron nanoparticles for environmental clean-up: Recent developments and future outlook. *Environmental Science: Processes & Impacts, 15*(1), 63–77.

Yang, Q., Li, Q., Li, H., & Li, F. (2021). pH-response quantum dots with orange–red emission for monitoring the residue, distribution, and variation of an organophosphorus pesticide in an agricultural crop. *Journal of Agricultural and Food Chemistry, 69*(9), 2689–2696.

Yin, J., Su, X., Yan, S., & Shen, J. (2023). Multifunctional nanoparticles and nanopesticides in agricultural application. *Nanomaterials, 13*(7), 1255.

Yu, B., Zeng, J., Gong, L., Zhang, M., Zhang, L., & Chen, X. (2007). Investigation of the photocatalytic degradation of organochlorine pesticides on a nano-TiO_2 coated film. *Talanta, 72*(5), 1667–1674.

Zand, A. D., Tabrizi, A. M., & Heir, A. V. (2020). Application of titanium dioxide nanoparticles to promote phytoremediation of Cd-polluted soil: Contribution of PGPR inoculation. *Bioremediation Journal, 24*, 171–189.

Zayed, A., Pilon-Smits, E., deSouza, M., Lin, Z. Q., & Terry, N. (2020). Remediation of selenium-polluted soils and waters by phytovolatilization. In *Phytoremediation of contaminated soil and water* (pp. 61–83). CRC Press.

Zhang, M., Wang, Y., Zhao, D., & Pan, G. (2010). Immobilization of arsenic in soils by stabilized nanoscale zero-valent iron, iron sulfide (FeS), and magnetite (Fe_3O_4) particles. *Chinese Science Bulletin, 55*, 365–372.

Zhang, L., He, F., & Guan, Y. (2022). Immobilization of hexavalent chromium in contaminated soil by nano-sized layered double hydroxide intercalated with diethyldithiocarbamate: Fraction distribution, plant growth, and microbial evolution. *Journal of Hazardous Materials, 430*, 128382.

Zheng, S., Hao, L., Zhang, L., Wang, K., Zheng, W., Wang, X., Zhou, X., Li, W., & Zhang, L. (2018). Tea polyphenols functionalized and reduced graphene oxide-ZnO composites for selective Pb^{2+} removal and enhanced antibacterial activity. *Journal of Biomedical Nanotechnology, 14*(7), 1263–1276.

Zheng, T., Ouyang, S., & Zhou, Q. (2023). Synthesis, characterization, safety design, and application of NPs@ BC for contaminated soil remediation and sustainable agriculture. *Biochar, 5*(1), 5.

Zhu, F., Li, L., Ma, S., & Shang, Z. (2016). Effect factors, kinetics and thermodynamics of remediation in the chromium contaminated soils by nanoscale zero valent Fe/cu bimetallic particles. *Chemical Engineering Journal, 302*, 663–669.

Zia, R., Taj, A., Younis, S., Bukhari, S. Z., Latif, F., Feroz, Y., Fatima, K., Imran, A., & Bajwa, S. Z. (2022). Application of nanosensors for pesticide detection. In *Nanosensors for smart agriculture* (pp. 259–302). Elsevier.

Nanotechnology in Waste Management: Pyrolysis-Driven Solutions for Transforming Challenges into Opportunities

M. Tamilmozhi and J. Krishnaraj

Abstract

This comprehensive review explores the transformative potential of nanotechnology-driven pyrolysis solutions in waste management. It investigates the composition, properties, and utilization pathways of waste tire, plastic, and biomass pyrolysis products, enhanced by nanotechnology interventions. Waste tire pyrolysis oil (WTPO), plastic pyrolysis products, and biomass-derived fuels are analyzed, with a focus on nanomaterial-based catalysts and purification methods to enhance product quality and yield. WTPO, rich in alkanes and aromatic hydrocarbons but hindered by impurities, benefits from nanocatalytic treatments for cleaner production and enhanced fuel properties. Plastic pyrolysis, yielding valuable products like liquid oil and chemicals, is augmented by nanomaterial catalysts to improve efficiency and selectivity. Biomass pyrolysis, offering bio-oils, gases, and char, is optimized through nanotechnology for increased yield and higher-value products. The chapter delves into past efforts and experimental results, discussing the role of nanomaterials in reactor design, catalyst development, and product refinement. It highlights the potential for nanotechnology-enabled pyrolysis processes to address global waste challenges and foster sustainable resource utilization. Co-pyrolysis of waste streams, enhanced by nanomaterial interventions, further enhances product quality and sustainability. The study underscores nanotechnology's pivotal role in advancing pyrolysis-driven solutions within the circular economy paradigm, emphasizing ongoing research trends, such as nanocatalyst design, reactor optimization, and product

M. Tamilmozhi
Department of Bioscience, Sri Krishna Arts and Science College, Coimbatore, Tamil Nadu, India

J. Krishnaraj (✉)
Department of Automobile Engineering, Hindusthan College of Engineering and Technology, Coimbatore, Tamil Nadu, India

D. K. Gupta et al. (eds.), *Nano-Ecosystems*,
https://doi.org/10.1007/978-981-95-5033-3_8

valorization. Experimental findings demonstrate the feasibility of using nanotechnology-enhanced pyrolysis products as alternative fuels and high-value chemicals, with promising results in energy efficiency and environmental sustainability. Overall, the chapter provides insights into the transformative potential of nanotechnology-driven pyrolysis solutions, offering pathways to convert waste challenges into sustainable opportunities.

Keywords

Nanotechnology · Waste management · Pyrolysis · Catalysts · Sustainable energy · Circular economy

1 Introduction

The exponential increase in the production of waste materials, spanning from discarded automobile tires and plastics to organic biomass, has catalyzed a global environmental crisis. Conventional disposal techniques like landfilling and incineration, once deemed effective, now stand as significant contributors to pollution and resource exhaustion. Consequently, the imperative to explore alternative avenues for waste management has never been more urgent (Xue et al., 2016; Zhao et al., 2019; Park et al., 2018; Iftikhar et al., 2019; Murthy et al., 2020). Amid this backdrop, pyrolysis emerges as a beacon of hope in the realm of sustainable waste treatment. At its core, pyrolysis is a transformative process that harnesses the power of controlled heat in the absence of oxygen to break down complex organic compounds. This controlled thermal decomposition leads to the creation of valuable by-products, including pyrolysis oil, gas, and char, each with its unique potential for reuse and repurposing (Campuzano & Brown, 2019; Park et al., 2019; Wong et al., 2020; Williams et al., 1998). The advent of pyrolysis technology signifies a paradigm shift in waste management philosophy, offering a pathway towards both environmental sustainability and economic viability. By converting waste materials into high-value products, pyrolysis not only mitigates the burden on landfills and reduces harmful emissions but also fosters the creation of new revenue streams from previously discarded resources. In essence, the adoption of pyrolysis represents a pivotal step forward in our collective efforts to combat the escalating challenges of waste proliferation and environmental degradation. Through innovation and strategic implementation, pyrolysis holds the promise of transforming waste from a liability into a valuable asset, thereby reshaping the trajectory of our planet's sustainability journey (Kar, 2011; Lucchesi & Maschio, 1983).

2 Composition of Feedstocks

The composition of feedstocks plays a crucial role in determining the efficiency and outcomes of the pyrolysis process. Each feedstock tires, plastics, and biomass possesses a distinct molecular makeup, which directly influences the pyrolysis conditions required and the resulting product yields (Galvagno et al., 2002). Tires, for instance, are predominantly composed of natural and synthetic rubbers, fortified with carbon black for reinforcement. These materials imbue tires with their characteristic elasticity and durability. During pyrolysis, the rubber components break down into smaller hydrocarbon molecules, while carbon black may yield valuable carbonaceous by-products. The presence of steel reinforcements in some tires also necessitates additional processing steps to recover metals effectively (Dogan et al., 2012; Fairburn et al., 1990; Demirbas, 2000). Plastics encompass a diverse range of materials, including polyethylene, polypropylene, polystyrene, and polyvinyl chloride, among others. These polymers consist primarily of carbon and hydrogen, with varying degrees of oxygen and other elements depending on their chemical structure. The heterogeneity of plastic feedstocks necessitates careful selection of pyrolysis conditions to optimize product yields. Furthermore, the presence of additives and contaminants in plastic waste streams may impact the pyrolysis process and require preprocessing or purification steps(Aguado et al., 2002; Cornelissen et al., 2008; Grace & Lim, 2013). Biomass materials, such as wood, agricultural residues, and organic wastes, comprise predominantly carbon, hydrogen, and oxygen. Cellulose, hemicellulose, and lignin are the primary constituents of lignocellulosic biomass, each with unique thermal degradation characteristics (Murugan et al., 2009; Leung et al., 2002; Banar et al., 2012; Aydın & Ilkılıç, 2012). Pyrolysis of biomass involves the decomposition of these complex polymers into simpler molecules, yielding a range of bio-oil, syngas, and biochar products. The composition of biomass feedstocks, influenced by factors like moisture content and lignin content, can significantly affect pyrolysis kinetics and product distribution.

Understanding the composition of feedstocks is essential for optimizing pyrolysis processes and maximizing the recovery of valuable products. Tailoring pyrolysis parameters to accommodate the specific characteristics of each feedstock enables efficient conversion of waste materials into valuable resources, thereby advancing the goals of sustainability and resource conservation.

3 Pyrolysis Process

Pyrolysis is a transformative process that harnesses the power of controlled heat in the absence of oxygen to break down complex organic compounds into valuable by-products. This thermal decomposition occurs through a series of chemical reactions, ultimately yielding products such as pyrolysis oil, gas, and char (Xue et al., 2016; Murthy et al., 2020; Williams et al., 1990). To execute the pyrolysis process, various reactor configurations are utilized, each with its unique advantages and challenges. The pyrolysis setup is shown in Fig. 1. Fixed-bed reactors, for instance, offer simplicity and ease of operation, making them suitable for small-scale applications. Fluidized bed reactors, on the other hand, facilitate better heat transfer and mixing, enhancing process efficiency and enabling continuous operation. Rotary kiln reactors provide versatility and scalability, accommodating a wide range of

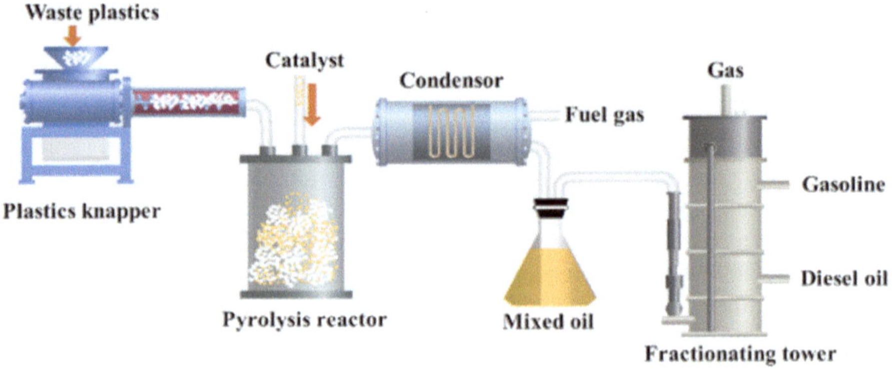

Fig. 1 pyrolysis setup (Maqsood et al., 2021)

feedstocks and operating conditions (Fairburn et al., 1990; Cornelissen et al., 2008; Maqsood et al., 2021).

The success of the pyrolysis process hinges upon carefully controlled parameters, including temperature, heating rate, and residence time. These factors dictate the kinetics of thermal decomposition and profoundly influence the distribution and properties of the resulting products. High temperatures promote rapid pyrolysis reactions, but excessive heat may lead to undesirable side reactions or product degradation. Figure 2 illustrates the analysis of waste-to-energy pyrolysis, including methods, kinetics, and output products. Optimal heating rates and residence times ensure efficient conversion of feedstock into desired products while minimizing energy consumption and process duration. By fine-tuning pyrolysis conditions and selecting appropriate reactor configurations, it is possible to tailor the process for specific feedstocks and desired product outcomes. This strategic approach not only maximizes resource utilization and product yields but also mitigates environmental impacts associated with traditional waste disposal methods. In essence, the pyrolysis process represents a versatile and sustainable solution for converting diverse waste streams into valuable resources, thereby contributing to the transition towards a circular economy and mitigating the global challenges of waste management and environmental degradation (Wang et al., 2021).

4 Types of Reactors Used in Pyrolysis

4.1 Fixed-Bed Reactor

A fixed-bed reactor is a type of pyrolysis system, as shown in Fig. 3, where Jin Q et al. investigated the stationary bed. The feedstock is placed in a stationary bed within the reactor chamber. This configuration offers simplicity and ease of operation, making it particularly suitable for small-scale operations and batch processing scenarios. In a fixed-bed reactor, the feedstock remains stationary while heat is

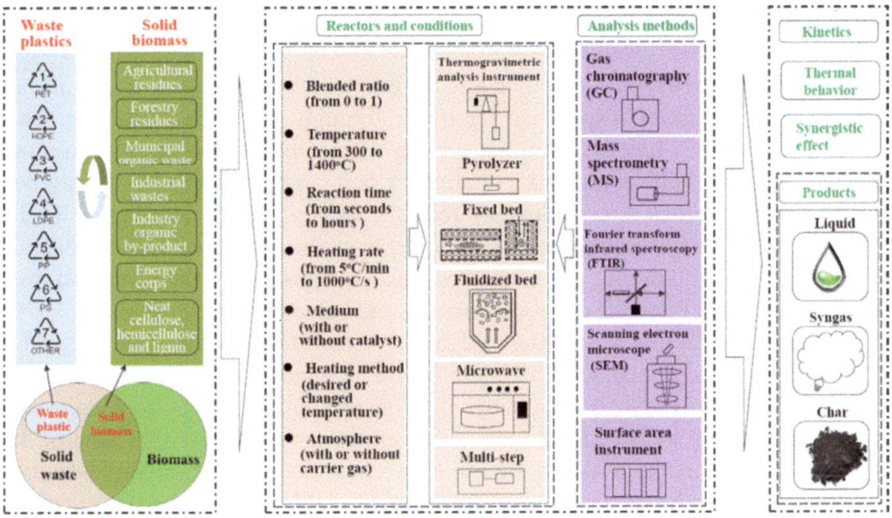

Fig. 2 Analysis of waste-to-energy pyrolysis: methods, kinetics, and output products (Wang et al., 2021)

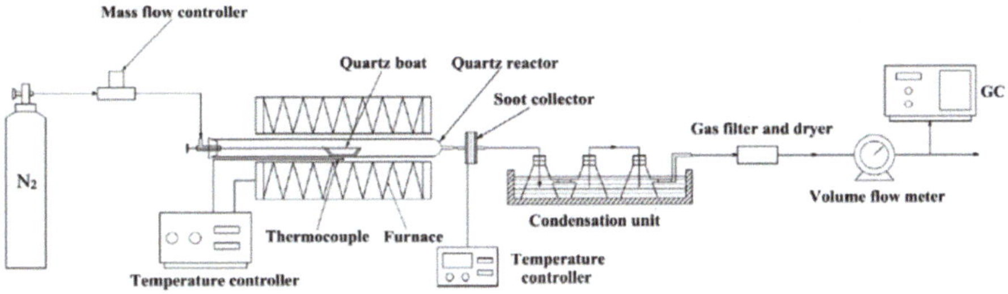

Fig. 3 Fixed-bed reactor (Jin et al., 2019)

applied to initiate the pyrolysis process. This stationary bed allows for straightforward loading and unloading of the feedstock, simplifying the operational procedures. Additionally, the stationary nature of the bed facilitates easy monitoring and control of the pyrolysis conditions. However, fixed-bed reactors may suffer from limitations such as poor heat transfer and uneven product distribution. Since the feedstock remains in a stationary position, heat transfer efficiency may be compromised, leading to non-uniform heating throughout the bed. This uneven heating can result in variations in product quality and distribution within the reactor. Despite these challenges, fixed-bed reactors remain a viable option for certain applications, especially those involving small-scale operations or batch processing requirements. Their simplicity and ease of operation make them well-suited for research and development purposes, as well as for niche applications where scalability is not a primary concern.

4.2 Moving Bed Reactors

Moving bed reactors represent a dynamic approach to pyrolysis, utilizing a mechanism to continuously transport the feedstock through the reactor chamber. This movement enhances heat transfer and mixing within the reactor, leading to more uniform product yields compared to stationary bed configurations. In moving bed reactors, the feedstock is continuously fed into one end of the reactor, while the pyrolysis products are simultaneously removed from the other end. This continuous movement ensures that the feedstock is exposed to consistent heat levels and residence times, optimizing the pyrolysis process and enhancing product quality.

The dynamic nature of moving bed reactors facilitates efficient heat transfer throughout the reactor bed, minimizing temperature gradients and ensuring uniform heating of the feedstock. Additionally, the constant movement helps to promote mixing, preventing localized hot spots and facilitating thorough conversion of the feedstock into valuable products. Due to their continuous operation and enhanced efficiency, moving bed reactors are particularly suitable for larger-scale pyrolysis plants where consistent production rates are essential. Their ability to handle a continuous feed of material makes them well-suited for industrial-scale applications, allowing for higher throughput and improved process control.

4.3 Fluidized Bed Reactor

The fluidized bed reactor, represented in Fig. 4, is a dynamic and highly efficient configuration for pyrolysis, wherein the feedstock particles are suspended in a fluidized state by blowing gas through the reactor chamber. In a fluidized bed reactor, a gas, typically air or steam, is introduced from the bottom of the reactor at a controlled flow rate. As the gas flows upward, it lifts and suspends the solid particles of the feedstock, creating a fluidized bed. This fluidized state allows for excellent heat transfer and mixing throughout the reactor bed, ensuring uniform temperatures and reaction conditions. The continuous movement of the feedstock particles in the fluidized bed reactor promote intimate contact with the surrounding gas, enhancing heat transfer efficiency and facilitating rapid pyrolysis reactions. This efficient heat transfer and mixing result in high pyrolysis efficiency and product yields compared to other reactor configurations.

One of the significant advantages of fluidized bed reactors is their versatility in processing a wide range of feedstock sizes and moisture contents. The fluidized state of the bed enables effective handling of both fine and coarse particles, as well as materials with varying moisture levels. This flexibility makes fluidized bed reactors well-suited for handling diverse feedstock streams, including biomass, plastics, and other waste materials. Additionally, fluidized bed reactors offer inherent advantages in terms of scalability and controllability, making them suitable for both pilot-scale research and large-scale industrial applications. Their ability to achieve high heat transfer rates and uniform reaction conditions makes them a preferred choice for industries seeking efficient and reliable pyrolysis processes.

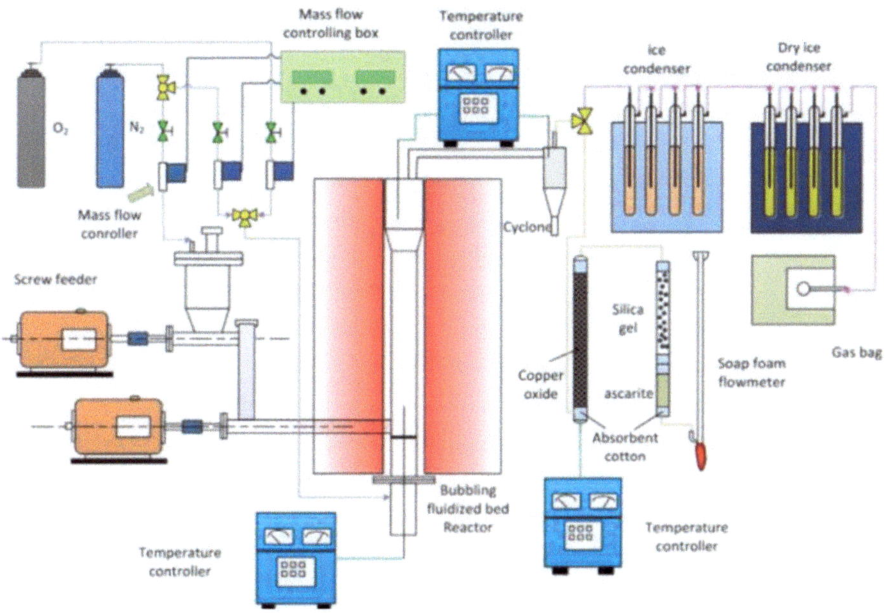

Fig. 4 Fluidized bed reactor (Zhang et al., 2015)

4.4 Rotating Bed Reactors

Rotating bed reactors, also known as rotary kiln reactors, employ a rotating drum or cylinder to agitate the feedstock and facilitate thermal decomposition. Figure 5 represents the layout of the rotating bed reactor. In these reactors, the feedstock is loaded into the rotating drum, which is heated externally to initiate the pyrolysis process. As the drum rotates, the feedstock is continuously agitated, ensuring uniform exposure to heat and promoting thorough thermal decomposition. This agitation helps to break down complex organic compounds and facilitate the release of volatile gases. One of the key advantages of rotating bed reactors is their flexibility in controlling residence time and heating rates. By adjusting the rotational speed of the drum and the feed rate of the material, operators can precisely control the time the feedstock spends in the reactor and the heating profile it experiences. This flexibility allows for optimization of the pyrolysis process to maximize product yields and quality. Rotating bed reactors are particularly suitable for continuous processing of homogeneous feedstocks. The continuous agitation provided by the rotating drum ensures consistent heat distribution and reaction conditions throughout the reactor bed, leading to uniform product yields. This makes them well-suited for applications where a consistent feedstock composition is maintained, such as in certain industrial processes or dedicated pyrolysis facilities.

Additionally, rotating bed reactors offer advantages in terms of scalability and adaptability to different feedstock types. They can accommodate a wide range of

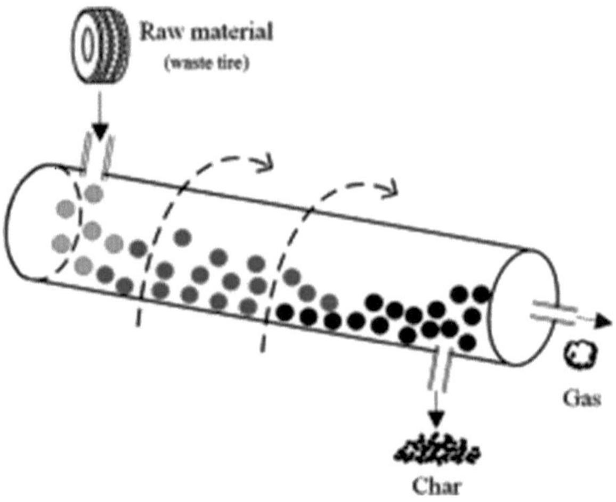

Fig. 5 Rotating bed reactor (Xu et al., 2020)

materials, including biomass, plastics, and tire-derived fuels, making them versatile for various pyrolysis applications.

Each reactor type has its advantages and considerations. Table 1 provides a comparison of reactor types, their operating conditions, and yield output. Fixed-bed reactors offer simplicity but may lack efficiency for large-scale operations. Moving bed reactors excel in heat transfer and yield consistency, suitable for continuous and large-scale pyrolysis. Fluidized bed reactors provide excellent heat transfer and versatility for diverse feedstocks, ideal for sustainable waste management. Rotating bed reactors offer flexibility and efficiency for continuous processing of homogeneous feedstocks. The choice depends on specific needs and objectives, ensuring optimal pyrolysis performance for waste management and resource utilization.

5 Yields of Pyrolysis Products

5.1 Liquid From Pyrolysis

The liquid fraction obtained from pyrolysis, often referred to as pyrolysis oil or bio-oil, constitutes a significant portion of the total product yield, typically ranging from 30% to 70%, depending on the feedstock composition and pyrolysis conditions. Pyrolysis oil is a complex mixture of organic compounds derived from the thermal decomposition of biomass, plastics, or other organic materials. It contains a diverse range of hydrocarbons, oxygenates, and other organic compounds, with varying chemical compositions and properties depending on the feedstock and pyrolysis parameters. The composition of pyrolysis oil can vary widely, with hydrocarbons such as alkanes, alkenes, and aromatics representing the predominant components. Additionally, oxygen-containing functional groups, including alcohols, ketones, and carboxylic acids, may also be present, depending on the feedstock's chemical structure and the pyrolysis conditions employed. Pyrolysis oil holds significant

Table 1 Types of reactors operating conditions and yield output

Type of reactor	Operating conditions	Maximum oil yield temperature (°C)	Oil (wt.%)	Char (wt.%)	Gas (wt.%)	References
Fixed bed	450–600 °C temperature, 5 °C/min heating rate, 3 kg	475 °C	58.2	37.3	4.5	Cunliffe and Williams (1998) and Williams and Brindle (2002)
Vacuum	450–600 °C temperature, batch, 100 g	550 °C	47.1	36.9	16	Roy et al. (1999)
Moving screw bed	600–800 °C temperature, 3.5–8 kg/h mass flow rate	600 °C	48.4	39.9	11.7	Aylón et al. (2010)
Fluidized bed	450–600 °C temperature, 220 g/h, tire granules	450 °C	55	42.5	2.5	Williams and Brindle (2002)
Rotary bed	550–680 °C temperature, 4.8 kg/h	550 °C	38.12	49	2.39	Galvagno et al. (2002)

potential as a valuable resource for various applications. It can be further refined through processes such as hydrotreating or fractional distillation to produce transportation fuels such as gasoline, diesel, or jet fuel (Williams & Brindle, 2002). Alternatively, it can serve as a feedstock for chemical synthesis, providing a renewable source of organic compounds for the production of specialty chemicals, polymers, or other high-value products.

However, the utilization of pyrolysis oil presents certain challenges, including its high oxygen content, low heating value, and tendency to undergo chemical degradation over time. Addressing these challenges requires the development of efficient upgrading and refining technologies to enhance the quality and suitability of pyrolysis oil for specific end uses.

5.2 Gases From Pyrolysis

The gases generated from pyrolysis constitute a significant portion of the total product yield, typically ranging from 10% to 50%, depending on the feedstock and pyrolysis conditions employed. These gases are primarily composed of hydrogen (H_2), methane (CH_4), carbon monoxide (CO), and light hydrocarbons such as ethylene (C_2H_4) and propylene (C_3H_6) (Janarthanan & Sivanandi, 2022).

Hydrogen is often one of the major components of the gas fraction obtained from pyrolysis. It is a valuable energy carrier and feedstock for various industrial processes, including hydrogenation reactions and ammonia synthesis. Methane, the primary component of natural gas, is another important constituent of pyrolysis gases. It can be utilized for heat and power generation through combustion or upgraded into higher-value products such as hydrogen or synthetic natural gas. Carbon monoxide is a versatile intermediate in chemical synthesis, serving as a precursor for the production of methanol, acetic acid, and other chemicals. Light hydrocarbons like ethylene and propylene are valuable feedstocks for the petrochemical industry, used in the production of plastics, synthetic fibers, and other materials.

The gases generated from pyrolysis can be utilized in various ways to maximize their value and contribute to a more sustainable energy system. They can be combusted directly for heat and power generation, providing an alternative to fossil fuels and reducing greenhouse gas emissions. Alternatively, they can be upgraded through processes such as steam reforming, water-gas shift, or methanation to produce higher-value products like hydrogen or synthetic natural gas. The utilization of gases from pyrolysis contributes to the circular economy by converting waste materials into valuable energy and chemical resources, thereby reducing dependence on finite fossil fuels and mitigating environmental impacts associated with waste disposal and energy production. Continued research and development efforts are essential to optimize the utilization of pyrolysis gases and unlock their full potential in the transition towards a more sustainable and resource-efficient future.

5.3 Char From Pyrolysis

Char, also known as biochar or carbon black, is a solid residue produced as a by-product of the pyrolysis process. It typically represents 10–50% of the total product yield, depending on factors such as the feedstock composition and pyrolysis conditions employed. Char consists primarily of carbonaceous material, with a highly porous structure and high surface area (Park et al., 2018; Wong et al., 2020; Banar et al., 2012). These unique physical and chemical characteristics make it suitable for various applications across different industries. In agriculture, biochar is commonly used as a soil amendment to improve soil health and fertility. Its porous structure provides habitat for beneficial microorganisms and helps retain moisture and nutrients in the soil, thereby enhancing crop yields and reducing the need for synthetic fertilizers. Additionally, biochar can contribute to carbon sequestration in soil, helping to mitigate climate change by storing carbon over the long term.

In addition to its agricultural applications, char finds use in environmental remediation and wastewater treatment. Its high surface area and adsorption capacity make it effective for removing contaminants such as heavy metals, organic pollutants, and nutrients from soil, water, and air. Furthermore, char can be utilized as a carbon source in metallurgical processes, such as iron and steel production. Its high carbon content and low impurity levels make it an attractive alternative to traditional carbonaceous materials, contributing to energy efficiency and reducing greenhouse gas emissions in metal smelting processes.

6 Pyrolytic Liquid Fuel Characteristics

Pyrolytic liquid fuel, commonly referred to as pyrolysis oil or bio-oil, derived from waste tires, plastics, and biomass, possesses properties reminiscent of conventional fossil fuels yet with distinct characteristics owing to its diverse chemical composition and production process.

1. *High energy density*: Pyrolytic liquid fuel typically exhibits a high energy density, making it a valuable energy source for various applications. This property stems from the presence of hydrocarbons and other organic compounds in the fuel, which can be efficiently converted into heat or electricity through combustion or other conversion processes.
2. *Diverse chemical composition*: The chemical composition of pyrolysis oil is diverse, reflecting the heterogeneity of the feedstock materials and the complex reactions occurring during pyrolysis. It may contain a mixture of hydrocarbons, oxygenated compounds (such as alcohols, ketones, and acids), nitrogen-containing compounds, and other organic constituents. This diverse composition can influence the fuel's properties, including its viscosity, stability, and combustion characteristics.
3. *Challenges with stability and compatibility*: Despite its similarities to conventional fossil fuels, pyrolytic liquid fuel may face challenges related to stability and compatibility due to the presence of oxygenated compounds and impurities. Oxygenates can contribute to increased acidity, viscosity, and susceptibility to degradation over time, potentially leading to issues such as phase separation, corrosion, and fouling in fuel storage and transportation systems. Additionally, impurities such as ash, metals, and sulfur compounds may affect combustion performance and emission characteristics.
4. *Importance of characterization*: Analyzing the characteristics of pyrolysis products is crucial for evaluating their potential applications, such as fuels, chemicals, and materials. Comprehensive characterization techniques, including spectroscopic analysis, chromatography, and calorimetry, can provide valuable insights into the fuel's composition, physical properties, and performance attributes. This information can inform decisions regarding fuel processing, refining, and utilization, as well as facilitate the development of tailored strategies for addressing stability and compatibility challenges.

In summary, pyrolytic liquid fuel derived from waste materials offers a promising avenue for renewable energy production and waste valorization. Table 2 lists the comparison of physical properties of tire, plastic, and biomass pyrolysis fuels. However, its unique properties and composition necessitate careful consideration and characterization to optimize its utilization and address potential challenges related to stability and compatibility. Through continued research and innovation, pyrolysis technologies can contribute to the transition towards a more sustainable and resource-efficient energy system.

Table 2 Comparison of physical properties of tire, plastic, and biomass pyrolysis fuels

S.no	Parameters	Diesel	Tire pyrolysis fuel (Williams & Brindle, 2002)	Plastic pyrolysis fuel (Krishnaraj et al., 2020)	Biomass pyrolysis fuel (Peng et al., 2022)
1.	Viscosity (sec)	2–4	6.3	4.3	2.3
2.	Calorific value (kcal/kg)	10191.20	10038.67	9201.72	9017.88
3.	Fire point (°C)	52	34	55	52
4.	Flash point (°C)	40	20	53	46
5.	Moisture (%)	0.05	0.46	0.21	0.2
6.	Sulfur (% wt)	0.05	0.6	0.0046	0.0012
7.	Specific gravity	0.85	0.85	0.8517	0.85
8.	Carbon (%)	0.035	2.1	0.044	0.016
9.	Density (kg/L)	0.8600	0.910	0.8654	0.86

7 Usage of Pyrolysis Products

The versatile products obtained through pyrolysis offer diverse applications across industries, providing opportunities to reduce dependency on nonrenewable resources and mitigate environmental impacts.

1. *Pyrolysis oil:* Pyrolysis oil, derived from various feedstocks including waste tires, plastics, and biomass, can serve as a sustainable substitute for fossil fuels in heating, power generation, and transportation. It can be combusted directly in boilers and furnaces to generate heat or electricity. Alternatively, it can be upgraded through processes like hydrotreating or fractional distillation to produce transportation fuels such as gasoline, diesel, or jet fuel. The utilization of pyrolysis oil helps reduce greenhouse gas emissions and fossil fuel consumption while promoting resource efficiency and waste valorization.
2. *Carbon black:* Carbon black, a solid residue produced from waste tires and plastics during pyrolysis, possesses valuable properties as a reinforcing agent in rubber products, plastics, and coatings. It enhances the mechanical strength, durability, and UV resistance of these materials, making them suitable for a wide range of applications. Additionally, carbon black can be utilized in the production of electrodes for batteries, conductive polymers, and pigment formulations. By incorporating carbon black derived from waste materials, industries can reduce their reliance on virgin carbon black and contribute to circular economic principles.

3. *Biochar:* Biochar, a carbon-rich residue obtained from the pyrolysis of biomass, holds significant potential for enhancing soil fertility and carbon sequestration. When applied to soil, biochar improves water retention, nutrient retention, and microbial activity, thereby enhancing crop yields and reducing the need for synthetic fertilizers. Moreover, biochar acts as a long-term carbon sink, sequestering carbon in soil and mitigating greenhouse gas emissions. By integrating biochar into agricultural practices, farmers can promote sustainable soil management and contribute to climate change mitigation efforts.

8 Pyrolysis Testing Methods

8.1 Proximate Analysis

Proximate analysis is a fundamental analytical technique used to determine the major components of a solid fuel or feedstock. It typically involves assessing the moisture content, volatile matter, fixed carbon, and ash content of the material. These parameters provide valuable insights into the combustibility, thermal behavior, and overall quality of the material, particularly in the context of pyrolysis and other thermal conversion processes.

1. *Moisture content*: Moisture content refers to the amount of water present in the material. High moisture content can significantly affect the efficiency of pyrolysis by requiring additional energy for water evaporation and reducing the available energy for thermal decomposition of organic compounds. Therefore, determining moisture content is crucial for optimizing pyrolysis conditions and maximizing process efficiency.
2. *Volatile matter*: Volatile matter represents the proportion of the material that vaporizes or combusts when heated to a certain temperature in the absence of oxygen. It includes organic compounds such as hydrocarbons, as well as moisture and other volatile substances. The volatile matter content is indicative of the material's combustibility and its potential to release gases during pyrolysis. Higher volatile matter content generally correlates with increased gas yields and faster pyrolysis rates.
3. *Fixed carbon*: Fixed carbon refers to the residual carbonaceous material remaining after volatile matter has been driven off from the feedstock. It represents the portion of the material that is nonvolatile and contributes to the formation of char during pyrolysis. Fixed carbon content is a key parameter influencing the yield and quality of char produced, as well as the overall energy content of the feedstock.
4. *Ash content*: Ash content refers to the inorganic residue left behind after combustion or pyrolysis of the material. It consists primarily of minerals and other inorganic compounds present in the feedstock. High ash content can lead to operational challenges during pyrolysis, such as increased slag formation, corro-

sion, and fouling of equipment. Therefore, assessing ash content is essential for evaluating the suitability of the feedstock for pyrolysis and predicting potential impacts on process performance.

8.2 Ultimate Analysis

Ultimate analysis is a comprehensive analytical technique used to quantify the elemental composition of a feedstock, providing insights into the distribution of carbon, hydrogen, nitrogen, sulfur, and oxygen within the material. This analysis is essential for understanding the chemical composition and properties of the feedstock, particularly in the context of pyrolysis and other thermal conversion processes.

1. *Carbon (C)*: Carbon is a fundamental element in organic materials and serves as the primary constituent of biomass, plastics, and other organic feedstocks. The carbon content of the feedstock influences the potential yields of pyrolysis products, including pyrolysis oil, gas, and char. Higher carbon content generally correlates with increased yields of carbonaceous products, such as char and hydrocarbons, during pyrolysis.
2. *Hydrogen (H)*: Hydrogen is another essential element in organic materials, contributing to their energy content and chemical properties. The hydrogen content of the feedstock affects the composition and properties of pyrolysis products, particularly their hydrogen-to-carbon ratio. Higher hydrogen content can lead to increased yields of gaseous products, such as hydrogen and hydrocarbons, with enhanced energy density and combustibility.
3. *Nitrogen (N)*: Nitrogen is commonly present in biomass and certain organic materials, albeit in smaller quantities compared to carbon and hydrogen. The nitrogen content of the feedstock influences the composition of pyrolysis products and can affect process performance and product quality. Nitrogen-containing compounds may contribute to the formation of nitrogenous gases, such as ammonia, during pyrolysis.
4. *Sulfur (S)*: Sulfur is a trace element present in some organic materials, particularly fossil-derived feedstocks such as coal and petroleum. The sulfur content of the feedstock is important for assessing its environmental impact and potential emissions during pyrolysis. Sulfur-containing compounds can lead to the formation of sulfur dioxide (SO2) and other sulfur oxides during combustion or pyrolysis, contributing to air pollution and acid rain.
5. *Oxygen (O)*: Oxygen is ubiquitous in organic materials, existing in various functional groups and chemical bonds. The oxygen content of the feedstock influences its energy content, reactivity, and stability during pyrolysis. Higher oxygen content may result in increased yields of oxygenated compounds, such as acids, alcohols, and ketones, in pyrolysis products like bio-oil.

8.3 Thermogravimetric Analysis (TGA)

Thermogravimetric Analysis (TGA) is a powerful analytical technique used to study the thermal behavior of materials as they undergo heating in a controlled atmosphere. During TGA, the weight of the sample is continuously monitored as a function of temperature or time, providing insights into its decomposition kinetics and thermal stability.

1. *Measurement of weight loss*: TGA involves subjecting the feedstock sample to a programmed temperature ramp in an inert or oxidizing atmosphere (such as nitrogen or air). As the sample is heated, it undergoes various physical and chemical transformations, leading to the release of volatile components and the decomposition of organic matter. The weight loss of the sample, resulting from the removal of volatile compounds and the conversion of organic matter to gases, is measured in real-time by the TGA instrument.
2. Insights into decomposition kinetics: By analyzing the weight loss data obtained from TGA, researchers can characterize the decomposition kinetics of the feedstock. This includes identifying key decomposition stages, determining onset temperatures for decomposition reactions, and quantifying reaction rates at different temperature ranges. TGA curves typically exhibit distinct peaks or inflection points corresponding to different decomposition processes, such as moisture evaporation, volatile release, and thermal degradation of organic compounds.
3. Assessment of thermal stability: TGA provides valuable information about the thermal stability of the feedstock, i.e., its resistance to decomposition or degradation at elevated temperatures. The onset temperature of weight loss, often referred to as the decomposition temperature or degradation temperature, indicates the temperature at which thermal decomposition reactions begin. Materials with higher decomposition temperatures are considered more thermally stable and may be better suited for high-temperature applications or processing methods.

8.4 Differential Scanning Calorimetry (DSC)

Differential Scanning Calorimetry (DSC) is a powerful analytical technique used to study the thermal behavior of materials by measuring the heat flow into or out of a sample as it undergoes controlled heating or cooling. DSC provides valuable insights into phase transitions, melting points, reaction kinetics, and thermal stability of the feedstock.

1. *Measurement of heat flow*: During a DSC experiment, the sample and a reference material are subjected to identical heating or cooling profiles, while the heat flow into or out of the sample is continuously monitored. Any heat absorbed or released by the sample during phase transitions, chemical reactions, or other thermal events results in a detectable change in the heat flow signal. The differ-

ence in heat flow between the sample and the reference material, known as the differential heat flow, provides information about the thermal behavior of the sample.

2. *Characterization of phase transitions*: DSC is particularly useful for characterizing phase transitions, such as melting, crystallization, glass transition, and solid-solid transitions. The appearance of peaks or inflection points in the DSC curve corresponds to these transitions, allowing researchers to determine the transition temperatures, enthalpies, and kinetics associated with each phase change. For example, the melting point of a crystalline material can be accurately determined from the onset or peak temperature of the endothermic melting peak observed in the DSC curve.
3. *Assessment of reaction kinetics*: DSC can also be used to study the kinetics of chemical reactions occurring in the feedstock during heating or cooling. By analyzing the shape, magnitude, and onset temperature of exothermic or endothermic peaks in the DSC curve, researchers can deduce information about reaction rates, activation energies, and reaction mechanisms. This information is valuable for understanding the thermal stability, decomposition pathways, and reactivity of the feedstock under specific temperature conditions.

8.5 Pyrolysis-Gas Chromatography/Mass Spectrometry (Py-GC/MS)

Pyrolysis-gas chromatography/mass spectrometry (Py-GC/MS) is a powerful analytical technique used to analyze the volatile pyrolysis products generated from a feedstock by heating it to high temperatures in a pyrolyzer. This technique combines two complementary analytical methods: gas chromatography (GC), which separates the individual components of a mixture based on their chemical properties, and mass spectrometry (MS), which identifies and quantifies these components based on their mass-to-charge ratios.

1. *Pyrolysis in a pyrolyzer*: The feedstock is introduced into a pyrolyzer, where it is rapidly heated to high temperatures (typically in the range of several hundred to over a thousand degrees Celsius) in an inert atmosphere. This thermal decomposition process, known as pyrolysis, breaks down the complex organic molecules in the feedstock into smaller, volatile compounds, including hydrocarbons, oxygenates, nitrogen-containing compounds, and other organic species.
2. *Separation by gas chromatography (GC)*: The volatile pyrolysis products generated in the pyrolyzer are then introduced into a gas chromatograph, where they are separated based on their chemical properties, such as molecular size, polarity, and volatility. The GC column separates the individual components of the mixture as they pass through it, with different compounds eluting from the column at different retention times.
3. *Identification and quantification by mass spectrometry (MS)*: As the separated compounds exit the GC column, they enter a mass spectrometer, where they are

ionized and fragmented into ions of different masses. The mass spectrometer analyzes these ions based on their mass-to-charge ratios (m/z) and generates mass spectra, which provide information about the molecular structure and composition of the compounds. By comparing the mass spectra of the detected ions to reference databases, the individual pyrolysis products can be identified and quantified.

4. *Detailed chemical composition analysis*: Py-GC/MS provides detailed information about the chemical composition of the volatile pyrolysis products, including the types of organic compounds present, their relative abundances, and their structural characteristics. This information helps researchers understand the decomposition pathways and mechanisms involved in the pyrolysis process, as well as identify potential product streams for further analysis or utilization.

8.6 Elemental Analysis

Elemental analysis is a crucial analytical technique used to determine the elemental composition of pyrolysis products, including pyrolysis oil, gas, and char. This analysis provides valuable information about the presence and distribution of key elements such as carbon, hydrogen, nitrogen, sulfur, and oxygen, which are essential for assessing the purity and quality of the pyrolysis products.

1. *Measurement of elemental composition*: Elemental analysis involves quantifying the relative amounts of different elements present in a sample of pyrolysis products. This is typically achieved using various analytical methods, such as combustion analysis, elemental combustion analysis, or elemental spectroscopy techniques. These methods involve subjecting the sample to controlled conditions, such as high temperature or combustion, to convert the elemental constituents into measurable forms (e.g., gases or compounds) that can be quantified using appropriate detectors or sensors.
2. *Assessment of purity and quality*: The elemental composition of pyrolysis products provides insights into their purity, composition, and quality.

   - *Carbon (c)*: Carbon content is a key indicator of the organic content and energy density of pyrolysis products. High carbon content is desirable for fuels and carbonaceous materials like char, indicating high energy content and purity.
   - *Hydrogen (h)*: Hydrogen content influences the energy content and reactivity of pyrolysis products. Higher hydrogen content may indicate the presence of hydrocarbons or other volatile compounds, which can contribute to the energy density and combustibility of the products.
   - *Nitrogen (n) and sulfur (s)*: Nitrogen and sulfur content are indicators of potential impurities or contaminants in pyrolysis products. Elevated levels of nitrogen or sulfur may indicate the presence of nitrogen-containing or sulfur-containing compounds, which can affect product quality and environmental compatibility.

- *Oxygen (o)*: Oxygen content provides insights into the presence of oxygenated compounds in pyrolysis products, such as alcohols, ketones, and acids. Oxygen content can influence the chemical properties, stability, and storage requirements of the products.

8.7 Rheological Testing

Rheological testing is a valuable analytical technique used to measure the flow properties and viscosity of fluids, including pyrolysis oil, at different temperatures and shear rates. This analysis provides crucial insights into the behavior of pyrolysis oil under various processing and handling conditions, guiding the selection of appropriate equipment and operational parameters for its utilization.

1. *Measurement of flow properties*: Rheological testing involves subjecting the pyrolysis oil sample to controlled flow conditions, such as varying temperatures and shear rates, while measuring its response to these changes. The key rheological properties assessed include viscosity, shear stress, shear rate, and flow behavior. These parameters describe how the fluid responds to applied forces and how its flow characteristics change under different conditions.
2. *Assessment of viscosity*: Viscosity is a critical rheological property that quantifies the resistance of fluid to flow. In the context of pyrolysis oil, viscosity measurement provides information about its thickness, fluidity, and resistance to flow at different temperatures and shear rates. Understanding the viscosity of pyrolysis oil is essential for predicting its behavior during transportation, storage, pumping, and processing operations.
3. *Selection of processing equipment*: Rheological testing helps guide the selection of appropriate processing and handling equipment for pyrolysis oil utilization. Equipment such as pumps, pipes, valves, and heat exchangers must be capable of handling fluids with specific viscosity profiles and flow behaviors. By characterizing the rheological properties of pyrolysis oil, engineers and operators can ensure the proper design and operation of equipment to minimize energy consumption, optimize process efficiency, and prevent operational issues such as clogging, fouling, or equipment damage.
4. *Optimization of process conditions*: Rheological testing also aids in optimizing process conditions for pyrolysis oil utilization. By understanding how viscosity changes with temperature and shear rate, operators can adjust operating parameters such as temperature, pressure, and flow rate to achieve desired flow behaviors and product characteristics. This optimization process can enhance the efficiency of pyrolysis oil processing, improve product quality, and reduce production costs.

8.8 Combustion Analysis

Combustion analysis is a fundamental analytical technique used to evaluate the calorific value, ash content, and combustion characteristics of pyrolysis products, including pyrolysis oil, gas, and char. This analysis provides essential information about the energy content, combustibility, and potential for energy recovery and utilization of these products as fuels.

1. *Calorific value*: Combustion analysis measures the calorific value, also known as the heating value or energy content, of pyrolysis products. This value quantifies the amount of energy released during the complete combustion of a unit mass or volume of the fuel. For pyrolysis products, the calorific value indicates their potential as energy sources for heating, power generation, or industrial processes. Higher calorific values indicate greater energy density and efficiency in converting the fuel into useful energy.
2. *Ash content*: Combustion analysis also assesses the ash content of pyrolysis products, particularly char. The ash content represents the inorganic residue remaining after combustion of the organic components. High ash content can affect combustion efficiency, equipment performance, and emissions during combustion processes. By quantifying the ash content, combustion analysis helps evaluate the purity and suitability of pyrolysis products for combustion applications and identifies potential challenges related to ash disposal or handling.
3. *Combustion characteristics*: Combustion analysis provides insights into the combustion characteristics of pyrolysis products, including their ignition temperature, burning rate, flame stability, and emissions profile. These characteristics influence the combustion behavior, efficiency, and environmental impact of using pyrolysis products as fuels. Understanding the combustion characteristics helps optimize combustion processes, design combustion equipment, and mitigate emissions of pollutants such as particulate matter, nitrogen oxides, and sulfur compounds.
4. *Energy recovery and utilization*: By assessing the calorific value and combustion characteristics of pyrolysis products, combustion analysis helps evaluate their potential for energy recovery and utilization in various applications. Pyrolysis oil, gas, and char can be utilized as renewable fuels for heating, power generation, industrial processes, and transportation. The information obtained from combustion analysis informs decisions regarding fuel selection, combustion system design, and operational parameters to maximize energy efficiency, minimize environmental impact, and ensure regulatory compliance.

8.9 Microscopic Analysis

Microscopic analysis, particularly techniques like scanning electron microscopy (SEM) and transmission electron microscopy (TEM), is a powerful tool for examining the morphology, structure, and surface properties of pyrolysis char. This analysis provides valuable insights into the physical and chemical characteristics of char, which are essential for understanding its properties and evaluating its suitability for various applications, including soil amendment and carbon sequestration.

1. *Scanning electron microscopy* (SEM): SEM is a microscopy technique that uses a focused beam of electrons to scan the surface of a sample, generating high-resolution images of its morphology and topography. SEM can reveal the microstructure, pore size distribution, particle size distribution, and surface features of pyrolysis char with excellent spatial resolution. By analyzing SEM images, researchers can characterize the physical structure and surface properties of char, including porosity, surface area, particle shape, and aggregation behavior. This information is crucial for understanding the adsorption capacity, reactivity, and mechanical properties of char in various applications.
2. *Transmission electron microscopy* (TEM): TEM is a microscopy technique that transmits a beam of electrons through a thin specimen, allowing for detailed imaging of its internal structure at the nanoscale. TEM provides insights into the crystalline structure, chemical composition, and nanoscale features of pyrolysis char, including the presence of carbonaceous nanostructures, such as graphene sheets, carbon nanotubes, and fullerenes. TEM analysis helps elucidate the formation mechanisms, graphitization degree, and heterogeneity of char, providing valuable information for understanding its chemical reactivity, electronic properties, and potential applications in nanotechnology and materials science.
3. *Insights for applications*: Microscopic analysis of pyrolysis char provides valuable insights into its suitability for various applications, including soil amendment, carbon sequestration, water treatment, energy storage, and catalysis. The morphology, structure, and surface properties of char influence its performance and effectiveness in these applications. For example, char with high porosity and surface area is desirable for soil amendment, as it enhances nutrient retention, water holding capacity, and microbial activity in soil. Similarly, char with well-defined nanostructures and active surface sites may exhibit superior adsorption capacity and catalytic activity for environmental remediation and energy conversion processes.

Overall, proximate analysis enables informed decision-making in pyrolysis system design, optimization, and operation by providing valuable insights into feedstock composition and properties. Ultimate analysis predicts pyrolysis product yields and aids in tailoring process parameters. TGA studies thermal behavior, while DSC examines thermal properties, guiding process optimization. Py-GC/MS characterizes volatile products, aiding in process optimization and product development. Elemental analysis ensures product quality and suitability for various applications. Rheological testing informs processing equipment selection. Combustion analysis assesses energy value and combustion characteristics. Microscopic analysis elucidates char properties, optimizing its utilization across industries. Together, these techniques advance sustainable pyrolysis technologies for waste valorization, energy production, and resource conservation.

9 Merits and Demerits of Pyrolysis on Environmental Concern

9.1 Merits

- Reduces greenhouse gas emissions compared to traditional waste disposal methods
- Recovers energy and valuable materials from waste streams, contributing to resource conservation and circular economy principles
- Provides renewable alternatives to fossil fuels and chemical feedstocks, mitigating environmental pollution and promoting sustainable development

9.2 Demerits

- Emissions of volatile organic compounds (VOCs), particulate matter, and other pollutants during pyrolysis require careful monitoring and control to minimize environmental impacts.
- Energy-intensive pyrolysis processes may result in carbon emissions and contribute to climate change if powered by fossil fuels.

9.3 Government Policies

- Many governments are implementing policies and regulations to promote waste management practices that prioritize resource recovery and recycling.
- Incentives such as tax credits, grants, and subsidies may be provided to support the development and adoption of pyrolysis technologies.
- Environmental standards and emission limits are enforced to ensure pyrolysis facilities operate in compliance with air and water quality regulations.

9.4 Opportunities for Scale-Up Models

- Collaboration between public and private sectors is essential to scale up pyrolysis technologies and infrastructure.
- Investment in research and development can lead to innovations in reactor design, process optimization, and product utilization.
- Public-private partnerships and funding initiatives can support the deployment of large-scale pyrolysis plants and integrated waste management systems.

10 Environmental Concerns Related to Pyrolysis

10.1 Air Pollution

Emissions of volatile organic compounds (VOCs), particulate matter (PM), and other pollutants during pyrolysis can contribute to air pollution. VOCs and PM can have adverse health effects on nearby communities and exacerbate respiratory problems.

10.2 Greenhouse Gas Emissions

While pyrolysis can reduce greenhouse gas emissions compared to traditional waste disposal methods, the process itself may emit carbon dioxide (CO2) and other greenhouse gases. The combustion of pyrolysis gases for energy recovery can release CO2 if derived from fossil-based feedstocks or operated with fossil fuel-derived energy.

10.3 Energy Intensity

Pyrolysis processes can be energy-intensive, especially if high temperatures and long residence times are required. Energy demands for heating reactors, pre-processing feedstocks, and gas purification can lead to increased carbon emissions if fossil fuels are used as the primary energy source.

10.4 Waste Management Impacts

Pyrolysis can generate solid residues such as ash and non-condensable gases that require proper management and disposal. Improper handling of pyrolysis by-products can lead to soil and water contamination, particularly if leachates containing harmful substances are not adequately treated.

10.5 Resource Depletion

While pyrolysis offers a means to recover energy and valuable materials from waste streams, it may also compete with other sustainable resource management strategies such as recycling and composting. Depending on feedstock availability and processing methods, pyrolysis could potentially divert organic materials away from soil amendment and composting, impacting soil health and agricultural productivity.

10.6 Technological Risks

Improperly designed or operated pyrolysis facilities may pose risks of fires, explosions, and chemical releases, endangering workers and surrounding communities. Safeguarding against equipment malfunctions, process upsets, and hazardous material handling is crucial to mitigate potential accidents and environmental incidents.

10.7 Land Use and Habitat Impact

Large-scale deployment of pyrolysis facilities may require significant land areas for sitting, potentially encroaching on natural habitats or agricultural lands. Clearing land for pyrolysis plant construction or feedstock cultivation could lead to habitat loss, deforestation, and biodiversity decline if not managed sustainably.

10.8 Mitigation Strategies

Implementing stringent emission controls, such as scrubbers and filters, to minimize air pollutants released during pyrolysis operations. Investing in renewable energy sources, such as solar or wind power, to reduce the carbon footprint of pyrolysis processes. Promoting integrated waste management approaches that prioritize waste reduction, recycling, and composting alongside pyrolysis for residual waste streams.

Enhancing regulatory oversight and enforcement to ensure pyrolysis facilities comply with environmental standards and best practices. Conducting comprehensive environmental impact assessments and stakeholder engagement processes prior to siting and permitting new pyrolysis projects. Encouraging research and innovation to develop cleaner, more efficient pyrolysis technologies and sustainable feedstock sourcing strategies.

11 Future Scope of Pyrolysis

1. *Technological advancements*: Continued research and development efforts to optimize pyrolysis processes, enhance reactor designs, and improve product yields and quality. Integration of advanced materials, catalysts, and process control technologies to enhance pyrolysis efficiency, scalability, and environmental performance.
2. *Feedstock diversification:* Exploration of novel feedstock sources, including agricultural residues, industrial by-products, and organic waste streams, to broaden the range of materials suitable for pyrolysis. Development of multi-feedstock processing systems capable of handling diverse feedstock blends and mixtures, maximizing resource utilization and waste diversion.
3. *Circular economy integration*: This involves the incorporation of pyrolysis into circular economy models emphasizing resource recovery, recycling, and sustain-

able waste management practices. Collaboration among pyrolysis operators, waste generators, and product manufacturers is crucial to establish closed-loop systems for material reuse and resource conservation. Table 3 showcases the top ten productive countries in pyrolysis research, which indicates a growing number of ongoing research initiatives focusing on the circular economy integration of the pyrolysis process, with an increasing number of innovations emerging each year.

4. *Renewable energy applications*: Expansion of pyrolysis-derived fuels and energy products in renewable energy markets, including biofuels, bioenergy, and biogas. Integration of pyrolysis oil, syngas, and biochar into decentralized energy systems, off-grid applications, and rural electrification initiatives.
5. *Sustainable agriculture solutions*: Utilization of pyrolysis-derived biochar as a soil amendment to improve soil fertility, water retention, and carbon sequestration in agricultural and forestry systems. Adoption of biochar-based soil management practices to enhance crop yields, reduce fertilizer requirements, and mitigate climate change impacts.
6. *Waste-to-value chains*: Development of integrated waste-to-value chains that leverage pyrolysis as a key technology for converting diverse waste streams into high-value products. Establishment of partnerships and collaborations across sectors, including waste management, manufacturing, agriculture, and energy, to create synergies and maximize economic and environmental benefits.
7. *Policy support and market incentives*: Implementation of supportive policy frameworks, regulations, and incentives to promote investment in pyrolysis technologies, infrastructure, and innovation. Alignment of government initiatives with sustainability goals, climate targets, and circular economy principles to accelerate the adoption and scale-up of pyrolysis solutions.
8. *Public awareness and education*: Increasing public awareness and understanding of pyrolysis technologies, their potential benefits, and their role in addressing environmental challenges. Education and outreach efforts to engage stakeholders, foster community participation, and build consensus around sustainable waste management practices and resource recovery solutions.

Table 3 Top ten productive countries in pyrolysis research (Armenise et al., 2021)

	Year						
Country	2001–2005	2006–2010	2011–2015	2016–2020	TP	TC	*h*-index
Spain	29	47	38	21	135	4605	41
China	1	9	15	74	99	1600	22
India	4	7	13	41	65	1270	17
S. Korea	10	6	8	31	55	731	16
England	11	11	10	17	49	3014	26
USA	5	1	6	28	40	1215	21
Japan	7	4	0	17	28	686	15
Pakistan	1	4	7	16	28	290	11
Popland	2	7	4	11	24	253	8
Malaysia	1	3	4	13	21	455	10

12 Conclusion

The exploration of pyrolysis as a transformative waste management solution elucidates its multifaceted potential in mitigating environmental impacts while fostering resource conservation and renewable energy generation. Pyrolysis exhibits commendable merits, including reduced greenhouse gas emissions and the recovery of energy and valuable materials from waste streams. However, it is imperative to address associated demerits such as emissions of volatile organic compounds and energy intensity to ensure sustainable implementation. Governmental policies play a pivotal role in incentivizing innovation and enforcing environmental standards, thereby accelerating the adoption of pyrolysis technologies. Collaborative efforts between public and private sectors are essential for scaling up pyrolysis infrastructure and advancing research and development initiatives. A comparative analysis of pyrolysis reactor types underscores the importance of tailoring technology selection to operational requirements and feedstock characteristics. Each reactor configuration offers distinct advantages, with fluidized bed reactors and moving bed reactors demonstrating competitive oil yields and efficiency. For instance, fluidized bed reactors can achieve remarkable oil yields of up to 55 wt.% at 450 °C, showcasing their efficiency in converting waste materials into valuable resources. Similarly, moving bed reactors exhibit competitive oil yields of approximately 48.4 wt.% at 600 °C, reaffirming their efficacy in continuous pyrolysis operations.

The characterization of pyrolysis products through analytical techniques provides valuable insights into their composition, properties, and suitability for various applications. Proximate analysis, elemental analysis, and combustion analysis offer critical data for process optimization and fuel selection. For instance, proximate analysis elucidates key parameters such as moisture content, volatile matter, fixed carbon, and ash content, influencing pyrolysis process efficiency and product quality. Meanwhile, ultimate analysis offers a comprehensive understanding of the elemental composition of feedstocks, guiding process optimization and facilitating informed decision-making in waste management and resource utilization. Furthermore, pyrolysis-gas chromatography/mass spectrometry elucidates the chemical diversity of volatile products, while rheological testing informs handling and processing of pyrolysis oil. Microscopic analysis delves into the nanostructural intricacies of char, informing its suitability for diverse applications. The integration of scientific insights and technical data underscores the interdisciplinary nature of pyrolysis research, driving innovation across chemistry, materials science, engineering, and environmental science domains. By leveraging the full potential of pyrolysis technologies and embracing sustainable waste management practices, stakeholders can navigate towards a circular economy and a greener, more resilient future.

In summary, the synthesis of scientific knowledge and analytical rigor empowers stakeholders to unlock the full potential of pyrolysis as a versatile and sustainable waste management solution. Through continued collaboration, innovation, and policy support, pyrolysis stands poised to revolutionize waste valorization, energy recovery, and resource conservation endeavors on a global scale.

References

Aguado, R., Olazar, M., Gaisan, B., Prieto, R., & Bilbao, J. (2002). Kinetic study of polyolefin pyrolysis in a conical spouted bed reactor. *Industrial & Engineering Chemistry Research, 41*, 4559–4566.

Armenise, S., Wong, S. L., Ramírez-Velásquez, J. M., Launay, F., Wuebben, D., Ngadi, N., Rams, J., & Muñoz, M. (2021). Plastic waste recycling via pyrolysis: A bibliometric survey and literature review. *Journal of Analytical and Applied Pyrolysis, 158*, 105265.

Aydın, H., & Ilkılıç, C. (2012). Optimization of fuel production from waste vehicle tires by pyrolysis and resembling to diesel fuel by various desulfurization methods. *Fuel, 102*, 605–612.

Aylón, E., Fernández-Colino, A., Murillo, R., Navarro, M. V., & García, T. (2010). Valorisation of waste tyre by pyrolysis in a moving bed reactor. *Waste Management, 30*(7), 1220–1224.

Banar, M., Akyıldız, V., Özkan, A., Çokaygil, Z., & Onay, Ö. (2012). Characterization of pyrolytic oil obtained from pyrolysis of TDF (Tire Derived Fuel). *Energy Conversion and Management, 62*, 22–30.

Campuzano, F., & Brown, R. C. (2019). Auger reactors for pyrolysis of biomass and wastes. *Renewable and Sustainable Energy Reviews, 102*, 372–409.

Cornelissen, T., Yperman, Y., Reggers, G., Schreurs, S., & Carleer, R. (2008). Flash co-pyrolysis of biomass with polylactic acid. Part 1: Influence on bio-oil yield and heating value. *Fuel, 87*, 1031–1041.

Cunliffe, A. M., & Williams, P. T. (1998). Composition of oils derived from the batch pyrolysis of tyres. *Journal of Analytical and Applied Pyrolysis, 44*(2), 131–152.

Demirbas, A. (2000). Recent advances in biomass conversion technologies. *Energy Education Science and Technology Part A: Energy Science and Research, 6*, 77–83.

Dogan, O., Celik, M. B., & Odalyan, B. (2012). The effect of tire derived fuel/diesel fuel blends utilization on diesel engine performance and emissions. *Fuel, 95*, 340–346.

Fairburn, J. A., Behie, L. A., & Svrcek, W. Y. (1990). Ultra pyrolysis of n-hexadecane in a novel micro-reactor. *Fuel, 69*, 1537–1545.

Galvagno, S., Casu, S., Casabianca, T., Calabrese, A., & Cornacchia, G. (2002). Pyrolysis process for the treatment of scrap tyres: Preliminary experimental results. *Waste Management, 22*(8), 917–923.

Grace, J. R., & Lim, C. J. (2013). Properties of circulating fluidized beds (CFB) relevant to combustion and gasification systems. In *Fluidized bed technologies for near-zero emission combustion and gasification* (pp. 147–176). Woodhead Publishing.

Iftikhar, H., Zeeshan, M., Iqbal, S., Muneer, B., & Razzaq, M. (2019). Co-pyrolysis of sugarcane bagasse and polystyrene with ex-situ catalytic bed of metal oxides/HZSM-5 with focus on liquid yield. *Bioresource Technology, 289*, 121647.

Janarthanan, K., & Sivanandi, P. (2022). Extraction and characterization of waste plastic pyrolysis oil for diesel engines. *Journal of Cleaner Production, 366*, 132924.

Jin, Q., Wang, X., Li, S., Mikulčič, H., Bešenič, T., Deng, S., et al. (2019). Synergistic effects during co-pyrolysis of biomass and plastic: Gas, tar, soot, char products and thermogravimetric study. *Journal of the Energy Institute, 92*, 108–117.

Kar, Y. (2011). Catalytic pyrolysis of car tire waste using expanded perlite. *Waste Management, 31*, 1772–1782.

Leung, D. Y. C., Yin, X. L., Zhao, Z. L., Xu, B. Y., & Chen, Y. (2002). Pyrolysis of tire powder: Influence of operation variables on the composition and yields of gaseous product. *Fuel Processing Technology, 79*, 141–155.

Lucchesi, G., & Maschio, G. (1983). Semi-active carbon and aromatics produced by pyrolysis of scrap tires. *Conservation & Recycling, 6*, 85–90.

Maqsood, T., Dai, J., Zhang, Y., Guang, M., & Li, B. (2021). Pyrolysis of plastic species: A review of resources and products. *Journal of Analytical and Applied Pyrolysis, 159*, 105295.

Murthy, K., Shetty, R. J., & Shiva, K. (2020). Plastic waste conversion to fuel: A review on pyrolysis process and influence of operating parameters. *Energy Sources, Part A: Recovery, Utilization, and Environmental Effects, 45*, 1–21.

Murugan, S., Ramaswamy, M. C., & Nagarajan, G. (2009). Assessment of pyrolysis oil as an energy source for diesel engines. *Fuel Processing Technology, 90*, 67–74.

Park, Y.-K., Lee, H. W., Lee, J. Y., & Kim, Y.-M. (2018). The use of high density polyethylene (HDPE) as a co-feeding feedstock on the catalytic pyrolysis of yellow poplar over Al-MCM-48 and Al-MSU-F. *Journal of Analytical and Applied Pyrolysis, 135*, 390–396.

Park, K.-B., Jeong, Y.-S., & Kim, J.-S. (2019). Activator-assisted pyrolysis of polypropylene. *Applied Energy, 253*, 113558.

Peng, B., Liu, Q., Li, X., Zhou, Z., Wu, C., & Zhang, H. (2022). Co-pyrolysis of industrial sludge and rice straw: Synergistic effects of biomass on reaction characteristics, biochar properties and heavy metals solidification. *Fuel Processing Technology, 230*, 107211.

Roy, C., Chaala, A., & Darmstadt, H. (1999). The vacuum pyrolysis of used tires: End-uses for oil and carbon black products. *Journal of Analytical and Applied Pyrolysis, 51*(1–2), 201–221.

Wang, Z., Burra, K. G., Lei, T., & Gupta, A. K. (2021). Co-pyrolysis of waste plastic and solid biomass for synergistic production of biofuels and chemicals-A review. *Progress in Energy and Combustion Science, 84*, 100899.

Williams, P. T., & Brindle, A. J. (2002). Catalytic pyrolysis of tyres: Influence of catalyst temperature. *Fuel, 81*(18), 2425–2434.

Williams, P. T., Besler, S., & Taylor, D. T. (1990). The pyrolysis of scrap automotive tyres. The influence of temperature and heating rate on product composition. *Fuel, 69*, 1474–1482.

Williams, P. T., Bottrill, R. P., & Cunliffe, A. M. (1998). Combustion of tyre pyrolysis oil. *Process Safety and Environmental Protection, 76*(4), 291–301.

Wong, S. L., Nyakuma, B. B., Nordin, A. H., Lee, C. T., Ngadi, N., Wong, K. Y., & Oladokun, O. (2020). Uncovering the dynamics in global carbon dioxide utilization research: A bibliometric analysis (1995–2019). *Environmental Science and Pollution Research, 28*, 13842–13860.

Xu, J., Yu, J., Xu, J., Sun, C., He, W., Huang, J., & Li, G. (2020). High-value utilization of waste tires: A review with focus on modified carbon black from pyrolysis. *Science of the Total Environment, 742*, 140235.

Xue, Y., Kelkar, A., & Bai, X. (2016). Catalytic co-pyrolysis of biomass and polyethylene in a tandem micropyrolyzer. *Fuel, 166*, 227–236.

Zhang, H., Xiao, R., Nie, J., Jin, B., Shao, S., & Xiao, G. (2015). Catalytic pyrolysis of black liquor lignin by co-feeding with different plastics in a fluidized bed reactor. *Bioresource Technology, 192*, 68–74.

Zhao, Y., Yang, X., Fu, Z., Li, R., & Wu, Y. (2019). Synergistic effect of catalytic co-pyrolysis of cellulose and polyethylene over HZSM-5. *Journal of Thermal Analysis and Calorimetry., 140*, 363.

Nano Ecosystems for Biodiversity Conservation and Habitat Restoration

M. Thenmozhi and S. T. Kumaravel

Abstract

An ecosystem is a community or group of living things that coexist in particular conditions and communicate with one another. Nano ecosystems, or nanotechnology-based solutions for biodiversity conservation and habitat restoration, hold promises for addressing various environmental challenges. Here are some potential applications and considerations: nanoparticle remediation, precision agriculture, nano-enhanced habitat structures, water purification, genetic monitoring and engineering, smart delivery systems for conservation agents, remote sensing, and monitoring. This chapter explores the emerging field of nano ecosystems and their potential applications in biodiversity conservation and habitat restoration. With ecosystems facing unprecedented threats, innovative solutions are imperative. Nano ecosystems offer a novel approach, leveraging nanotechnology to address ecological challenges at the molecular level. Through an analysis of recent research and case studies, this chapter elucidates the diverse ways in which nano-scale technologies can contribute to conservation efforts. From the remediation of contaminated habitats to the targeted delivery of nutrients, nano ecosystems demonstrate versatility and efficacy. However, their adoption raises questions regarding safety, ethics, and long-term ecological impacts, necessitating careful consideration. By synthesizing current knowledge and identifying future directions, this chapter aims to catalyze further research and facilitate informed decision-making in the pursuit of sustainable biodiversity conservation and habitat restoration.

M. Thenmozhi (✉)
Department of Biotechnology, Selvam College of Technology, Namakkal, Tamil Nadu, India

S. T. Kumaravel
Department of Mechanical Engineering, Paavai Engineering College, Namakkal, Tamil Nadu, India

D. K. Gupta et al. (eds.), *Nano-Ecosystems*,
https://doi.org/10.1007/978-981-95-5033-3_9

Keywords

Nano ecosystems · Facilitate · Habitat restoration · Adoption · Safety · Ecological impacts

1 Introduction to Nano Ecosystems

Nano ecosystems are an initiative to restore the polluted habitat, and it is an idea of restoring habitat using nanoscale materials. Nano means it should be in nanoscale between 1 and 100 nanometers. This review aims to restore the habitat using biotechnological approaches and ideas using biomaterials in restoring nature. The greater reactivity resulting from the larger surface area could be one plausible reason for the observed change. It is not unusual for attributes within nanomaterials to vary, in addition to the differences in characteristics between bulk and nanoscale materials. One example is the demonstration of distinct absorption wavelengths for nanoparticles (NPs) made of the same material but with varying sizes. The parameter influencing the absorption performance was also discovered to be the size of the NPs. The shape of the NPs also controls their other characteristics, such as reactivity. This chapter deals with the incorporation of nanotechnology in environmental sustainability using biomaterials to restore the habitation to conserve nature for the next generation (Bakshi, 2020) (Fig. 1).

1.1 Role in Environment Monitoring

Nanosensors are being used to monitor the pollution levels in the air, water, and soil. Nanoparticles detect pollutants in earlier stages, thereby aiding in the restoration of wildlife. This sensor alerts us to identify changes in earlier stages, which helps us restore the system earlier. Nanomembranes help in detecting water pollutants in earlier stages (Chee et al., 2022).

1.2 Nanotechnology for Controlling Pollutants

Nanomaterials and nanocomposites remove toxins from the environment by adsorbing toxins and breaking down the hazards into harmless components.

1.3 Nanotechnology in Resource Management

Nanofertilizers and nanocoating materials help in nutrient delivery to the plants and their growth with less water consumption. It helps with well-planned irrigation systems.

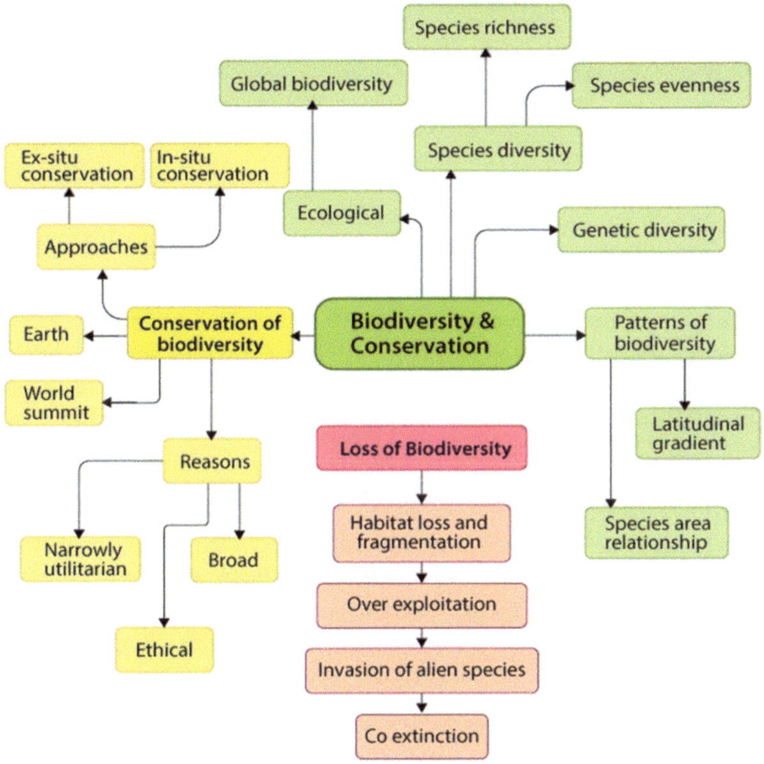

Fig. 1 Illustration of biodiversity habitat restoration possibilities

1.4 Nanotechnology in Habitat Restoration

Nanotechnology helps with soil fertility restoration, coral reef rehabilitation, and supporting plants in challenging environmental conditions.

1.5 In Restoring Climatic Changes

Nanotechnology helps by absorbing carbon and other pollutants, which are the major cause of our increased atmospheric temperature as well as pressure. There are some constraints while implementing tested or invented technology in real life due to the following considerations: environmental safety, public health, and socioeconomic equity. Based on the regulatory framework, stakeholder feedback plays a major role in sustainable outcomes. In summary, environmental resources diminished and were polluted due to the increased number of industries, increased population, and advancements in technology. It might be a good sign of advancement in technology, but upon considering future generations, it is supposed to find an innovative solution to conserve biodiversity. Nanotechnology is a promising

Table 1 Habitat restoration and creation

S. no	Interventions	Steps for habitat restoration
1.	Likely to be beneficial	1. Creation of artificial caves for bats and other species 2. Creating artificial water resources 3. Creating wetlands
2.	Unknown effectiveness	1. Creating grasslands 2. Creating artificial holes in trees for birds' habitat formation

solution to conserve biodiversity due to its size (Bajpai et al., 2023). Environmentalists can improve species preservation, habitat restoration, and pollution management by using nanoparticles. These uses cover the treatment of soil, air, and water and offer a long-term solution for environmental control. This means that the combination of biodiversity and nanotechnology tackles major issues in wildlife preservation and improves the effectiveness of conservation (Fernanda et al., 2018) (Table 1).

Integrating nanotechnology along with traditional conservation methods helps in enhancing the effectiveness and sustainability in soil and water remediation by increasing nutrient delivery and pollutant degradation. Nanotechnology offers to preserve the endangered species by improving habitat conditions or supporting genetic diversity. Nanotechnology offers to promote community involvement through innovative educational tools and demonstrations. Various workshops and training programs are offered to educate everyone on the importance of biodiversity and its habitat restoration.

2 Understanding Biodiversity Conservation Challenges

The following key challenges are threatening the biodiversity of life on earth.

2.1 Habitat Loss

Deforestation leads to many species becoming extinct, and wildlife animals, birds, and other living things becoming extinct due to their habitat loss. The increased fragmentation of their habitats makes it smaller for them to survive, which disrupts their ecosystem connectivity and also diminishes their resources, which disrupts the biodiversity of life because of deforestation (Martin et al. 2022).

2.2 Climate Changes

Climate change is due to increased earth's surface temperature, which alters habitats and also affects species growth. Extreme weather conditions threaten the habitats.

2.3 Contaminants

Chemical pollutants affect the air quality, soil, and water, and they disrupt the ecosystem and habitats. Due to the increased usage of plastics, they accumulate in the soil as well as in marine sources, which affects the quality of soil fertility as well as the lives of marine species.

2.4 Invasive Species

Invasive species disrupt the food and alter the ecosystems. Because of invasive species, new pathogens and diseases will affect the ecosystem, leading to a declining population.

2.5 Illegal Wildlife Trade

Due to the monetary benefit and for other uses, many people are involved in animal organ trading and hunting them for medicine as well as for food. Hunting animals, which has less reproduction rate, leads to the extinction of the same animal.

2.6 Lack of Awareness

Lack of awareness among people leads to the loss of support to restore the polluted and contaminated habitat.

2.7 Lack of Financial Support

Limited and inadequate funding slows down the monitoring and restoration process. Government policy and support due to the stringent government policies lead to environmental loss. Due to the fragmented governance delayed the conservation planning and implementation. Smart materials that are capable of detecting and eliminating hazardous chemical pollutants in the environment are being developed, and nanotechnology is essential to this process (Bansal et al., 2023). Cleaning, monitoring, and pollution control are some of the ways that nanotechnology is being used in conservation. Because nanoparticles may be designed to target particular pollutants, remediation efforts can be more concentrated and accurate while having less adverse effect on species that are not the intended target. By minimizing unintentional harm to ecosystems, this precision increases the efficacy of conservation efforts. Monitoring biodiversity is being revolutionized by cutting-edge instruments like unmanned aerial systems (UAS), automated sensory devices, remote sensing applications, and environmental DNA (eDNA) research (Schulz et al., 2023). Conservationists may now collect extensive data in real time, thanks to these

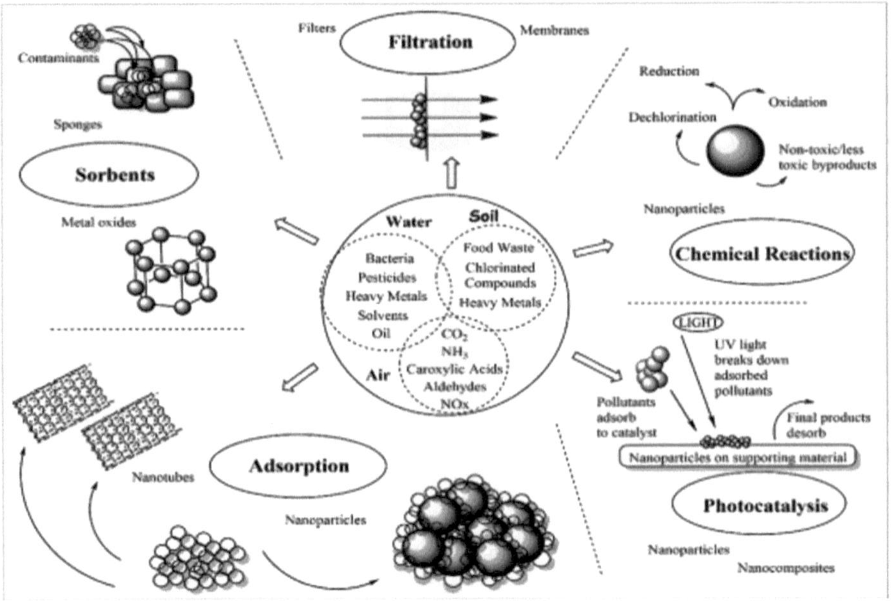

Fig. 2 Environmental remediation approaches. (Source: Fernanda et al., 2018, Adeleye et al. 2016)

technologies, artificial intelligence, and machine learning. This enables adaptive management and better-informed conservation policies (Kerry et al., 2022) (Fig. 2).

3 Case Studies

3.1 Application of Nanotechnology in Restoring Marine Resources

3.1.1 In Coral Health

Many researchers work toward preserving coral health. Scientists and researchers developed titanium dioxide nanoparticles, which help to increase the rate of photosynthesis in corals, thereby increasing their growth rates. The prepared nanoparticles help the corals to live in extreme weather changes, as well as the salinity changes of the seawater. Application of titanium nanoparticles helps in restoring the reef (Yafei et al., 2023).

3.1.2 In the Case of Oil Spills

Oil spills in oceans are a major threat to marine resources and marine life. There are many microbes available to convert the oil spills and other pollutants into harmless products, but it takes over years to process. By applying magnetic nanomaterials, such as functionalized super-paramagnetic iron oxide nanoparticles (SPIONs), magnetic nanocomposites, and magnetic carbon nanotubes (CNTs), nanotechnology proves to be a useful tool in cleaning up these oil spills or leaks in the water. Advanced properties, including increased adsorption capacity and high oil removal efficiency, are achieved by functionalizing nanoparticles with various functional moieties, such as organic non-polymeric and inorganic compounds, polymers, and polysaccharides (Amber et al., 2021; Bhat, 2022).

3.1.3 In the Case of Harmful Algal Growth

The silver nanoparticles help in controlling algal bloom, which threatens the marine biodiversity, and in habitat health.

3.1.4 In Mangrove Restoration

Biochar nanoparticles help in mangrove restoration by improving soil quality through nitrogen fixation and help in nutrient retention. This way biochar helps in mangrove rehabilitation efforts (Be et al., 2021). Nanotechnology offers better and more innovative solutions for restoring habitats in marine systems.

3.2 Application of Nanotechnology in Restoring Agricultural Resources

3.2.1 Nano Fertilizer

It is anticipated that the application of nanoparticles in the delivery of fertilizers and pesticides will lower dosages and provide regulated, gradual delivery. Applying nanoparticles to stabilize biocontrol preparations is expected to be a major contribution, as it will significantly reduce environmental hazards. Nanofertilizers are a promising tool to meet the food demand in the future because by 2050 overall population is expected to exceed 9 billion.

3.2.2 Nanosensors

Nanosensors are used for various purposes like soil health monitoring, water quality management, pest detection and management, crop health assessment, nutrient monitoring in growing plants, etc. Nanosensors that can identify diseases at concentrations as low as parts per billion have been created via nanotechnology by taking advantage of the special qualities of nanomaterials. In addition to detection, persistent compounds can be broken down into harmless and occasionally beneficial components using nanotechnology. Utilizing the instruments of nanotechnology, the pressing problems of pollution and environmental preservation can be tackled. The technologies now employed in environmental detection, sensing, and cleanup can be radically streamlined and provided by nanotechnology (Fincheira et al., 2023; Manjunatha et al., 2019; Zhao et al. 2019).

3.2.3 Nanopesticides

Nanoencapsulated controlled-release pesticides developed in India to improve the crop yield to avoid frequent usage of pesticides (Chaud et al., 2021). Recent studies showed that silver nanoparticles help in reducing the aphid populations to limit the usage of chemical fertilizers with improved plant health (Bhattacharya et al., 2016).

3.2.4 Nanocarriers

Nanocarriers are intended to deliver nutrients to the plant in a controlled manner. Nanocoating also helps in preventing seeds from pathogen attacks and environmental stresses and helps in improving plant growth and yield.

3.3 Application of Nanotechnology in the Field of Medicine

Nanotechnology in medicine plays a wonderful role in the field of medicine. Targeting cancer cells, DNA-based nanobots are also being developed. To carry a molecular payload, for example, scientists at Harvard Medical School in the USA recently described in science how they created a DNA "origami nanorobot." The barrel-shaped nanobot is able to transport compounds that instruct cells on how to behave. The group effectively demonstrates in their work how to deliver chemicals that cause lymphoma and leukemia cells to commit suicide (Gonzalez-Bulnes & Hashem, 2022).

There are other plans to produce nanobots composed of various materials. For example, Northwestern University scientists utilize gold to create "nanostars," which are straightforward, specialized, star-shaped nanoparticles that have the ability to transport medications straight to cancer cells' nuclei. A recent article published in the journal *ACS Nano* describes how drug-loaded nanostars operate as tiny hitchhikers, depositing their payload directly into the nuclei of human cervical and ovarian cancer cells after becoming attracted to an overexpressed protein on the cell surface (Sweta et al., 2023). Liposomal-formulated nanomaterials are used to treat breast cancer as well as ovarian cancer with reduced side effects) (Soundararajan et al. 2009; Barenholz, 2012). Pegylated liposomal doxorubicin (doxorubicin HCl liposome injection; Doxil or Caelyx) is a liposomal formulation of doxorubicin, which is a promising drug in cancer treatment by the reticuloendothelial system. This drug stabilizes itself by anchoring itself in the lipid layer with the help of its polyethylene glycol polymers, which result in better tumor uptake (Gabizon et al., 2003). Iron oxide nanoparticles are helping in imaging cancer cells. Quantum dots are nanocrystals, which help in imaging cancer cells with improved accuracy (Shestovskaya et al., 2023).

Lipid nanoparticles are used to deliver mRNA vaccines, which will produce fewer side effects and produce a better immune response in biological systems (Zhang et al., 2020). Filomicelles-shaped nanoparticles produced better and prolonged release in cancer treatment as well as cardiovascular diseases than nanospheres (Geng et al., 2007). Nanoparticles loaded with polyglycolic acid and polylactic acid target the malignant tumor cells since these have a high affinity toward the tumor cells. This kind of targeted drug delivery helps in effective drug delivery and also helps to prevent the off-target toxicity by decreasing exposure to the healthy cells (Human Ghazal et al., 2024). Various plants are used to produce nanoparticles, which serve as antimicrobial agents as well as vehicles to carry the particular antimicrobial components (Thenmozhi et al., 2023). Nanotechnology has revolutionized the field of medicine by enabling precise diagnostics, targeted therapies, and innovative drug delivery systems.

4 Ethical Considerations

Although nanotechnology presents novel answers for environmental problems, its application presents a number of difficulties and moral dilemmas. The possible hazards that nanoparticles could cause to the environment and human health are a major worry. Nanoparticles can quickly permeate biological systems because of their small size, which can have unforeseen effects (Jingyi et al., 2022). Before a substance is widely used in the environment, it is necessary to conduct extensive research and risk assessments in order to understand the long-term effects and potential hazards of toxicity or bioaccumulation (Inshyna et al., 2022). The relationships between business, research, and public acceptance of nanotechnology also give rise to ethical questions. Industry sponsorship has a big impact on research integrity, and it can skew studies in favor of positive results while ignoring possible hazards. In order to gain public acceptability, nanotechnology's advantages and risks must be openly communicated. Maintaining ethical standards requires careful consideration of consent procedures for nano-research, privacy concerns, and the equitable distribution of nanomedical remedies.

These difficulties highlight the requirement for an international regulatory framework to properly control the application of nanotechnology in environmental contexts. Damage to the environment and to people could happen in the absence of consistent legislation.Uneven distribution of access to nanomedical breakthroughs could lead to the emergence of social inequities. In order to solve these problems and make sure that the public is aware of the safety and ethical ramifications of nanotechnology, education is essential. We may encourage ethical and responsible advancement in nanotechnology for environmental applications by tackling these issues. Synthesized nanoparticles should not disrupt the ecological cycle, nor should they affect local species (Thenmozhi et al., 2023). Increased risk of affecting genetic integrity while restoring habitat using nanotechnology (Ghosh et al., 2018). Maintaining equity is important because of the adaptability of all the communities to the use of this system by marginalized groups. Technology and its important applications should be informed by localities (Kuhlbusch et al., 2011). There should be interactions between nanotechnologists and ecologists to bridge the gap between the theoretical approach and the practical requirements and possibilities (Sandeep et al., 2023) (Table 2).

Table 2 Summary of ethical considerations

S. no.	Considerations	Descriptions	Example reference
1	Environment impact	Assessing long-term effects on ecosystems	Guerra et al. (2018)
2	Biodiversity concerns	Risk of disrupting native species	Ghosh et al. (2018)
3	Equity and justice	Ensuring fair access to benefits	Kuhlbusch et al. (2011)
4	Regulatory challenges	Need for updated regulatory frameworks	Letro et al. (2025)
5	Public perception	Importance of transparency and trust	Siegrist et al. (2013)
6	Technological dependence	Risk of overshadowing traditional methods	Corlett, (2017)
7	Interdisciplinary collaboration	Need for collaboration across fields	Sandeep et al. (2023)

5 Future Directions

With great commercialization potential in a number of industries, such as energy, environmental management, and medical, conservation has a bright future. By making it possible to collect data more precisely and effectively, the integration of nanotechnology with other fields, like drug delivery and diagnostic imaging, can improve conservation. This results in better methods for maintaining ecosystems and safeguarding threatened species (Bajpai et al., 2023). Commercially, feasible nanodevices specifically designed for conservation efforts could be developed as a result of the potential to improve product dependability and computational efficiency. Technological developments in biodegradable goods, high-performance computing, and nano-based sensors present promising opportunities for environmental restoration and biodiversity monitoring (Schulz et al., 2023). Nanotechnology will have a major impact on global biodiversity initiatives as it develops, opening up new avenues for addressing conservation issues (Aarju et al., 2023).

The development of nanosensors involves real-time monitoring of biodiversity and ecosystem health and integrates with AI tools to analyze ecological data (Sharma et al., 2020; Khan et al., 2021). Developing nanomaterials that mimic natural processes for effective habitat recovery (Patel et al., 2021). There is interdisciplinary research required to promote research in sustainable development and restoring habitats. An increased number of funding schemes should be available for this emerging field.

6 Conclusion

Finally, a viable route toward the advancement of sustainable biodiversity conservation is the integration of nano ecosystems. Through the application of nanotechnology's efficiency and accuracy, we may improve our comprehension of ecosystems and carry out focused interventions that support ecological equilibrium. By making it easier to monitor biodiversity, restore habitats, and lessen environmental stressors, these advances help vulnerable ecosystems become more resilient. As we move forward, it is essential to adopt a multidisciplinary approach that combines scientific research, policymaking, and community engagement. This collaborative effort will ensure that nano ecosystems are utilized effectively, promoting sustainable practices that safeguard biodiversity for future generations. Embracing this technology can help us create a harmonious relationship between human activities and the natural world, paving the way for a more sustainable and biodiverse planet. Nanomaterials help in solving potential issues, which are as follows: contaminated water and polluted soil and air, using conservative methods. Nanoremediation has a greater impact in cleaning up large, contaminated sites and eliminating the contaminants. Nanoparticles and nanotubes are used as sensors in detecting contaminating substances in the environment, as well as in other resources. The development of these sensors uses cheaper raw materials with less energy. It is also understood little what happens to the materials once they have been used for pollutant degradation or capture, despite the fact that the methods by which the various nanotechnologies are applied are well established. Certain materials appear to lose their usefulness after a certain point, even though their capacity to be recycled has been documented. Determining the destiny of these materials once they are introduced to the environment is therefore vital for remediation purposes to prevent the chance that these materials may end up becoming a cause of environmental contamination.

References

Aarju, et al. (2023). Enabling technologies for wildlife conservation. *IEEE Devices for Integrated Circuit*. https://doi.org/10.1109/DevIC57758.2023.10134561

Adeleye, A. S., Conway, J. R., Garner, K., Huang, Y., Su, Y., & Keller, A. A. (2016). Engineered nanomaterials for water treatment and remediation: Coats, benefits and applicability. *Journal of Chemical Engineering, 286*, 640–662.

Amber, J., Bhuvnesh, B., & Michael, G. B. (2021). Nano-enhanced bioremediation for oil spills: A review. *ACS ES&T Engineering Journal, 1*, 928–946.

Bajpai, S., et al. (2023). Nanotechnology for remediating current environmental problems. *International Journal for Multidisciplinary Research*. https://doi.org/10.36948/ijfmr.2023.v05i01.1565

Bakshi, M. S. (2020). Impact of nanomaterials on ecosystems: Mechanistic aspects invivo. *Environmental Research, 182*, 1–20.

Bansal, A., et al. (2023). Nanotechnology in environmental clean-up. *Nanobiomaterials*. https://doi.org/10.21741/9781644902370-11

Barenholz, Y. (2012). Doxil® — the first FDA-approved nano-drug: lessons learned. *Journal ofControlled Release, 160*(2), 117–134.

Be, S., Vinitnantharat, S., & Pinisakul, A. (2021). Effect of mangrove biochar residue amended shrimp pond sediment on nitrogen adsorption and leaching. *Sustainability, 13*(13), 7230. https://doi.org/10.3390/su13137230

Bhat, I. A. (2022). Nanotechnology in reproduction, breeding, and conservation of fish biodiversity: Current status and future potential. *Reviews in Aquaculture*. https://doi.org/10.1111/raq.12736

Bhattacharya, A., Prasad, R., et al. (2016). One-pot fabrication and characterization of silver nanoparticles using *solanum lycopersicum*: An eco-friendly and potent control tool against rose Aphid, Macrosiphum rosae. *Journal of Nanoscience, 1*, 1–7.

Chee, P. L., Toh, W. L., Yew, P. Y., Peng, S., & Kai, D. (2022). In Z. Li, J. Zheng, & E. Ye (Eds.), *Sustainable nanotechnology* (Vol. 1, pp. 1–32). The Royal Society of Chemistry.

Chen, M., Chen, X., & Wu, D. (2021). *Solar thermal conversion of plasmonic nanofluids: Fundamentals and applications*. IntechOpen. https://doi.org/10.5772/intechopen.9699

Corlett, R. T. (2017). A bigger toolbox: biotechnology in biodiversity conservation. *Trends in Biotechnology, 35*(1), 55–65. https://doi.org/10.1016/j.tibtech.2016.06.009

Fernanda, D. G., Mohamed, F. A., Daniel, C. W., & Frank, A. (2018). Nanotechnology for environmental remediation: Materials and applications. *Molecules, 23*, 1760.

Fincheira, P., et al. (2023). Eco-efficient systems based on nanocarriers for the controlled release of fertilizers and pesticides: Toward smart agriculture. *Nanomaterials (Basel)*. https://doi.org/10.3390/nano13131978

Gabizon, A., Shmeeda, H., & Barenholz, Y. (2003). Pharmacokinetics of pegylated liposomal doxorubicin: Review of animal and human studies. *Clinical Pharmacokinetics, 42*(5), 419–436. https://doi.org/10.2165/00003088-200342050-00002

Geng, Y., Dalhaimer, P., Cai, S., Tsai, R., Tewari, M., Minko, T., & Discher, D. E. (2007). Shape effects of filaments versus spherical particles in flow and drug delivery. *Nature Nanotechnology, 2*(4), 249–255. https://doi.org/10.1038/nnano.2007.70. Epub 2007 Mar 25. PMID: 18654271; PMCID: PMC2740330.

Ghazal, H., Waqar, A., Yaseen, F., Shahid, M., Sultana, M., Tariq, M., Bashir, M. K., Tahseen, H., Raza, T., & Ahmad, F. (2024). Role of nanoparticles in enhancing chemotherapy efficacy for cancer treatment. *Next Materials, 2*, 100128.

Ghosh, S., et al. (2018). Biodiversity and nanotechnology: The double-edged sword. *Nature Sustainability, 1*(3): 154–156.

Gonzalez-Bulnes, A., & Hashem, N. M. (2022). Nanotechnology in animal science. *Animals*. https://doi.org/10.3390/books978-3-0365-5946-9

Guerra, F. D., Attia, M. F., Whitehead, D. C., & Alexis, F. (2018). Nanotechnology for environmental remediation: Materials and applications. *Molecules, 23*(7), 1760. https://doi.org/10.3390/molecules23071760

Inshyna, et al. (2022). Ethical and societal aspects of nanotechnology applications in medicine. In *12th International conference nanomaterials: Applications & properties*. https://doi.org/10.1109/NAP55339.2022.9934298

Jingyi, S., et al. (2022). Ethics of nanomedicine. *Nanomedicine*. https://doi.org/10.1007/978-981-13-9374-7_22-1

Kerry, R. G., et al. (2022). An overview of remote monitoring methods in biodiversity conservation. *Environmental Science and Pollution Research*. https://doi.org/10.1007/s11356-022-23242-y

Khan, Y., et al. (2021). AI and nanotechnology in ecological data analysis. *Ecological Informatics, 62*, 101253.

Kuhlbusch, T. A. J., et al. (2011). Nanoparticle exposure at nanotechnology workplaces: A review. *Particle and Fibre Technology, 8*(22), 1–18.

Lee, J., et al. (2019). Nanomaterials for climate resilience. *Environmental Research Letters, 14*(9), 095003.

Letro, L., Aier, A., & Singh, A. (2025). Regulatory Framework for Nanomaterials in Environmental Pollution Control. In: Singh, A. K., Singh, A. K., Susan, M. A. B. H. (Eds.), Advances in Nanomaterials for Detection, Control, and Removal of Environmental Pollutants. Engineering Materials. Springer, Cham. https://doi.org/10.1007/978-3-031-87409-3_16

Manjunatha, R. L., Dhananjay, N., & Usharani, K. V. (2019). Nanotechnology application in agriculture: A review. *Journal of Pharmacognosy and Phytochemistry, 8*(3), 1073–1083.

Martin, T., et al. (2022). Designing connectivity for wildlife. *Ecological Applications, 32*(4), e2548.

Patel, S., et al. (2021). Bio-inspired nanomaterials for ecological applications. *Nano Today, 38*, 101143.

Sandeep, B., Alka, N., Anand, K., & Madhu, M. (2023). Nanotechnology for remediating current environmental problems. *International Journal of Multidisciplinary Research, 5*(1), 1–8.
Schulz, A. K., et al. (2023). Conservation tools: The next generation of engineering-biology collaborations. *Journal of the Royal Society Interface*. https://doi.org/10.1098/rsif.2023.0232
Sharma, A., et al. (2020). Nano-sensors for environmental monitoring. *Environmental Science: Nano, 7*(5), 1234–1245.
Shestovskaya, M. V., Luss, A. L., Bezborodova, O. A., Makarov, V. V., & Keskinov, A. A. (2023). Iron oxide nanoparticles in cancer treatment: Cell responses and the potency to improve Radiosensitivity. *Pharmaceutics, 15*(10), 2406. https://doi.org/10.3390/pharmaceutics15102406. PMID: 37896166; PMCID: PMC10610190.
Siegrist, M., et al. (2013). Public acceptance of nanotechnology foods and food packaging: the influence of affect and trust. *Appetite, 9*(2), 459–466. https://doi.org/10.1016/j.appet.2007.03.002
Soundararajan, A., Bao, A., Phillips, W. T., Perez, R., 3rd, & Goins, B. A. (2009). [(186) Re]Liposomal doxorubicin (Doxil): In vitro stability, pharmacokinetics, imaging and biodistribution in a head and neck squamous cell carcinoma xenograft model. *Nuclear Medicine and Biology, 36*(5), 515–524. https://doi.org/10.1016/j.nucmedbio.2009.02.004. Epub 2009 May 7. PMID: 19520292; PMCID: PMC2696057.
Sweta, G., et al. (2023). Toxicology related to nanoparticles – Challenges and future prospects. In *Therapeutic nanocarriers in cancer treatment: Challenges and future perspective*. https://doi.org/10.2174/9789815080506123010014
Thenmozhi et al. (2023) Green synthesis of copper oxid nanoparticles using Allmania nodiflora leaf extract and its characterization. *BME Horizon, 1*(6), 69–75. https://doi.org/10.37155/2972-449X
Yafei, D., Yang, Y., Zhang, Z., Xing, Y., & Li, H. (2023). Toxicity of titanium oxide nanoparticles on the histology, liver physiological and metabolism, and intestinal microbiota of grouper. *Marine Pollution Bulletin, 87*. https://doi.org/10.1016/j.marpolbul.2023.114600
Zhang, Y., et al. (2020). Sustainable nano-fertilizers for agriculture. *Nature Sustainability, 3*(4), 289–298.
Zhao, X., et al. (2019). Monitoring pollutants with nanosensors. *Sensors & Actuators: B. Chemical, 283*, 45–52.

Nanorobotics: A Futuristic Approach to Microplastic Removal

R. Yuvarajan and R. Karthika

Abstract

An enormous amount of microplastics accumulates in the environment, leading to a major global environmental problem. It can cause a negative impact on the living marine ecosystems and human health. Due to their occurrence in edible aquamarine organisms, they can enter into the food web and also affect food security. In previous research, various nanomotors were constructed efficiently to remove and degrade soluble organic pollutants. Recent research focused on design rationally and surface functionalization to achieve nanomotors capable of capturing, transporting, and releasing microplastics of different shapes and chemical structures. Self-propelled nanorobots have been used to degrade microplastics from contaminated water with a combination of physiochemical properties of nanomaterials and active motion. This chapter revealed that the most advanced strategies used to degrade microplastics using autonomous nanorobots. Fenton, Fenton-like, photo-Fenton reactions, and heterogeneous photocatalysis are the most recurrent advanced oxidation processes used for remediation processes by nanorobots. The ability to propel themselves has given micro- and nanomaterials a new engineering dimension, opening the door to the creation of intelligent swarms of small-scale robots that cooperate to complete specific tasks and move in reaction to outside stimuli. Nanorobots have shown promise for water-remediation applications, where the effectiveness and speed of the purification process are critical. This is because of the way that material design and active motion work together to create programmable pollutant removal degrada-

R. Yuvarajan
PG & Research Department, Department of Biotechnology, Mahendra Arts and Science College (Autonomous), Kallipatti, Namakkal, Tamil Nadu, India

R. Karthika (✉)
Department of Biotechnology, K. S. Rangasamy College of Arts and Science (Autonomous), Tiruchengode, Namakkal, Tamil Nadu, India

D. K. Gupta et al. (eds.), *Nano-Ecosystems*,
https://doi.org/10.1007/978-981-95-5033-3_10

tion mechanisms. Numerous obstacles still need to be overcome before nanorobots may be used in the real world, despite their broad range of efficacy against pollutants that vary in nature and size (from millimeter to atomic). To meet commercial demands, however, the translation of this technology into practical applications would require the collaborative efforts of scientists from diverse fields, such as engineers, physicists, chemists, and biologists.

Keywords

Microplastics · Self-propelled nanorobots · Photocatalysis · Dimension

1 Introduction

Synthetic polymers, or plastics, are composed of hundreds to thousands of monomers—organic subunits—that are bonded together by powerful covalent bonds. They are frequently utilized due to their exceptional qualities, which include strong chemical and thermal stability, and they are also very difficult to remove. Consequently, there is an ongoing global increase in the quantity of plastic waste that is being generated. Impressive photos show aquatic ecosystems teeming with plastic bottles, bags, and other items. But the true threat posed by plastic pollution comes from its slow disintegration into tiny pieces as a result of weathering and water action. Microplastics are bits of plastic that are smaller than 5 mm. These have the potential to become more harmful by acting as a substrate for the adsorption of additional contaminants or the development of bacterial biofilms. The food chain is already contaminated with microplastics. They pose serious risks to the health of all living things because they can be consumed by animals like fish, spread through the food chain and end up on our table, or directly contaminate drinking water systems. Regretfully, microplastics have the ability to degrade into nanoplastics—smaller, more dangerous particles (<1 μm). Because of their small weight and the marine currents that carry them around, nanoplastics stay suspended in the water while microplastics tend to settle on the seafloor. Because they have a higher surface-to-volume ratio than micro/nanoplastics (MNPs), they also permeate tissues and adsorb more contaminants and microorganisms (Ostle et al., 2019; Lönnstedt & Eklöv, 2016). These have the potential to become more harmful by acting as a substrate for the adsorption of additional contaminants or the development of bacterial biofilms. The food chain is already contaminated with microplastics. They pose serious risks to the health of all living things because they can be consumed by animals like fish, spread through the food chain and end up on our table, or directly contaminate drinking water systems. Regretfully, microplastics have the ability to degrade into nanoplastics—smaller, more dangerous particles (<1 μm). Because of their small weight and the marine currents that carry them around, nanoplastics stay suspended in the water while microplastics tend to settle on the seafloor (González-Fernández et al., 2021; Galloway et al., 2017) (Fig. 1).

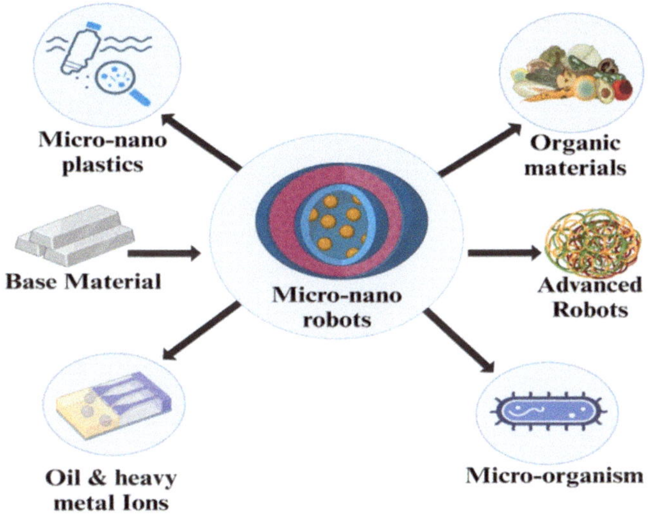

Fig. 1 Removal of pollutants by micro/nanorobot

MNP use by aquatic species can result in several health problems, including digestive tract blockages and physical injury (Mamun et al., 2023; Vethaak & Legler, 2021). Moreover, MNPs have the potential to negatively impact aquatic creatures' behavior and reproductive health (Gao et al., 2022). Current research suggests that metal nanoparticles (MNPs) can enter human tissues through the food chain and accumulate there, where they may cause a variety of health issues, such as cancer, genetic damage, and inflammation, even though the full impact of MNPs on human health is still unknown (Wright & Kelly, 2017; Borrelle et al., 2020; Bank et al., 2022). There are several obstacles that need to be taken into account in order to properly treat a sizable amount of MNP-contaminated water. These issues include efficiently removing ultrasmall particles, achieving high removal efficiency for low environmental MNP concentrations, and handling huge volumes of sewage at a low cost. To address these issues, several conventional technologies have been employed, including chemical treatments, physical adsorption and filtration (Cole et al., 2013; Talvitie et al., 2017), and biological elimination by ingestion (Shimao, 2001; Dawson et al., 2018; Auta et al., 2017; Li et al., 2018). Nevertheless, it's important to thoroughly consider each option's benefits and drawbacks in practice. Although filtration has high removal efficiency, large-scale treatment is hampered by its energy consumption and loss of nanoplastic filtration. Although chemical flocculation is a low-energy option, it may result in secondary pollutants and has a limited capacity for removal (Lee et al., 2012). Furthermore, biological therapy is an eco-friendly method, but its elimination rate is slow, and it is only appropriate in certain situations (Singh et al., 2022; Padervand et al., 2020). Therefore, an optimum solution is needed to balance the trade-offs between removal efficiency, cost, and environmental impact.

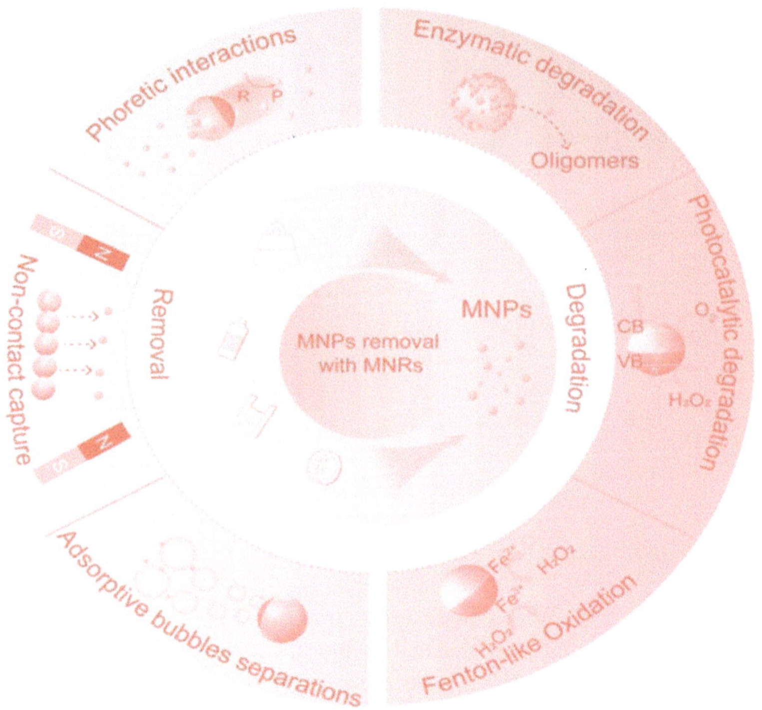

Fig. 2 Removal of MNPs by MNRs

Micro/nanorobots (MNRs) have attracted a lot of attention lately for the purpose of environmental remediation because of their remarkable ability to move autonomously and precisely (Iyare et al., 2020; Urso & Pumera, 2022; Wang et al., 2021; Hermanová & Pumera, 2022). These minuscule micro/nanomachines can operate at the micro/nanoscale in intricate surroundings by transforming a variety of energy sources, including heat, magnetic, light, electricity, ultrasound, and light energy, into mechanical motion (Urso et al., 2023). MNRs are a useful technique for remediating water pollution because of their autonomous motion, which can enhance the likelihood of coming into contact with contaminants and speed up their removal. In the meantime, MNRs' small size makes it possible for them to fit through narrow gaps and pass through areas that larger devices would not be able to reach (Chen et al., 2021) (Fig. 2).

2 Photocatalytic MNRs for MNPs Removal

The remarkable photochemical activity of photocatalytic materials has led to their broad application in the development of light-driven micro/nanorobots (LMNRs). Several methods, including self-electrophoresis, self-diffusiophoresis, and bubble propulsion, can power photocatalytic LMNRs. Spatial separation of the anodic and cathodic processes produces an unbalanced charge distribution and a propulsive electric field for self-electrophoresis. Diffusiophoresis, on the other hand, depends on anodic and cathodic processes that overlap spatially and produce distinct ion species near the same side of the MNR. A concentration gradient can provide an electric field to drive the MNR as cations and anions migrate outward at varying rates (Chen et al., 2017).

Bubble propulsion is a method of propelling LMNRs by the generation of bubble reactions, like H2O2 splitting. Further degrees of freedom can be obtained by using incident light as an encoding signal and a controllable energy source. Owing to their biocompatibility, controllability, and efficiency, LMNRs have been widely used in smart assembly, environmental cleanup, and biomedical devices. In the field of environmental remediation (Xu et al., 2017), LMNRs have been the focus of much research. For example, it has been discovered that Janus Au/WO3@C microrobots have increased motility due to the photocatalytic degradation of organic dyes in the presence of UV light (Zhang et al., 2017). AgCl microstar-based microrobots, on the other hand, can be powered by light and actively break down organic compounds in addition to eliminating bacteria (Escherichia coli) (Simmchen et al., 2017). A crucial part of this bactericidal process is played by silver nanoparticles, and microrobots' independent mobility helps to disperse these particles and creates more possibilities for them to come into contact with bacteria, which in turn increases their bactericidal efficacy. It is possible to logically design the phoretic interactions of LMNRs to produce special benefits in cargo transit, adsorption, and clustering. BiVO4 microrobots powered by photocatalysis have been proven to be active in loading and moving both living microbes and passive particles. They produce highly oxidative reactive oxygen species (ROS) through photocatalytic processes, which eradicate bacteria. These multipurpose MNRs have bright futures for a variety of uses, such as the removal of microbiological water pollution and the transportation of freight (Villa et al., 2019).

This clustering and capturing tendency is very promising for the removal of microplastics since large-scale aggregation and adsorption may also greatly increase the efficacy of MNP collection. Wang et al. developed a unique Janus Au@Ni@TiO_2 microrobot for removing microplastics. By employing the electrophoresis process in the presence of UV radiation, this microrobot was able to move in both water and H2O2 solution. The Au@Ni@TiO2 microrobot was able to successfully collect microplastic particles from samples of Baltic saltwater and personal care items through phoretic contact. Large-scale cleanup of microplastics would also be possible since the microrobots could move independently in reaction to UV light and join into a chain when exposed to an external magnetic field. The remarkable MNPs removal performance of this microrobot was 77% in H2O2 and 67% in pure water, respectively. This suggests a stable and material-independent method that opens the door for an intelligent MNPs "sweeper." Furthermore, to remove microplastics effectively, Chattopadhyay et al. employed self-asymmetric Pac-Man TiO_2 microrobots, which have the ability to both collect microplastics and promote photocatalytic breakdown. Recently, when exposed to UV light, Martin's group's incredible multipurpose microrobot, which is based on g-Fe2O3/Pt/TiO_2 created from MXene, can capture nanoplastics in a three-dimensional (3D) space (Urso et al., 2022). When these microrobots were subjected to UV light, they showed robust 3D mobility and negative photogravitactic activity. Furthermore, the microrobots' pH-programmable surface charge greatly shortened the preconcentration period, which concurrently improved the electrochemical sensor's sensitivity and the nanoplastics' capture efficiency. The clearance efficiency of 50-nm PS nanoparticles was 97% in just 1 min, which is a major improvement over the static MXene treatment. Since nanoplastics can be easily separated and collected by magnetic force after treatment, these microrobots are also an ideal alternative for removing nanoplastics from water treatment applications.

3 Magnetically Driven MNRs for MNPs Removal

Due to its great versatility, magnetic microrobots can be operated in multiple ways through the use of external magnetic control systems. For example, they can be made to roll by applying a magnetic field that rotates. Conversely, microrobots can be guided in a straight path using a gradient magnetic field while keeping a portion of their structure or their entire body in a fixed position. When rotational and gradient magnetic fields are connected, an alternating magnetic field that can power microrobots by asymmetric reciprocating switching is produced (Yang et al., 2020). Swarm manipulation, biological therapy, and environmental remediation are just a few of the many uses for magnetic manipulation technology, which allows for accurate, wireless, and distant manipulation. For example, a "hedgehog" magnetic robot based on sunflower pollen grains (SPGs) was created by Sun et al. and has demonstrated significant promise in eliminating microplastic and oil pollution. The microrobot can be configured to roll, rotate, or swing in three different motion modes using a rotating magnetic field.

Microrobots are able to transport 100-mm PS microspheres while actively assembling into chains in the rolling mode. The joined chains can revolve along their long axes when the swinging mode is engaged, creating a stable, symmetrical fluid flow field that is appropriate for the transportation of cargo. Cargo can be released by modifying the fluid flow field and motion mode after transportation. Therefore, the microrobot can effectively gather, move, and release microplastics by combining the rolling and swinging modes. The removal efficiencies of microplastics were 75% for the facial cleanser samples and 70% for the contaminated seawater, respectively (Won et al., 2022). A ternary nanocomposite made of polydimethylsiloxane, magnetic particles, and a nanoporous carbon nanotube yarn (CNTY) framework was used to construct the microrobot. This microrobot can swim in a pulsed quadrupole electromagnetic field in three different directions: linearly, rotationally, and oscillatorily. Upon assembling numerous CNTY microrobots, they can produce an intensified forced vortex in the rotating mode that attracts the microbeads in their direction. The collective microrobots create a clockwise vortex that gradually catches microplastics, enabling the removal of microplastics in 150 s. On the other hand, the fact that a single CNTY microrobot cannot produce hundreds of microbeads highlights how important the swarming effect is to the successful removal of microplastics. MNRs are powered by ion exchange to get rid of MNPs. Ion-exchange-based MNRs have found widespread use because of their notable intelligent assembly and swarming effects (Niu et al., 2018). For instance, a variety of modular swimming microrobots constructed with cation exchange resins were examined by Niu et al. (Wu et al., 2021). It's interesting to note that while the exchange resin particles themselves do not move actively, active propulsion is produced by self-assembly that happens when the resin particles are mixed with extra inert particles. When impurity ions in the solution interact with the resin particles, a localized diffusion electric field is produced that is directed inward. The different diffusion coefficients of the exchanged protons and cations explain why this happens. Inert particles are drawn to the area around the cation exchange resin by this field's induction of an electro-osmotic solvent flow around the negatively charged substrate. But in the absence of inert particles, the homogeneous fluid flow generated by the resin particles was unable to propel them in any particular direction. The asymmetric adsorption of inert particles finally disrupts the uniformly distributed flow and accelerates the autonomous mobility of the resin particles. As a result, the ion-exchange energy propels ion-exchange MNRs forward with the help of an autonomously constructed asymmetric structure. Ion-exchange-induced chemical communication greatly increased the reactivity and mobility of SPS microbeads and ZnO nanorods, forming active swarm complexes. Macroscopic and microscopic behaviors like quorum decision-making and phase segregation are accomplished.

A flexible technique that may be utilized to create intricate interactions between active and passive particles is the perspective ion-exchange-based mechanism. This technique opens up new possibilities for the development of responsive active materials and systems. It is essential to take into account the necessary energy input for MNR-based MNP elimination. External control devices and chemical fuels should ideally be avoided. Fortunately, energy from the ion-exchange process may be collected by the ion-exchange driven mechanism and turned into a propulsion force without requiring complex external manipulation equipment or additional chemical fuels. Because of this, the ion-exchange process is especially advantageous for practical, large-scale applications. Our team has recently created a self-driven magneto robot (SMR) that uses an ion-exchange mechanism to extract MNPs from nonmarine waters in a scalable and recyclable manner (Li et al., 2022). The SMR is made up of Fe3O4 nanoparticle-functionalized ion-exchange resin microspheres, which facilitate facile magnetic collection and separation. This allows for accurate magnetically regulated adsorption as well as cyclic operation. The use of SMRs for controlled MNP release, magnetic separation, and MNP removal demonstrates the adaptability and potency of SMRs as a scalable solution.

Moreover, the MNP particles' nonuniform adsorption creates an initially asymmetric flow field that drives the SMRs and permits autonomous motion. We found that as the degree of asymmetry brought about by adsorption grew, so did the initial motion speed. Notably, SMRs don't need an asymmetric structure to be predesigned, which significantly lowers preparation effort and expense. The attraction range and propulsion efficiency are determined by the ion-exchange capacity and the diffusion difference of the exchanged ions. High diffusivity ions, such H+ and OH, are excellent for enhancing the SMR system's functionality. OH-SMR creates a long-range pH gradient field and a 3D adsorption field that extends over 100 mm to draw negatively charged MNPs. It accomplishes this by releasing OH and absorbing impurity anions from the surrounding aqueous medium. This greatly improves the system's removal efficiency. It has been extensively shown that SMRs have the ability to remove microplastic particles from a variety of nonmarine fluids and from particles with diverse compositions, sizes, and forms. SMRs had exceptionally high removal efficiency of more than 90% for distinct MNPs present in a range of nonmarine water samples. SMRs' removal efficiency was compared with conventional techniques at low concentrations of 10 and 20 ppm, which are around ten times lower than usual environmental concentrations, in order to show how practical they are. The findings show that SMRs outperform conventional filtration and coagulation techniques, with a removal efficiency of over 90%. Moreover, following 100 treatment cycles, the SMRs showed outstanding cycling stability, with a sustained removal effectiveness of over 90%. This quality is essential for long-term recyclable materials.

To illustrate the large-scale decontamination of MNP-contaminated water, a closed-loop miniplant was created. It consists of four units: MNP adsorption by SMRs, MNP desorption, SMR regeneration, and SMR rinsing with deionized (DI) water. After using SMRs, the colorful MNPs were totally eliminated. SMRs gathered the MNPs and moved them into prospective desorption cells, releasing the adsorption cell's treated wastewater and replenishing it with fresh, tainted water for further cycles.

The desorption cells' color changed after five consecutive cycles, which could be seen. SMR removal efficiency, SMR regeneration rate, and MNP desorption rate were all above 90% in this miniplant. Many benefits are available with the SMR-based dynamic adsorption technology proposed in this work, including an easy-to-use structure, low cost, the capacity to operate at an industrial scale, and no need for chemical fuel input. This offers a method for eliminating and segregating MNPs that shows great promise for a range of applications. To effectively remove microplastics, seawater's high ionic strength still poses a major challenge; therefore, further developments in salt resistance are essential to extending the use of ion-exchange-based MNRs. However, this technology offers a promising approach to addressing MNP pollution, and more research and development are needed to maximize its utility in practical contexts.

Manganese dioxide (MnO_2)-based catalyst microrobots have been successfully employed to remove a variety of microplastics utilizing H2O2 bubbles as fuel (Zeng et al., 2019). A common method for removing solid particles, droplets, or other suspended debris from liquids is called adsorptive bubble separation, or ABS (Villa et al., 2018). Using MnO_2-based microrobots, which produce a lot of bubbles when they are propelled due to H2O2 breakdown, presents a viable method for removing pollutants using a bubble separation process. Ye et al. used scalable hydrothermal techniques to create magnetic Fe3O4-MnO2 and Fe2O3-MnO2 microrobots with a core-shell configuration in a recent work. These microrobots were proven to be effective at dynamically removing microplastics via the ABS mechanism at a concentration of 5% H2O2. The H_2O_2 breakdown produced microbubbles that captured microplastics throughout the microrobots' motion. These microbubbles then gathered on the solution's surface foam layer, which made them easier to separate. After 2 h of treatment, the ABS mechanism's microplastics removal efficiency was 10%. Furthermore, the utilization of light-thermal-driven MNRs for MNP removal has been investigated by researchers. Through the use of photothermal materials in Janus structures, MNRs can be generated by using the photothermal effect to create a temperature gradient. This microrobot consists of a polyurethane (PU) matrix functionalized with sulfonated graphene (SG) modified by b-cyclodextrin (b-CD). Through hydrogen-bonding interactions, b-CD can be included into a polyurethane matrix to generate a gradient nanostructure that balances stiffness, improves mechanical characteristics, and increases the material's use and longevity. This is particularly advantageous for applications like soft robotics. Under near-infrared (NIR) light irradiation, the soft robot showed high actuation speed, several actuation modes, and arbitrary directional mobility. The PU and SG thermal expansion coefficients differ by an order of magnitude, which results in an imbalance in the volume change between the two sides of the film, causing bending and straightening motions. The fish-shaped PU/CD-SG microrobot deforms at the irradiated region when NIR light is applied to different parts of it. This periodic switching of NIR light also causes the microrobot to move reciprocally. The PU/CD-SG microrobot exhibited remarkable adsorption capabilities in relation to microplastics. The robot that adsorbed microplastics had a bumpy surface that could be reused up to ten times, in contrast to its original smooth and level surface.

4 Degradation of MNPs by MNRs

Degradation measurement: Understanding the degrading effect of MNPs requires effective detection. Many techniques can be used to investigate the surface morphology and elemental composition of microplastics, including optical microscopy and scanning electron microscopy with energy-dispersive X-ray spectroscopy (SEM-EDS). The degree of oxidation can be determined using X-ray photoelectron spectroscopy (XPS); however, its poor penetration depth and surface-only sensitivity are its main limitations (Dong et al., 2020). FTIR, or Fourier transform infrared spectroscopy, is the method most frequently employed to find MNPs. Plastic particles are exposed to infrared radiation, which produces spectra that show the vibrations of different atoms' chemical bonds. The spectra were then compared to reference spectra stored in a library to ascertain the composition of the plastic particles (Velimirovic et al., 2021). Raman spectroscopy is another helpful technique for categorizing various types of polymers, despite the fact that it might be challenging to obtain information on differences in the polymer surface composition.

On the other hand, changes in the functional groups on plastic surfaces can be detected using FTIR in combination with Raman spectroscopy. Mass spectrometry

(MS) is a more effective technique for giving complete characterization in low-concentration nanoparticles when analyzing MNPs in the submicrometer range (Anderson et al., 2016). Increasingly, thermal decomposition in conjunction with MS is being used as a supplementary technique to spectroscopic approaches (m-FTIR and m-Raman) for the bulk identification of MNPs in extremely complicated environments (Montaño et al., 2016). Pyrolysis-gas chromatography-mass spectrometry (Pyr-GC-MS), for instance, allows both qualitative and quantitative.

However, in low-mass areas, the intensity of fingerprint peaks is significantly increased by thermal fragmentation utilizing matrix-assisted laser desorption/ionization-time-of-flight mass spectrometry (MALDI-TOF-MS). As a result, there are many opportunities for the technology to detect MNP degradation. MNPs are photocatalytically degraded by light-driven MNRs. The very effective and eco-friendly process of photocatalytic degradation is used to purify MNPs (Zhang et al., 2021). Photocatalysts use light to produce electrons and holes, which are then used to decompose MNPs into inorganic molecules like CO_2 and H_2O. In order to build LMNRs, photocatalytic materials, including zinc oxide (ZnO) and titanium dioxide (TiO2), have been used extensively (Turchi & Ollis, 1990). These materials have shown outstanding performance in the destruction of MNPs from wastewater. The application of photocatalytically functionalized MNRs to hasten microplastic degradation has garnered increasing attention in recent times. Photocatalytic MNRs were used in a number of research, reported by the Pumera group, to induce MNP degradation. Bi2WO6-based visible-light-driven microrobots were created to break down textile fibers. Under visible light stimulation, microrobots may move on their own without the need for fuel or at very low fuel concentrations (0.025 wt% H_2O_2). Due to electrophoretic contacts and penetration caused by the chemical gradient concentration, the microrobot swarms collectively moved and gathered on the textile fiber surface. After a period of light-induced degradation, the fiber network architecture collapsed, and surface damage and pore formation with a diameter of 0.3–0.6 mm were seen by scanning electron microscopy, indicating partial breakdown of the fiber. The generation of reactive species, such as hydroxyl radicals, during the photocatalytic reaction, explains this phenomenon. These species desorb from the catalyst and attack the fiber surface. The formation of these pores makes the inside surface of the fiber more vulnerable to further degradation, which may ultimately lead to total failure. An efficient visible-light photocatalyst, bismuth vanadate ($BiVO_4$), has been developed as a useful microrobot for ensnaring and breaking down MNPs. This microrobot is made of BiVO4 that has been enhanced with Fe_3O_4 nanoparticles. Its naturally asymmetric geometric structure allows it to move independently under visible light irradiation at a low H_2O_2 concentration of 0.1 wt%. The self-propelled BiVO4-based microrobots are able to break down organic short molecules and oligomers by attaching themselves to microplastics with varying polymer compositions, including polylactic acid (PLA), polycaprolactone (PCL), polyethylene terephthalate (PET), and polypropylene (PP). Furthermore, external magnetic navigation and separation are made possible by the presence of magnetic nanoparticles. The effectiveness of removing microplastics might vary from 20% to 70%, depending on how hydrophobic they are (Lam et al., 2021). This is due to the

Fe_3O_4/$BiVO_4$ microrobot's ability to adhere strongly to the surface of microplastics, which enables the collection of both free and microrobot-loaded microplastics using a magnet. The degree of deterioration was then determined by weighing the microplastics after they had been treated for 1, 3, and 7 days. Following a 7-day course of therapy, 2%–3% of PCL and PLA were lost in mass. To further illustrate the Fe_3O_4/$BiVO_4$ microrobot's efficacy in microplastic degradation, the surface morphologies of microplastics were compared before and after photocatalytic destruction. The magnetic/photocatalytic bifunctional hybrid motion displayed by the same Fe_3O_4/$BiVO_4$ microrobot may be useful in the extraction and degradation of cellulose acetate microfilaments from cigarette filters (Mayorga-Burrezo et al., 2023). Because of their cooperative character, the microrobots were able to migrate collectively over and along the PP ultrafine fiber networks.

This led to improved physical contact time and active interfacial interaction. Because the 1D microrobots have different electrostatic properties from the PP microfibers, they can stick to the fiber surface either via electrostatic or physical attraction. Following a 30-h light exposure, the morphological characteristics of the PP filters were compared. The raw filter membrane showed a uniformly smooth surface of microfiber networks, suggesting that substantial deterioration cannot be caused by light irradiation. On the other hand, the PP filter membrane treated with the 1D microrobot cluster exhibited notable variations, exhibiting 100–500 nm-diameter fissures and voids across the PP microfiber network. Surface area and porosity measurements, along with the degree of morphological degradation indicated by the corresponding physical deformation of cracks and voids, were used to quantitatively characterize these differences. The changes in the chemical composition of the photocatalytic PP filter membrane are confirmed by Fourier transform infrared spectroscopy.

Further evidence of photocatalytic degradation was obtained by scanning electron microscopy with energy-dispersive X-ray spectroscopy (SEM-EDX) and solid-phase micro-extraction gas chromatography MS (SPME-GC-MS). MNP breakdown with enzymes, biological catalysts that have been well researched for their capacity to degrade MNPs, catalyze the natural process of biodegradation, which converts materials into ecologically benign byproducts (Othman et al., 2021). Immobilizing enzymes on microrobots altered by bio-based material binders or adhesion promoters is a novel strategy for degrading MNPs. Enzyme interactions with MNPs are improved by this technique, and microrobot mobility can increase MNP degradation efficiency even more. A magnetic field-driven mussel-inspired microrobot (MagRobots) attaches itself to microplastics and breaks them down enzymatically (Zhou et al., 2021). It was simple to build the magnetic microrobot inspired by mussels by coating Fe_3O_4 nanoparticles with the lipase enzyme on a polydopamine (PDA) layer. Because of the high adhesion of PDA, PDA@Fe_3O_4/lipase stuck to microplastics under a transversal rotating magnetic field produced by triaxial coils. Following a 24-h enzymatic degradation period, MagRobots containing microplastics were gathered, and optical microscopy was used to determine the structural alterations in the microplastics prior to and following enzymatic degradation. This process demonstrated the partial decomposition of microplastics. When compared to conventional approaches, the utilization of enzyme-based microrobots for MNP degradation has various advantages. Firstly, the enzymatic degradation process is extremely selective, which guarantees that MNPs are effectively broken down without endangering the environment (Azevedo & Reis, 2005). Secondly, MNPs can be targeted and localized to be degraded by microrobots, which will limit their diffusion and spread as pollutants. Thirdly, continuous and prolonged degradation is made possible by the integration of enzymes with microrobots, leading to more effective and durable remediation. Enzyme-based microrobots, therefore, have the ability to lessen the effects of MNPs.

Pollution that occurs later: There is a lot of promise for reducing plastic pollution through the use of MNRs to eliminate MNPs. But it's imperative to make sure MNRs don't start producing secondary pollutants. To lessen this task, two strategies might be taken into account. First, after their duty is done, MNRs can be built using biodegradable materials, such as biological-based materials, which will dissolve into the environment without causing any harm. Second, we think that utilizing dependable recovery techniques and integrating the MNRs' treatment process into already-existing wastewater treatment facilities are workable and sensible ways to reduce the risk of MNRs overflow and stop secondary pollution. Effective filtration systems can be used to cascade MNRs in order to collect undesired byproducts that are produced throughout the removal process. Cutting-edge nanocomposite materials may be essential in the future to stop secondary pollution. For example, the performance and safety of MNRs in gathering and eliminating microplastics can be enhanced by the incorporation of smart materials, such as responsive hydrogels. The secondary pollutants and carbon footprint connected to the installation and operation of MNRs could also be reduced by optimizing their energy sources through the use of green and sustainable energy. The mobility of electric-field-driven MNRs is hampered by tolerance for ions' high-salt solutions, which is a significant barrier to their practical use in numerous industries (Ceylan et al., 2017).

Using naturally salt-resistant materials, like polyelectrolytes or very porous materials, salt-resistant MNRs can be built to get around this problem (Lipczynska-Kochany, 1991; Kim & Vogelpohl, 1998). Complex macromolecules called polyelectrolytes, which have strongly charged groups, can improve surface conductivity and offset the inhibiting effects of high salt concentrations by functioning as an artificial electric double layer. Additionally, polyelectrolytes can improve MNRs' mechanical strength and stability, which will improve their performance in high-salt settings. Combining MNRs with extremely porous materials is an additional way to improve their resistance to salt and allow them to move in saline liquids (Peng et al., 2021). Through the utilization of the distinct characteristics of porous materials, MNRs are able to withstand the impact of salts and retain their performance, thus increasing the range of possible applications.

Moving toward scalable engineering design: Because there is so much MNP-contaminated water, the practical use of MNRs for the removal of MNPs necessitates a commitment to large-scale industry. MNRs require careful engineering design and economic study in order to reach their full potential. However, before these technologies can be fully incorporated into industrial applications, a number of important issues need to be resolved. The MNRs' removal efficiency is one of the primary obstacles (Wang et al., 2022a).

Effective deployment of MNRs is challenging due to their restricted propulsion capability and the intricate fluidic environment in which they function. Furthermore, MNPs are very difficult to remove due to their small size and complex structure. The creation of cutting-edge MNR propulsion systems and the incorporation of innovative materials and surface alterations to improve removal efficiency are potential future research areas. One more difficulty is scaling. Although MNRs have proven to have high clearance rates in lab environments, scaling these technologies for large-scale commercial applications presents a number of difficulties. Larger-scale high removal rates will necessitate the development of more reliable, recyclable, and effective propulsion systems in addition to power sources. Because MNRs can be harmed by chemicals, physical abrasion, and other environmental factors, durability is also an essential concern in industrial wastewater situations. To achieve long-term operation in industrial applications, it is imperative to develop more robust MNRs that can survive these circumstances. The expense of creating and executing MNRs to eliminate MNPs continues to be a major obstacle to their use in industrial settings (Wang et al., 2022b).

5 Conclusion

In order to overcome this obstacle, more affordable materials, practical propulsion, and affordable power sources must be developed. New business models and financing sources must also be found in order to enable the large-scale implementation of these technologies. In the future, MNRs will be able to eliminate MNPs from real-world surroundings more effectively thanks to the development of sophisticated sensing and control capabilities. Additionally, MNRs will be more effective in tackling the MNPs pollution issue if they are integrated with other environmental technologies, such as waste management and water treatment systems.

References

Anderson, A. G., Grose, J., Pahl, S., Thompson, R. C., & Wyles, K. J. (2016). Microplastics in personal care products: Exploring perceptions of environmentalists, beauticians and students. *Marine Pollution Bulletin, 113*, 454–460. https://doi.org/10.1016/j.marpolbul.2016.10.048

Auta, H. S., Emenike, C. U., & Fauziah, S. H. (2017). Screening of Bacillus strains isolated from mangrove ecosystems in Peninsular Malaysia for microplastic degradation. *Environmental Pollution, 231*, 1552–1559. https://doi.org/10.1016/j.envpol.2017.09.043

Azevedo, H. S., & Reis, R. L. (2005). *Understanding the enzymatic degradation of biodegradable polymers and strategies to control their degradation rate*. https://doi.org/10.1201/9780203491232.ch12

Bank, M. S., Mitrano, D. M., Rillig, M. C., Lin, C. S. K., & Ok, Y. S. (2022). Embrace complexity to understand microplastic pollution. *Nature Reviews Earth and Environment, 3*, 736–737. https://doi.org/10.1038/s43017-022-00365-x

Borrelle, S. B., Ringma, J., Law, K. L., Monnahan, C. C., Lebreton, L., McGivern, A., Murphy, E., Jambeck, J., Leonard, G. H., Hilleary, M. A., et al. (2020). Predicted growth in plastic waste exceeds efforts to mitigate plastic pollution. *Science, 369*, 1515–1518. https://doi.org/10.1126/science.aba3656

Ceylan, H., Giltinan, J., Kozielski, K., & Sitti, M. (2017). Mobile microrobots for bioengineering applications. *Lab on a Chip, 17*, 1705–1724. https://doi.org/10.1039/C7LC00064B

Chen, C., Mou, F., Xu, L., Wang, S., Guan, J., Feng, Z., Wang, Q., Kong, L., Li, W., Wang, J., & Zhang, Q. (2017). Light-steered isotropic semiconductor micromotors. *Advanced Materials, 29*, 1603374. https://doi.org/10.1002/adma.201603374

Chen, M., Lin, Z., Xuan, M., Lin, X., Yang, M., Dai, L., & He, Q. (2021). Programmable dynamic shapes with a swarm of LightPowered colloidal motors. *Angewandte Chemie, International Edition, 60*, 16674–16679. https://doi.org/10.1002/anie.202105746

Cole, M., Lindeque, P., Fileman, E., Halsband, C., Goodhead, R., Moger, J., & Galloway, T. S. (2013). Microplastic ingestion by zooplankton. *Environmental Science & Technology, 47*, 6646–6655. https://doi.org/10.1021/es400663f

Dawson, A. L., Kawaguchi, S., King, C. K., Townsend, K. A., King, R., Huston, W. M., & Bengtson Nash, S. M. (2018). Turning microplastics into nanoplastics through digestive fragmentation by Antarctic krill. *Nature Communications, 9*, 1001. https://doi.org/10.1038/s41467-018-03465-9

Dong, M., Zhang, Q., Xing, X., Chen, W., She, Z., & Luo, Z. (2020). Raman spectra and surface changes of microplastics weathered under natural environments. *Science of the Total Environment, 739*, 139990. https://doi.org/10.1016/j.scitotenv.2020.139990

Galloway, T. S., Cole, M., & Lewis, C. (2017). Interactions of microplastic debris throughout the marine ecosystem. *Nature Ecology & Evolution, 1*, 0116. https://doi.org/10.1038/s41559-017-0116

Gao, D., Liu, X., Junaid, M., Liao, H., Chen, G., Wu, Y., & Wang, J. (2022). Toxicological impacts of micro(nano)plastics in the benthic environment. *Science of the Total Environment, 836*, 155620. https://doi.org/10.1016/j.scitotenv.2022.155620

González-Fernández, D., Cózar, A., Hanke, G., Viejo, J., Morales-Caselles, C., Bakiu, R., Barceló, D., Bessa, F., Bruge, A., Cabrera, M., et al. (2021). Floating macrolitter leaked from Europe into the ocean. *Nature Sustainability, 4*, 474–483. https://doi.org/10.1038/s41893-021-00722-6

Hermanová, S., & Pumera, M. (2022). Micromachines for microplastics treatment. *ACS Nanoscience Au, 2*, 225–232. https://doi.org/10.1021/acsnanoscienceau.1c00058

Iyare, P. U., Ouki, S. K., & Bond, T. (2020). Microplastics removal in wastewater treatment plants: A critical review. *Environmental Science, Water Research & Technology, 6*, 2664–2675. https://doi.org/10.1039/D0EW00397B

Kim, S. M., & Vogelpohl, A. (1998). Degradation of organic pollutants by the photo-Fenton-process. *Chemical Engineering & Technology, 21*, 187–191. https://doi.org/10.1002/(SICI)1521-4125(199802)21:23.0.CO;2-H

Lam, S.-M., Sin, J.-C., Zeng, H., Lin, H., Li, H., Chai, Y.-Y., Choong, M.-K., & Mohamed, A. R. (2021). Green synthesis of Fe-ZnO nanoparticles with improved sunlight photocatalytic performance for polyethylene film deterioration and bacterial inactivation. *Materials Science in Semiconductor Processing, 123*, 105574. https://doi.org/10.1016/j.mssp.2020.105574

Lee, K. E., Morad, N., Teng, T. T., & Poh, B. T. (2012). Development, characterization and the application of hybrid materials in coagulation/flocculation of wastewater: A review. *Chemical Engineering Journal, 203*, 370–386. https://doi.org/10.1016/j.cej.2012.06.109

Li, L., Xu, G., Yu, H., & Xing, J. (2018). Dynamic membrane for micro-particle removal in wastewater treatment: Performance and influencing factors. *Science of the Total Environment, 627*, 332–340. https://doi.org/10.1016/j.scitotenv.2018.01.239

Li, W., Wu, C., Xiong, Z., Liang, C., Li, Z., Liu, B., Cao, Q., Wang, J., Tang, J., & Li, D. (2022). Self-driven magnetorobots for recyclable and scalable micro/nanoplastic removal from nonmarine waters. *Science Advances, 8*, eade1731. https://doi.org/10.1126/sciadv.ade1731

Lipczynska-Kochany, E. (1991). Degradation of aqueous nitrophenols and nitrobenzene by means of the Fenton reaction. *Chemosphere, 22*, 529–536. https://doi.org/10.1016/0045-6535(91)90064-K

Lönnstedt, O. M., & Eklöv, P. (2016). Environmentally relevant concentrations of microplastic particles influence larval fish ecology. *Science, 352*, 1213–1216. https://doi.org/10.1126/science.aad8828

Mamun, A. A., Prasetya, T. A. E., Dewi, I. R., & Ahmad, M. (2023). Microplastics in human food chains: Food becoming a threat to health safety. *Science of the Total Environment, 858*, 159834. https://doi.org/10.1016/j.scitotenv.2022.159834

Mayorga-Burrezo, P., Mayorga-Martinez, C. C., & Pumera, M. (2023). Photocatalysis dramatically influences motion of magnetic microrobots: Application to removal of microplastics and dyes. *Journal of Colloid and Interface Science, 643*, 447–454. https://doi.org/10.1016/j.jcis.2023.04.019

Montaño, M. D., Olesik, J. W., Barber, A. G., Challis, K., & Ranville, J. F. (2016). Single particle ICP-MS: Advances toward routine analysis of nanomaterials. *Analytical and Bioanalytical Chemistry, 408*, 5053–5074. https://doi.org/10.1007/s00216-016-9676-8

Niu, R., Fischer, A., Palberg, T., & Speck, T. (2018). Dynamics of binary active clusters driven by ion-exchange particles. *ACS Nano, 12*, 10932–10938. https://doi.org/10.1021/acsnano.8b04221

Ostle, C., Thompson, R. C., Broughton, D., Gregory, L., Wootton, M., & Johns, D. G. (2019). The rise in ocean plastics evidenced from a 60-year time series. *Nature Communications, 10*, 1622. https://doi.org/10.1038/s41467-019-09506-1

Othman, A. R., Hasan, H. A., Muhamad, M. H., Ismail, N., & Abdullah, S. R. S. (2021). Microbial degradation of microplastics by enzymatic processes: A review. *Environmental Chemistry Letters, 19*, 3057–3073. https://doi.org/10.1007/s10311-021-01197-9

Padervand, M., Lichtfouse, E., Robert, D., & Wang, C. (2020). Removal of microplastics from the environment. A review. *Environmental Chemistry Letters, 18*, 807–828. https://doi.org/10.1007/s10311-020-00983-1

Peng, X., Urso, M., & Pumera, M. (2021). Photo-Fenton degradation of nitroaromatic explosives by light-powered hematite microrobots: When higher speed is not what we go for. *Small Methods, 5*, 2100617. https://doi.org/10.1002/smtd.202100617

Shimao, M. (2001). Biodegradation of plastics. *Current Opinion in Biotechnology, 12*, 242–247. https://doi.org/10.1016/S0958-1669(00)00206-8

Simmchen, J., Baeza, A., Miguel-Lopez, A., Stanton, M. M., Vallet-Regi, M., Ruiz-Molina, D., & Sa'nchez, S. (2017). Dynamics of novel photoactive AgCl microstars and their environmental applications. *ChemNanoMat, 3*, 65–71. https://doi.org/10.1002/cnma.201600300

Singh, S., Kumar Naik, T. S. S., Anil, A. G., Dhiman, J., Kumar, V., Dhanjal, D. S., AguilarMarcelino, L., Singh, J., & Ramamurthy, P. C. (2022). Micro (nano) plastics in wastewater: A critical review on toxicity risk assessment, behaviour, environmental impact and challenges. *Chemosphere, 290*, 133169. https://doi.org/10.1016/j.chemosphere.2021.133169

Talvitie, J., Mikola, A., Koistinen, A., & Setälä, O. (2017). Solutions to microplastic pollution – Removal of microplastics from wastewater effluent with advanced wastewater treatment technologies. *Water Research, 123*, 401–407. https://doi.org/10.1016/j.watres.2017.07.005

Turchi, C., & Ollis, D. F. (1990). Photocatalytic degradation of organic water contaminants: Mechanisms involving hydroxyl radical attack. *Journal of Catalysis, 122*, 178–192. https://doi.org/10.1016/0021-9517(90)90269-P

Urso, M., & Pumera, M. (2022). Nano/microplastics capture and degradation by autonomous nano/microrobots: A perspective. *Advanced Functional Materials, 32*, 2112120. https://doi.org/10.1002/adfm.202112120

Urso, M., Ussia, M., Novotny´, F., & Pumera, M. (2022). Trapping and detecting nanoplastics by MXene-derived oxide microrobots. *Nature Communications, 13*, 3573. https://doi.org/10.1038/s41467-022-31161-2

Urso, M., Ussia, M., & Pumera, M. (2023). Smart micro- and nanorobots for water purification. *Nature Reviews Bioengineering, 1*, 236–251. https://doi.org/10.1038/s44222-023-00025-9

Velimirovic, M., Tirez, K., Verstraelen, S., Frijns, E., Remy, S., Koppen, G., Rotander, A., Bolea-Fernandez, E., & Vanhaecke, F. (2021). Mass spectrometry as a powerful analytical tool for the characterization of indoor airborne microplastics and nanoplastics. *Journal of Analytical Atomic Spectrometry, 36*, 695–705. https://doi.org/10.1039/D1JA00036E

Vethaak, A. D., & Legler, J. (2021). Microplastics and human health. *Science, 371*, 672–674. https://doi.org/10.1126/science.abe5041

Villa, K., Parmar, J., Vilela, D., & Sa'nchez, S. (2018). Metal-oxide-based microjets for the simultaneous removal of organic pollutants and heavy metals. *ACS Applied Materials & Interfaces, 10*, 20478–20486. https://doi.org/10.1021/acsami.8b04353

Villa, K., Novotny´, F., Zelenka, J., Browne, M. P., Ruml, T., & Pumera, M. (2019). VisibleLight-driven single-component BiVO4 micromotors with the autonomous ability for capturing microorganisms. *ACS Nano, 13*, 8135–8145. https://doi.org/10.1021/acsnano.9b03184

Wang, L., Haq, K. T., & Tereshchenko, L. G. (2021). Self-propelled micro/nanomotors for removal of insoluble water contaminants: Microplastics and oil spills. *ESC Heart Failure, 8*, 3440–3441. https://doi.org/10.1039/D1EN00663K

Wang, K., Li, Y., Xie, L.-H., Li, X., & Li, J.-R. (2022a). Construction and application of basestable MOFs: A critical review. *Chemical Society Reviews, 51*, 6417–6441. https://doi.org/10.1039/D1CS00891A.161

Wang, K., Li, Y., Xie, L.-H., Li, X., & Li, J.-R. (2022b). Construction and application of basestable MOFs: A critical review. *Chemical Society Reviews, 51*, 6417–6441. https://doi.org/10.1039/D1CS00891A

Won, S., Lee, H. E., Cho, Y. S., Yang, K., Park, J. E., Yang, S. J., & Wie, J. J. (2022). Multimodal collective swimming of magnetically articulated modular nanocomposite robots. *Nature Communications, 13*, 6750. https://doi.org/10.1038/s41467-022-34430-2

Wright, S. L., & Kelly, F. J. (2017). Threat to human health from environmental plastics. *BMJ, 358*, j4334. https://doi.org/10.1136/bmj.j4334

Wu, C., Dai, J., Li, X., Gao, L., Wang, J., Liu, J., Zheng, J., Zhan, X., Chen, J., Cheng, X., et al. (2021). Ion-exchange enabled synthetic swarm. *Nature Nanotechnology, 16*, 288–295. https://doi.org/10.1038/s41565-020-00825-9

Xu, L., Mou, F., Gong, H., Luo, M., & Guan, J. (2017). Light-driven micro/nanomotors: From fundamentals to applications. *Chemical Society Reviews, 46*, 6905–6926. https://doi.org/10.1039/C7CS00516D

Yang, X., Shang, W., Lu, H., Liu, Y., Yang, L., Tan, R., Wu, X., & Shen, Y. (2020). An agglutinate magnetic spray transforms inanimate objects into millirobots for biomedical applications. *Science Robotics, 5*, eabc8191. https://doi.org/10.1126/scirobotics.abc8191

Zeng, M., Yuan, S., Huang, D., & Cheng, Z. (2019). Accelerated design of catalytic watercleaning nanomotors via machine learning. *ACS Applied Materials & Interfaces, 11*, 40099–40106. https://doi.org/10.1021/acsami.9b14792

Zhang, Q., Dong, R., Wu, Y., Gao, W., He, Z., & Ren, B. (2017). Light-driven Au-WO3@ C Janus micromotors for rapid photodegradation of dye pollutants. *ACS Applied Materials & Interfaces, 9*, 4674–4683. https://doi.org/10.1021/acsami.6b12081

Zhang, M., Yang, Y., An, X., & Hou, L.-A. (2021). A critical review of g-C3N4-based photocatalytic membrane for water purification. *Chemical Engineering Journal, 412*, 128663. https://doi.org/10.1016/j.cej.2021.128663

Zhou, H., Mayorga-Martinez, C. C., & Pumera, M. (2021). Microplastic removal and degradation by mussel-inspired adhesive magnetic/enzymatic microrobots. *Small Methods, 5*, 2100230. https://doi.org/10.1002/smtd.202100230

Integration of Artificial Intelligence in Nano-ecosystems

Kiruthika Balasubramanian

Abstract

Nanotechnology refers to the use of atoms and molecules at the nanoscale for solving a wide range of problems. Artificial Intelligence (AI) can process large amounts of data, identify unique patterns, and attain impossible insights for the human mind. The advancement in AI, when integrated with nanotechnology, gives us an enhanced insight into the behavior of the materials at the nanoscale. One of the most powerful applications of AI in nanotechnology is the evolution of nanosensors. These AI-integrated nanosensors play a vital role in many fields. In agriculture, it can monitor soil conditions such as nutrients, pH, pest outbreaks, and moisture content and provide precise information, helping farmers make informed decisions on irrigation, harvesting, and disease management. Pollution is a crucial problem in our environment. Nanoparticles integrated with AI algorithms help to optimize the deployment of materials to effectively clean the contaminants from the soil, air, and water. It also enables smart sorting and provides efficient recycling methods in waste management. These AI-integrated nanoparticles can improve energy efficiency by enhancing the performance of solar cells, batteries, and catalysts, which aids in reducing energy consumption. It also supports identifying the precise locations for installing wind turbines and solar panels and assists in generating renewable energy to its fullest. Another pressing problem in today's world is climate change. Artificial nanosystems can help us predict extreme weather events and minimize the impact of natural disasters.

K. Balasubramanian (✉)
New Jersey Academy of Science, Kean University, Union, NJ, USA

D. K. Gupta et al. (eds.), *Nano-Ecosystems*,
https://doi.org/10.1007/978-981-95-5033-3_11

Keywords

Nanosensors · AI algorithm · Energy conservation · Climate change · Agriculture · Waste management

1 Introduction

Everything is made up of atoms—the food we eat, the clothes we wear, and even the air we breathe. Imagine a world where we can just control atoms and molecules. The materials at the nanoscale behave uniquely with characteristics such as massive strength, lightweight, vast control of the light spectrum, and increased chemical reactivity than their larger-scale materials (https://www.nano.gov/nanotech-101/what/definition). When we are studying and dealing with things at the nanoscale, the volume of data we generate is enormous. It needs the right tool to analyze the data at real-time speed to fully understand. With the help of artificial intelligence, data analytics and data mining are easier and can be done beyond the human mind. AI is imperative, as the amount of data generated is mountainous, and a human's ability to handle, interpret, and make the right decision on time is impossible, and the pace at which AI is performing is breathtaking (Zohuri & Mossavar, 2020). AI is a subdivision of computer science that emphasizes the development of superintelligent machines with high performance. It simulates human intelligence and has an analytical and problem-solving capacity. AI encompasses machine learning and deep learning, which involves the development of AI algorithms, simulating the decision-making processes of the human brain which can utilize the available data and make precise decisions at the right time (Wang et al., 2023). AI can complete any task in an unsupervised manner without any human intervention (Hu et al., 2023).

AI algorithms, such as deep learning and neural networks, can analyze data meticulously, identifying hidden correlations and predicting material properties, and assisting in improving their performance. These capabilities significantly led to the discovery and development of nanomaterials and nanodevices without any trial or error, whereas the traditional method of designing materials was time-consuming and limited in scope. AI-integrated nanofabrication systems can boost control over the manufacturing procedure with the least defects and high-quality nanomaterials and nanodevices. These nanofabrication systems can adapt to changing conditions and perform better in real time (How AI Is Transforming the Nanotechnology Field | by Eye Of Unity | Medium). Nanotechnology collaborates with the knowledge of physics, chemistry, and engineering, while artificial intelligence has its basis in the biological inspiration to develop neural networks or evolutionary algorithms. Converging nanotechnology and AI can provide an unimaginable generation of information and communication, which can solve numerous problems in our society (Sacha & Varona, 2013). AI shelters a wide array of technologies, such as deep learning (LeCun et al., 2015), computer vision (Voulodimos et al., 2018), natural language processing (Chowdhary, 2020), and more. These technologies play an important role in analyzing data and making decisions.

2 Artificial Intelligence

AI is a field of study in computer science that makes machines perform tasks that require human intelligence. It requires human skills such as recognizing speech, decision-making, and understanding languages. AI has two subfields too (machine learning and deep learning) (understanding the differences between AI, machine learning, and deep learning (neurosnap.ai)). Machine learning depends on the patterns from raw data and makes decisions based on that learning. It's like teaching machines to learn from their experience. Deep learning is a subset of ML, which tries to imitate the human brain in analyzing data and creating patterns to make decisions. It uses a neural network, just like the human brain uses neurons. It uses large amounts of data and computational power and employs complex architectures like recurrent neural networks and convolutional neural networks.

(https://medium.com/iotforall/the-difference-between-artificial-intelligence-machine-learning-and-deep-learning-3aa67bff5991)

3 Relationship Among AI, ML, and DL

- Both ML and DL are encompassed by AI
- ML mainly focuses on algorithm development that enables machines to learn from data, and it is a subset of AI
- DL is characterized by the utilization of deep neural networks, which is further a subset of ML, as illustrated by Fig.1.

AI can be categorized as general AI and narrow AI. General AI includes all the characteristics of human intelligence, including planning, language understanding, object recognition, sound learning, and the capacity to solve problems. Narrow AI is good at recognizing images and can perform a few facets of human intelligence well, but not all.

4 Applications of AI in Nanomedicine

4.1 Nanomedicine

Traditional methods of treatment assume one size fits all, but later, it is clear that variations in genetics, environment, and lifestyle always have an impact on an individual's response to treatment. This awareness sparked the idea of personalized medicine and an approach to each individual, based on their unique characteristics. The road map for a personalized treatment plan lies in the identification of

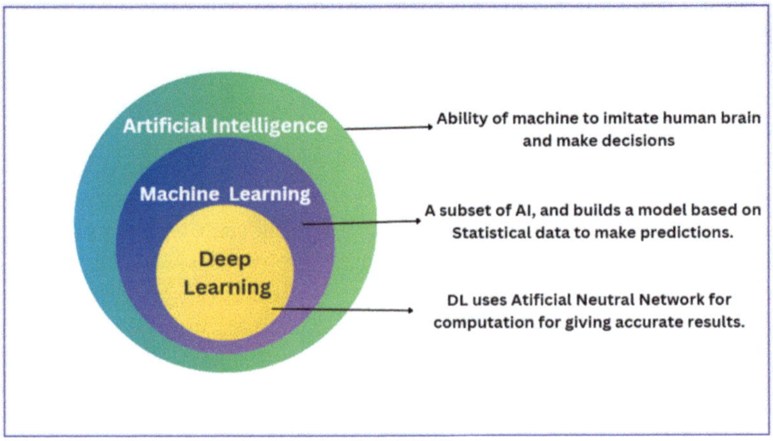

Fig. 1 The interconnection of AL, ML, and DL

disease-relevant biomarkers. Personalized medicine gets its full potential through the integration of AI. AI algorithms can process excessive patient data, including genetic information, medical records, and lifestyle parameters, to plan personalized treatment. By analyzing patterns within this data, AI can identify the most precious hidden insights regarding treatment. Every patient has a unique molecular structure and other differences, such as age, gender, height, and blood groups. This uniqueness leads to differences in response to the same drug among individuals. This is why a lot of research focuses now on personalized medicine.

AI integration in personalized medicine can analyze large amounts of data in a short duration, leading to the discovery of new biomarkers, therapeutic targets, and genetic variants. Nanomedicine is a combination of nanotechnology, biomedical, and pharmaceutical sciences, and it incorporates nanopharmaceutical sciences, theranostics, and nano-imaging techniques. Nanodrug formulations can bestow many physical and biological benefits, such as enhanced solubility and pharmacokinetics, improved efficiency, less toxicity, and a wide range of tissue selectivity compared to conventional medicines. The biochemical, electronic, magnetic, and/or optical properties of a drug formulation can be changed using nanoparticles and can be applied for various therapeutic purposes (Ventola, 2017). In the medical field, integrating nanotechnology and AI paves a great hope for personalized and targeted therapies. The drug delivery system using nanorobots, which can be controlled by AI, will directly target the cells and reduce the side effects caused to normal cells (Peer et al., 2007). AI-powered nanosensors for disease biomarker detection assist in detecting any disease in the early stages and save lives (Ventola, 2017). Cells will behave differently under different stress or environmental conditions. AI systems can easily predict the patterns and the cellular activity or interactions between themselves. Nanomaterials can be easily designed using AI and Nanotechnology. Large-scale biomedical data generated using medications with therapeutic efficiency will be used by AI to predict the efficiency and the cellular

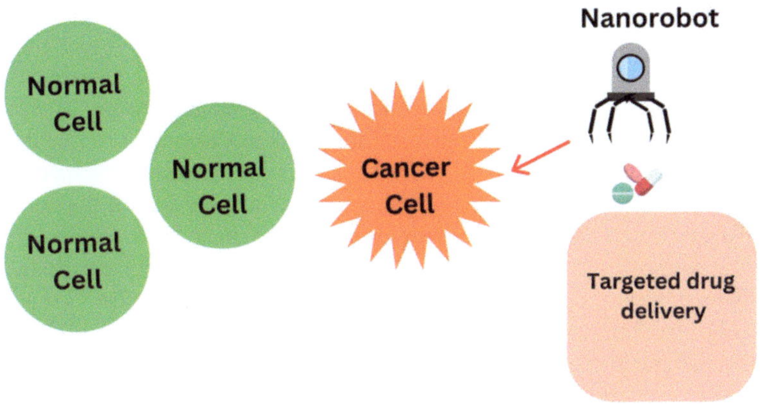

Fig. 2 The role of AI and nanotechnology in drug delivery

response. This helps in reducing the time and expense required for drug development (Vijayan et al., 2022).

Structure-based drug discovery is very much possible by predicting the 3D structures of the lead molecules and by performing drug design optimization. AI has the efficiency to assist with drug design and discovery. It also helps to process the clinical studies data and identify the target precisely (Adir et al., 2020; You et al., 2018). It assists in foreseeing the transport efficiency of the nano-drug. Cancers are very difficult to detect because it is not a single disease. But when the cancer is not detected earlier, it can metastasize to different organs, which is then very challenging to cure. Detecting cancer at an early stage can save people's lives. Some cancers are difficult to detect because of their asymptomatic nature and indistinctive signs on scans and mammograms. AI is more accurate than statistical analysis and multivariate analysis, especially when AI combines with any of the bioinformatics tools, which can significantly improve diagnostic, prognostic, and predictive accuracies (Huang et al., 2020) (Fig. 2).

5 Targeted Drug-Delivery System

AI helps in enhanced image analysis by analyzing the subtle anatomical and pathological changes that can be easily missed by human radiologists (Esteva et al., 2017). AI can discover and validate new biomarkers for various diseases by mining large datasets of clinical and molecular information (Huang et al., 2020). Machine learning models analyze patient data to predict disease risk and enable early intervention, enhancing the chances of successful treatment (Sujatha et al., 2021). By mining large datasets of clinical and molecular information, AI can discover and validate new biomarkers. The early diagnosis of disease and intervention is achievable using machine learning analysis (Huang et al., 2020).

6 Agriculture

In today's world, agriculture faces critical challenges, such as water scarcity, poor environmental conditions, and dependence on conventional energy. The lives of humans and animals are at risk because of intensive transformation and landscaping (Betts et al., 2017., Qi et al., 2018). The rising population, which is expected to be 10 billion by 2050, is placing an enormous amount of pressure on the agriculture sector. To maximize agricultural yields, we should expand land use and perform large-scale farming or adapt to innovative technologies to enhance productivity. Precision agriculture, which is considered to be the third agricultural revolution, mainly relies on digital technology, such as IoT, Airdate analytics, cloud computing, and blockchain technology. This type of farming helps to reduce the cost of production and keep a check on the environmental effects (Stafford, 2000) (Gubbi et al., 2013).

The data related to soil nutrients, fertilizers, climate change, pest infestation, and water requirements are collected using IoT-based smart sensors. Satellite images and robots are used in precision agriculture for identifying diseases in plants and overall monitoring of the field (Nex & Remondino, 2014). The data collected by nanosensors are processed by an AI algorithm to fine-tune the fertilization and irrigation schedules, to provide optimum conditions for the crops. The water and the nutrients will be released to the plants by the nano-enabled smart delivery systems according to the AI recommendation based on data analysis. This decreases the budget by 30% and increases the crop yield by 20%, boosting agricultural productivity and also helping in conserving resources (10 Ways AI Can Be Integrated With Nanotechnology [2024]—DigitalDefynd). Precision agriculture, combined with AI, helps farmers to maximize crop growth with fewer resources. It helps to monitor crops in real-time and easily understand the needs for irrigation, fertilization, or pesticide treatment.

The primary objective of precision agriculture is to select the crops that yield the highest yield with reduced water and fertilizer use without affecting the environment. Automating agricultural processes, like driverless tractors, smart irrigation, IoT-powered agricultural drones, fertilization systems, and AI-based harvesting robots. AI-driven tools are more efficient than humans. There are several tools, such as GIS, GPS, IoT, proximate sensors, and remote sensors, being used. In addition to that, precision irrigation systems, yield monitoring and mapping, variable rate technology, auto guidance equipment, different hardware and software in computers, along with AI, would completely open a new level of agriculture, fulfilling the needs of agriculture for the growing population (Mathur et al., 2023). Potential leaks and anomalies can be identified by analyzing data. Changes in water flow or pressure can be recognized by machine learning models. Along with the crop water requirements data, it incorporates the weather data and helps to identify the areas with excessive water usage (AI in Agriculture and Farming: Revolutionizing Crop Growth—Intellias) (Fig. 3).

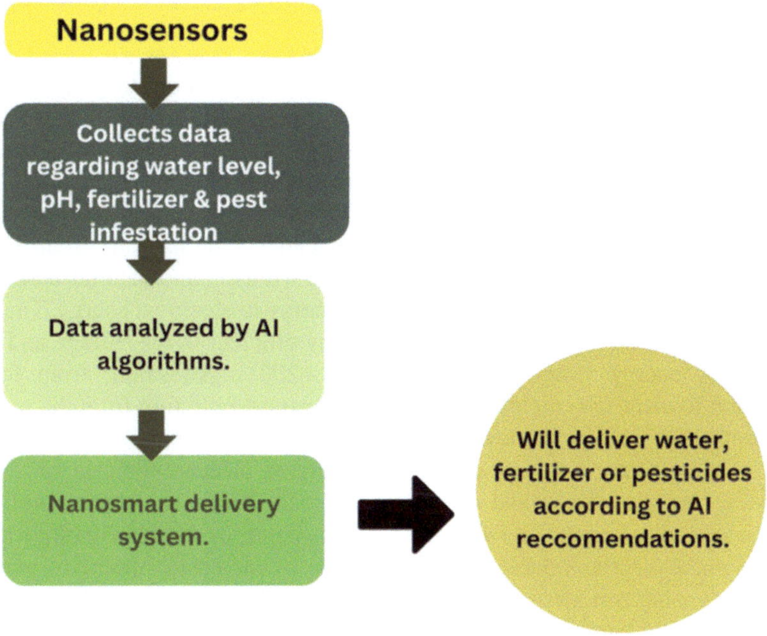

Fig. 3 Role of AI and nanotechnology in agriculture

7 Waste Management

Waste management is a colossal environmental issue, which can contaminate the water, air, and soil. We accumulate lots of waste that is produced through pollution, industrial production, construction, and an increasing population (Chen et al., 2023; Ukaogo et al., 2020). Nanotechnology and AI play an important role in waste management. AI helps in waste sorting and waste-to-energy conversion, and it also helps to enhance efficiency and cost reduction. AI can use its sensor and data analysis capacity, which helps to reduce transportation distance, save costs, and enhance public health in smart cities (Fang et al., 2023). Nanoparticles and nanomaterials can remove contaminants at the molecular level, including heavy metals and pathogens from wastewater, as they can be used in a powerful filtration system. The valuable metals and industrial by-products can be easily recovered from electronic waste through the adsorption process using nanomaterials. They also help with efficient and cost-effective recycling. AI helps in planning and operating the recovery operations and helps to predict waste stream compositions. The nanosensor data helps in optimizing the crucial parameters for the recycling processes. Nanosensors can detect pollutants even at low concentrations in the air, water, and soil. The datasets from nanosensors can be used to detect patterns and trends in population events. Response action can be automated using AI, like activating treatment systems or notifying authorities regarding hazardous conditions.

8 Energy Conservation

The reliance on fossil fuels, which are the major contributors to greenhouse gas emissions and air pollution, can be reduced by integrating hybrid power sources, such as solar energy and wind energy. The energy production and management by AI helps to optimize energy generation, leading to well-planned usage of resources and to low pollution levels. This system contributes to a cleaner environment and reduces pollution (Hseiki et al., 2022). The dependence on natural resources, such as oil, coal, gas, or uranium, can be reduced if we use nanomaterial-based solar cells. Photovoltaic solar cells convert sunlight energy to electrical energy. The semiconductor nanomaterial solar cells can be highly efficient in the storage and transportation of energy (Somrita Mondal et al., 2021). Switching from internal combustion to electric vehicles could vastly reduce the pollution from the vehicles, and also, there is no dependency on oil. Currently, we are depending on fossil fuels for most of our transportation needs. Switching to electric vehicles can open a brand-new source of power and energy management. The main problem with electric vehicles is the capacity and the lifetime of the batteries. Nanotechnology has the ability to make electric vehicles more accessible by increasing the performance of the batteries using nanoparticles and nanocomposite materials as electrolytes.

In lithium-ion batteries, carbon nanotubes elevate electron transport, electrical conductivity, and mechanical stability, while their high surface area improves lithium-ion diffusion, which in turn increases charge transfer. This leads to higher capacity, better cycling stability, and optimized rate performance thus it helps in improving the battery life (https://www.azonano.com/article.aspx?ArticleID=6753).

9 Waste Water Management

The freshwater level on the Earth is less than 1%, and the increasing pollution and population are decreasing its level to an unexpectedly low level. The micro and nanoscale robots, which are self-driven, help to overcome the diffusion-limited reactions, allowing for effective engagement with the target pollutants and helping to clean water efficiently (Preetam, 2024). AI, due to its independent nature, can work without human intervention with high efficiency and speed. AI has its spot in wastewater treatment, too. This technology helps to monitor the performance of wastewater plants by checking biological oxygen demand (BOD), chemical oxygen demand (COD) determination, uptake of contaminants, elimination of nitrogen and sulfur, prediction of turbidity and hardness, etc. The three models under AI predominantly used in the wastewater sector are artificial neural networks (ANN), fuzzy logic algorithms (FL), and genetic algorithms (GA) (Malviya & Jaspal, 2021).

Nanomaterials like nanocomposites, nanoparticles, carbon nanotubes (CNT), and zinc, iron, or titanium oxide nanoparticles have been tremendously used in wastewater and groundwater treatments because of their excessive adsorption capability of pollutants from water due to their broad spectrum of surface area with high reactivity (Singha et al., 2021). This nanotechnology controlled by AI helps to

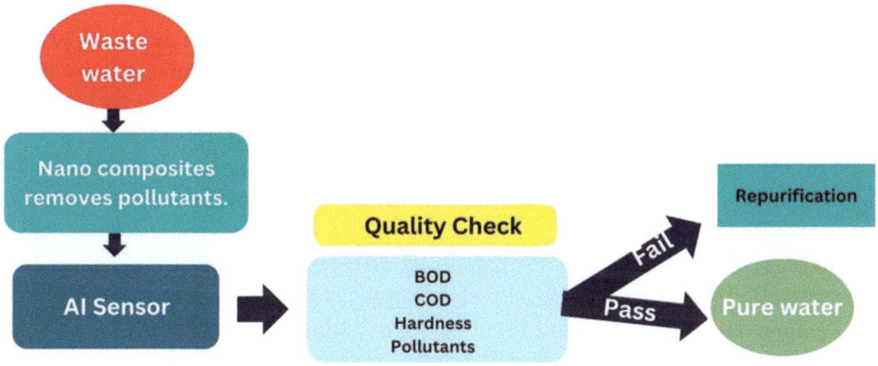

Fig. 4 Role of AI and nanotechnology in wastewater treatment

precisely record the parameters and efficiency of the water plant and also to identify any defect in prior hand in order for the plant to run without any hindrance. Industry 4.0 refers to the monitoring of the wastewater plant using IoT devices and sensors. Nanocomposites are used in different forms, such as membranes, adsorbents, and catalysts, to remove contaminants. Industry 4.0 allows the scalability of the advanced water treatment systems, making them suitable for large- and small-scale plants. Automated control systems improved water cleaning efficiency and reduced human intervention (Uwamungu et al., 2022) (Fig. 4).

10 Climatic Change

Carbon dioxide emissions due to industrial production and the use of fossil fuels in manufacturing processes are leading to climate change, which is one of the biggest problems for mankind. The rise of sea level, the reduced level of crop production, the frequent show up of natural disasters, and the deprivation of biodiversity are all the effects of climate change (Shivanna, 2022).

Global warming is also one of the principal causes of climate change, which needs immediate attention. The Earth's average temperature has been increasing steadily for the past few decades. NOAA's (National Oceanic and Atmospheric Administration, US) data records showed that 2020 was the warmest year (Chausali et al., 2023).

Carbon sequestration and storage are crucial elements of climate change reduction strategies (Liu et al., 2022; Osman et al., 2022; Yang et al., 2023). There are sustainable CO_2 management materials that emphasize amine-based materials, metal-organic frameworks (MOFs), porous organic polymers (POPs), hybrid nanomaterials, and carbon-based adsorbents. These are very powerful as the absorbents can enhance CO_2 capture, with high surface area, and can provide special CO_2 capture capabilities. Integration of AI helps in optimizing the algorithm and supports modeling the CO_2 capture systems with higher capabilities. It also helps in analyzing the material performance, optimizing the synthesis conditions, and outlining the materials with optimal properties for efficient CO_2 capture (Zhang et al., 2022). These approaches help create more effective and measurable solutions for managing CO_2 emissions.

AI helps to measure the melting of the iceberg faster than humans. Scientists can easily understand the amount of iceberg meltwater entering the ocean, a process that accelerates climate change and warms the atmosphere. AI can map the satellite images of the Antarctic icebergs in just one-hundredth of a second, which is so important in determining climate change. AI also helps to track the deforestation levels in the landscape and gives data regarding the stored carbon levels. AI can also predict the accurate location and the time when a climate disaster will occur. This helps the business and the government to better prepare for the challenges caused by disasters. AI, along with drones, can disperse 180 seed capsules per minute, which is 100 times faster than the traditional method of reforestation. Reforestation plays a vital role in controlling climate change (nine ways AI is being deployed to fight climate change | World Economic Forum (weforum.org))

11 Material Science

The combination of nanotechnology and AI is helping in the material revolution that impacts various industries, such as electronics, health care, and manufacturing. Manipulation of materials at the atomic and molecular scale is possible through nanotechnology, while AI provides algorithms for designing, discovering, and optimizing the materials. The material revolution involves the major changes taking place in material science. This revolution involves the use of improved materials to create products. From the Stone, Bronze, and Iron Ages, humans are pushing their limits, and today, material science is highly based on nanotechnology and artificial intelligence.

The development of next-generation batteries with high energy densities, longer life spans, and faster charging times. The prediction of battery failures or degradation can be analyzed by AI from battery usage data. The improvisation of battery reliability and life span extension can be achieved by predictive capability, allowing proactive maintenance and management (https://datafloq.com/read/nanotechnology-ai-impact-materials-revolution/).

12 Conclusion

The integration of AI and nanotechnology represents a cutting-edge, transformative technique that can solve tons of problems. The data from nanosensors and nanodevices are analyzed to tailor personalized treatment plans. AI can design and optimize the nanocarriers for drug delivery, which deliver drugs only to the target cell and not to the normal cells. It also helps in identifying the new biomarkers even at low concentrations for various diseases, which can detect the disease at an early stage. Minimal invasive surgery is also possible by targeting cancer cells at the cellular level. Sensitive detectors can easily monitor the pollution level with AI to get real-time monitoring and help take precautionary measures. The unexpected climatic changes can also be accurately predicted, and the data analysis also provides us with the safest area for people to relocate during extreme climatic conditions. Waste treatment can also be monitored and controlled by AI. Nanomaterials are also effective in recycling waste, and AI helps in sorting the waste efficiently. It also optimizes the design and function of nanomaterials to purify water. The water parameters, like COD, BOD, and hardness, are efficiently monitored by AI, which helps to recycle water to the maximum. The nanomaterials in batteries help to increase the life of batteries and, at the same time, reduce the cost of batteries. This opens up a new platform in transportation, where conventional fuel-based vehicles are replaced by electric cars. This can, in turn, reduce pollution. AI-integrated nanosensors monitor soil conditions, like pH, water, nutrient levels, and pest infestation, and provide a clear idea about the plant conditions. The AI processes these data, and it recommends the amount of water, nutrients, and fertilizer to the plants that the small delivery systems will deliver.

AI helps in restoring the climate by precisely monitoring the melting of the iceberg, which plays an inevitable role in climate change. Deforestation is also a major factor in climate change. AI-installed drones help in reforestation by sowing seeds at a higher speed compared to traditional methods. The seeds can be easily shown in inaccessible areas beyond human reach. Nanotechnology and AI combination also helps in predicting the climatic disaster, which alerts the government and the residents to be prepared to face the challenges. AI usually requires high-quality and precise data to make decisions, which is quite challenging. Integrating AI with nanotechnology raises ethical issues, privacy, and any other potential misuse. Ensuring the safe use of nanomaterials requires proper regulatory frameworks and strict monitoring.

References

Adir, O., Poley, M., Chen, G., Froim, S., Krinsky, N., Shklover, J., Shainsky-Roitman, J., Lammers, T., & Schroeder, A. (2020). Integrating artificial intelligence and nanotechnology for precision cancer medicine. *Advanced Materials, 32*, 1901989. [Crossref], [Google Scholar], [Publisher].

Betts, M. G., et al. (2017). Global forest loss disproportionately erodes biodiversity in intact landscapes. *Nature, 547*(7664), 441–444.

Chausali, N., Saxena, J., & Prasad, R. (2023). Nanotechnology as a sustainable approach for combating the environmental effects of climate change. *Journal of Agriculture and Food Research, 12*, 100541. https://doi.org/10.1016/j.jafr.2023.100541

Chen, L., Huang, L., Hua, J., Chen, Z., Wei, L., Osman, A. I., Fawzy, S., Rooney, D. W., Dong, L., & Yap, P.-S. (2023). Green construction for low-carbon cities: A review. *Environmental Chemistry Letters*. https://doi.org/10.1007/s10311-022-01544-4

Chowdhary, K. R. (2020). Natural language processing. In *Fundamentals of artificial intelligence* (pp. 603–649). Springer. https://doi.org/10.1007/978-81-322-3972-7_19

Digital Defynd, 10 Ways AI Can Be Integrated With Nanotechnology. Available at: https://digitaldefynd.com

Environmental News Service, New Nanotech Batteries Allow Super-Fast Charging.
Environmental Protection Agency, Lithium-ion Batteries and Nanotechnology for Electric Vehicles: A Life Cycle Assessment.
Esteva, A., Kuprel, B., Novoa, R. A., Ko, J., Swetter, S. M., Blau, H. M., & Thrun, S. (2017). Dermatologist-level classification of skin cancer with deep neural networks. *Nature, 542*, 115.
Eye Of Unity, How AI is Transforming the Nanotechnology Field. Medium. Available at: https://www.nano.gov/nanotech-101/what/definition
Eye Of Unity, Understanding the Differences between AI, Machine Learning, and Deep Learning. Available at: https://medium.com/iotforall/the-difference-between-artificial-intelligence-machine-learning-and-deep-learning-3aa67bff5991
Fang, B., Yu, J., Chen, Z., et al. (2023). Artificial intelligence for waste management in smart cities: A review. *Environmental Chemistry Letters, 21*, 1959–1989. https://doi.org/10.1007/s10311-023-01604-3
Gubbi, J., Buyya, R., Marusic, S., & Palaniswami, M. (2013). Internet of things (IoT): A vision, architectural elements, and future directions. *Future Generation Computer Systems, 29*(7), 1645–1660.
Hseiki, H., Bazzi, H., Kassem, R., El-hajj, A., & Haidar, A. (2022). AI to preserve energy and environment. In *2022 international conference on smart systems and power management (IC2SPM)* (pp. 29–34). https://doi.org/10.1109/IC2SPM56638.2022.9988899
Hu, Y., Yang, J., Chen, L., Li, K., Sima, C., Zhu, X., Chai, S., Du, S., Lin, T., Wang, W., et al. (2023). Planning-oriented autonomous driving. In *Proceedings of the 2023 IEEE/CVF conference on computer vision and pattern recognition (CVPR)* (pp. 17853–17862). IEEE.
Huang, S., Yang, J., Fong, S., & Zhao, Q. (2020). Artificial intelligence in cancer diagnosis and prognosis: Opportunities and challenges. *Cancer Letters, 471*, 61. [Crossref], [Google Scholar], [Publisher].
LeCun, Y., Bengio, Y., & Hinton, G. (2015). Deep learning. *Nature, 521*(7553), 436–444. https://doi.org/10.1038/nature14539
Liu, T., Chen, L., Yang, M., Sandanayake, M., Miao, P., Shi, Y., & Yap, P.-S. (2022). Sustainability considerations of green buildings: A detailed overview on current advancements and future considerations. *Sustainability, 14*, 14393. https://doi.org/10.3390/su142114393
Malviya, A., & Jaspal, D. (2021). Artificial intelligence as an upcoming technology in wastewater treatment: A comprehensive review. *Environmental Technology Reviews, 10*(1), 177–187. https://doi.org/10.1080/21622515.2021.1913242
Mathur, R., Katyal, R., Gunwal, I., & Chandra, S. (2023). Artificial Intelligence in Sustainable Agriculture. *International Journal for Research in Applied Science & Engineering Technology (IJRASET), 11*(6), 4047.
Mondal, S., Manna, A. S., & Maiti, D. K. (2021). Nanotools and devices in solar power energy. In S. Devasahayam & C. M. Hussain (Eds.), *Nano tools and devices for enhanced renewable energy, Micro and Nano Technologies* (pp. 429–446). Elsevier. https://doi.org/10.1016/B978-0-12-821709-2.00012-8
Nandipati, M., Fatoki, O., & Desai, S. (2024). Bridging Nanomanufacturing and artificial intelligence—A comprehensive review. *Materials, 17*, 1621. https://doi.org/10.3390/ma1707162
Nex, F., & Remondino, F. (2014). UAV for 3D mapping applications: A review. *Applied Geomatics, 6*(1), 1–15.
Osman, A. I., Chen, L., Yang, M., Msigwa, G., Farghali, M., Fawzy, S., Rooney, D. W., & Yap, P.-S. (2022). Cost, environmental impact, and resilience of renewable energy under a changing climate: A review. *Environmental Chemistry Letters, 21*, 741–764. https://doi.org/10.1007/s10311-022-01532-8
Peer, D., Karp, J. M., Hong, S., Farokhzad, O. C., Margalit, R., et al. (2007). Nanocarriers as an emerging platform for cancer therapy. *Nature Nanotechnology, 2*(12), 751–760.
Preetam, S. (2024). Nano revolution: Pioneering the future of water reclamation with micro-/nano-robots. *Nanoscale Advances, 6*(10), 2569–2581. https://doi.org/10.1039/D3NA01106B

Qi, X., Fu, Y., Wang, R. Y., Ng, C. N., Dang, H., & He, Y. (2018). Improving the sustainability of agricultural land use: An integrated framework for the conflict between food security and environmental deterioration. *Applied Geography, 90*, 214–223.

Sacha, G. M., & Varona, P. (2013). Artificial intelligence in nanotechnology. In *Grupo de NeurocomputaciónBiológica, Escuela Politécnica Superior, Universidad Autónoma de Madrid.* Cantoblanco. Spain E-28049.

Shivanna, K. R. (2022). Climate change and its impact on biodiversity and human welfare. *Proceedings of the Indian National Science Academy, 88*, 160–171. https://doi.org/10.1007/s43538-022-00073-6

Singha, K., Maity, S., Pandit, P., Maiti, S., & Lakshmanan, S. O. (2021). Nanotechnologies for wastewater treatment. In S. Senthilkannan Muthu (Ed.), *Sustainable Technologies for Textile Wastewater Treatments, the textile institute book series* (pp. 1–12). Woodhead Publishing. https://doi.org/10.1016/B978-0-323-85829-8.00009-2

Stafford, J. V. (2000). Implementing precision agriculture in the 21st century. *Journal of Agricultural Engineering Research, 76*(3), 267–275.

Sujatha, R., Chatterjee, J. M., Jhanjhi, N. Z., & Brohi, S. N. (2021). Performance of deep learning vs machine learning in plant leaf disease detection. *Microprocessors and Microsystems, 80*, 103615. [Crossref], [Google Scholar], [Publisher].

Ukaogo, P. O., Ewuzie, U., & Onwuka, C. V. (2020). Environmental pollution: Causes, effects, and the remedies. In P. Chowdhary, A. Raj, D. Verma, & Y. Akhter (Eds.), *Microorganisms for sustainable environment and health* (pp. 419–429). Elsevier. https://doi.org/10.1016/B978-0-12-819001-2.00021-8

Uwamungu, J. Y., Kumar, P., Alkhayyat, A., Younas, T., Capangpangan, R. Y., Alguno, A. C., & Ofori, I. (2022). Management by industry 4.0 integrated nanocomposite manufacturing. https://doi.org/10.1155/2022/5316228

Ventola, C. L. (2017). Progress in nanomedicine: Approved and investigational nanodrugs. *P&T: A Peer-Reviewed Journal for Formulary Management, 42*(12), 742–755.

Vijayan, R. S. K., Kihlberg, J., Cross, J. B., & Poongavanam, V. (2022). Enhancing preclinical drug discovery with artificial intelligence. *Drug Discovery Today, 27*, 967. [Crossref], [Google Scholar], [Publisher].

Voulodimos, A., Doulamis, N., Doulamis, A., et al. (2018). Deep learning for computer vision: A brief review. *Computational Intelligence and Neuroscience, 2018*, 7068349. https://doi.org/10.1155/2018/7068349

Wang, H., Fu, T., Du, Y., Gao, W., Huang, K., Liu, Z., Chandak, P., Liu, S., Van Katwyk, P., Deac, A., et al. (2023). Scientific discovery in the age of artificial intelligence. *Nature, 620*, 47–60. [CrossRef] [PubMed].

Yang, M., Chen, L., Wang, J., Msigwa, G., Osman, A. I., Fawzy, S., Rooney, D. W., & Yap, P.-S. (2023). Circular economy strategies for combating climate change and other environmental issues. *Environmental Chemistry Letters, 21*, 55–80. https://doi.org/10.1007/s10311-022-01499-6

You, J., Liu, B., Ying, Z., Pande, V., & Leskovec, J. (2018). Graph convolutional policy network for goal directed molecular graph generation. *Advances in Neural Information Processing Systems, 31*. [Google scholar], [Publisher].

Zhang, Z., Zheng, Y., Qian, L., Luo, D., Dou, H., Wen, G., Yu, A., & Chen, Z. (2022). Emerging trends in sustainable CO2-management materials. *Advanced Materials, 34*(29), 2201547. https://doi.org/10.1002/adma.202201547

Zohuri, B., & Mossavar, F. R. (2020). Artificial intelligence versus human intelligence: A new technological race. *Acta Scientific Pharmaceutical Sciences, 4*(5), 50–58.

An Overview of Multiscale Modeling and Simulation of Nano-ecosystems

K. Sheela

Abstract

It is essential to model and simulate the changes in shape and structure of polymer blends at the nanoscale that do not mix well during the process of injection molding. This is important for accurately forecasting and adjusting the performance of the final products. This chapter examines the latest advancements in the modeling and simulation of injection molding of polymer blends at many scales. A comprehensive assessment was conducted on the technological advancements in macroscale injection molding simulation. The discussion focused on several models that explain the evolution of morphology at a mesoscale level during the process of injection molding. An analysis and comparison were conducted on the existing strategies that bridge the gap between macroscopic mold-filling flow and mesoscopic morphological evolution. The advantages and disadvantages of these solutions were also examined. Ultimately, a thorough overview of the aforementioned models is provided, along with the prospects for future investigation in this domain.

Keywords

Molecular dynamics · FRP composites · 3D printer · Nanostructure

1 Introduction

Polymer materials that have been strengthened with nanoparticles, such as nanospheres, nanotubes, and clay platelets, have garnered significant interest in scientific and industrial circles. This is because they exhibit exceptional improvements in

K. Sheela (✉)
PG & Research Department of Biochemistry, Tirupattur, Tamil Nadu, India
e-mail: sheela@shctpt.edu

D. K. Gupta et al. (eds.), *Nano-Ecosystems*,
https://doi.org/10.1007/978-981-95-5033-3_12

their properties (Yu et al., 2014). Nevertheless, when considering the experimental aspect, it is indeed a significant undertaking to analyze the composition and control the production of polymer nanocomposites (Tam et al., 2017). The development of such materials is currently based primarily on empirical methods, and achieving a more precise control over their properties has not yet been possible. Consequently, simulation and modeling using computers will increasingly influence the prediction and design of material characteristics, as well as direct experimental endeavors, such as synthesis and characterization. Computer modeling and simulation are very valuable in tackling the fundamental difficulties related to polymer nanocomposites (Tam et al., 2017).

This chapter explores the application of modeling and simulation techniques to the study of polymer nanocomposites. This is a broad field that includes methods addressing diverse length and time scales, as well as multiple characteristics of these materials. The discussion is organized as follows.

It begins with a concise summary of the computational approaches commonly employed in this area. These techniques are grouped into three primary categories based on their scale: molecular-scale methods (e.g., molecular dynamics and Monte Carlo simulations), microscale methods, and mesoscale/macroscale methods. The intention is not to provide a detailed theoretical account of each method but rather to outline their core principles, discuss their respective strengths and limitations, and indicate their potential applications.

Progress in nano-engineering over recent decades has enabled detailed investigation of complex phenomena in materials at the nanoscale. These advances have facilitated the creation of stronger materials with unique properties. As construction materials are deeply integrated into daily life, their sustainability has become a growing concern. Consequently, the mechanical performance and long-term durability of these materials have been a major research focus. It has been observed that some construction materials may perform adequately initially but undergo substantial deformation or failure over time when exposed to expected service loads. This is often a result of material degradation or adverse interactions with the environment. Since material deformation originates at the atomic level, it is often difficult to observe using conventional experimental methods or continuum theories (Allen & Tildesley, 1989).

The advancement of construction materials has faced considerable challenges in recent decades. However, the introduction of nano-engineering concepts, particularly through simulation, has emerged as a powerful tool with the potential to transform research in this field (Ritschl et al., 2002). In response to the need for construction materials with enhanced performance and greater resistance to degradation, simulation techniques have been rapidly developed. The use of nano-engineering allows researchers to pinpoint the fundamental failure mechanisms within material systems (Sun, 1995), providing critical insights for structural design that are applicable beyond civil engineering.

The crucial significance of multiple-scale modeling and simulation techniques for comprehending and forecasting the hierarchical structures and characteristics that result from the combination of polymers and nanoparticles is one of the main themes. We evaluate the current state of various multiscale methods as applied to polymer nanocomposites (Van Duin et al., 2001).

2 Modeling and Simulation Techniques

2.1 Molecular Scale Methods

At the molecular level, the fundamental building blocks of modeling and simulation methods are usually atoms, molecules, or clusters. Monte Carlo (MC) simulation and molecular mechanics (MM) are the most commonly used techniques in this area (Vandamme & Ulm, 2009). The thermodynamics and kinetics of polymer nanocomposites' creation, as well as intricate molecular structures and interactions, are the main topics of modeling at this level.

In molecular dynamics (MD) simulations, the material is represented as an assembly of atoms. These atoms are treated as classical particles, meaning their behavior follows Newton's laws of motion rather than quantum mechanics (Shahsavari et al., 2016a). This approach generates a trajectory for each atom, revealing how the material's nanostructure evolves over time under specific external conditions. The core algorithm involves the iterative calculation of three key parameters like position, velocity, and acceleration, to numerically solve Newton's second law. To make these repetitive calculations more efficient, several numerical integration methods have been developed, such as Verlet, leapfrog, velocity Verlet, and predictor-corrector algorithms (Shahsavari & Chen, 2015). Each of these methods has its own strengths and weaknesses, making them suitable for different material types.

The accuracy of an MD simulation is heavily dependent on the force field, which defines the interatomic interactions. Numerous specialized force fields exist for different material classes, such as distinct ones for inorganic or organic polymers. However, a single force field directly describing generic construction materials is not available. Instead, force fields for these complex systems often rely on common functional forms for bonded and nonbonded energies, allowing the same interaction models to be applied across structurally similar components. When simulating material systems where no chemical bonds are formed or broken, nonreactive force fields are used to represent the interactions between the various constituents (Shahsavari et al., 2016a, b).

Under specific conditions, such as when atoms are released and new chemical phases form, the connections between components can change. In these scenarios, which involve chemical reactions, nonreactive force fields are no longer adequate (Wu et al., 2011). For simulating such processes, where chemical bonds break and form, the ReaxFF reactive force field can be used. This capability of reactive molecular dynamics (MD) enables the investigation of chemical reaction mechanisms, including catalysis, pyrolysis, and combustion (Alex et al., 2015).

Refinements to force fields for various interactions are essential to accurately describe not only the interactions between different material components but also those between materials and their external environment, such as water or other solvent molecules. These interactions are responsible for structural changes in construction materials and can have long-term consequences for their properties (Wu et al., 2016).

The outcomes of MD simulations can offer important new information about the molecular-level characteristics of common building materials, a topic that has received little attention lately. However, applying MD to building materials presents significant challenges. A primary obstacle is extracting meaningful information from nanoscale physical and chemical processes and, crucially, linking this nanoscale data to the material's overall macroscale performance (Mishra et al., 2013). Although MD simulations have seen growing use in construction materials research, the technique is inherently constrained by its limited time and length scales. Operating typically at the nanosecond (10^{-9} s) and nanometer (10^{-9} m) scale, MD is not directly suitable for studying bulk construction materials, which are far larger and whose long-term performance is key for sustainable design (Manzano et al., 2015).

A proposed strategy to overcome these limitations in scale is the adoption of multiscale modeling for construction materials (Fig. 1). This approach seeks to bridge the gap, connecting MD results with microscopic observations to comprehensively explain material behaviors. Recent advances in computational power and methods have begun to make it possible to capture material behaviors on millisecond timescales. These very limitations of MD serve as a further incentive for integrating nanoengineering principles into the development of new construction materials (Huang et al., 2015).

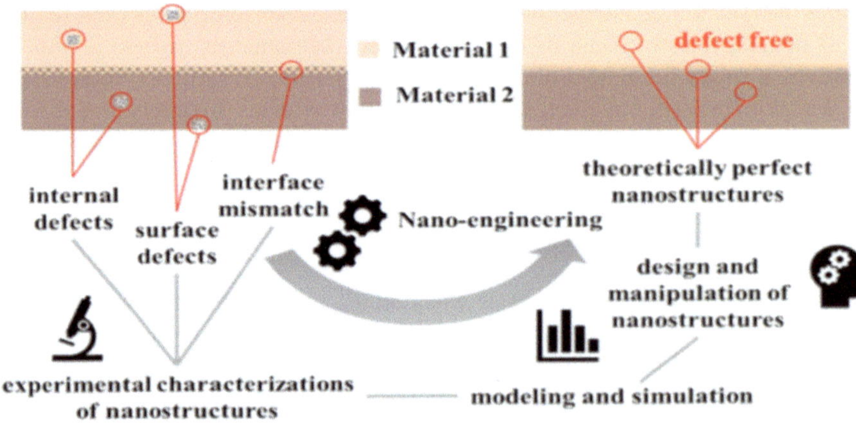

Fig. 1 Proposed implementation of nano-engineering processes for enhancing material properties

MD simulations are employed to reveal the fundamental physical and chemical processes that govern the performance of materials at the macroscopic level. Additionally, progress in bioinspired materials provides a new direction for future research in construction materials. The evolution of these materials should be guided by actively learning from the inherent properties of biomaterials. Also, significant challenges associated with using MD simulations in nano-engineering need to be assessed which are crucial for guiding future research and development efforts (Padilla Espinosa et al., 2017).

3 Fiber-Reinforced Polymer

Conventional construction materials like concrete and wood often undergo unwanted deformation and deterioration when subjected to mechanical loads and varying environmental conditions. Fiber-reinforced polymer (FRP) composites have emerged as a preferred solution for strengthening and rehabilitating such structures to prolong their service life. Their suitability stems from a combination of high strength-to-weight and stiffness-to-weight ratios, corrosion resistance, and the practicality of installation and customization.

The mechanical performance of both FRP composites and the systems where they are bonded to other materials is fundamentally governed by the integrity of two key interfaces: the one between the fiber and the polymer matrix within the composite and the one between the FRP layer and the underlying substrate (Wang et al., 2015). While these interfacial bonds have been investigated through various experiments and numerical models, questions remain regarding the long-term durability and resilience of these systems—critical factors for design. Conventional macroscale observational techniques are insufficient for uncovering the profound influence of fiber-matrix interface characteristics or the fundamental mechanisms behind deformation. This knowledge gap appears to limit the broader practical application of FRP composites in construction (Zhao et al., 2017).

In this context, nano-engineering methods based on molecular modeling have arisen as a powerful alternative, offering deeper insight into processes at the nanoscale. The capacity to model and manipulate the atomic structure of FRP, study inter-component interactions, and extract molecular-level data that correlates with macro-properties through MD simulations enables a fresh understanding of nano-mechanical behavior. This paves the way for designing FRPs with tailored functionalities. In a bonded FRP system, the integrity of the interface is paramount, especially when considering long-term durability under environmental stressors. Initially, research in this area relied heavily on experimental methods paired with empirical models (Manzano et al., 2011). While this approach can help quantify degradation processes, it often fails to identify the root cause of the deterioration.

A significant milestone was reached in 2011 with a publication that introduced MD simulation to the field of structural mechanics and engineering. This work aimed to provide a fundamental understanding of atomic-scale behavior and demonstrate its relevance to structural engineering applications (Cygan et al., 2004). It presented foundational information on MD and examined an organic-inorganic bilayer system to accurately characterize the concrete-epoxy interface in the

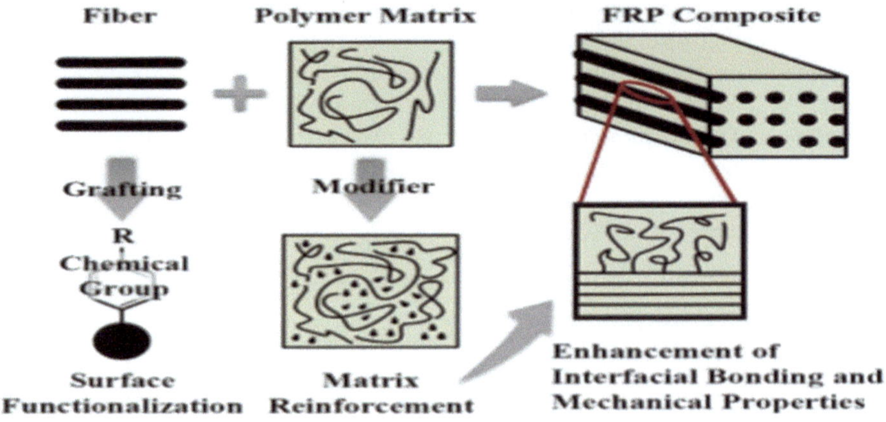

Fig. 2 Schematic layout of FRP composites containing two main components: fiber and matrix

presence of moisture. Predicting the behavior of this interface is crucial for ensuring structural design safety and for giving engineers confidence in using multi-material solutions like FRP-strengthened concrete.

This was followed by a 2012 study on a multiscale modeling approach, which demonstrated a seamless transfer of information from the atomistic to the continuum scale. It achieved this through a hierarchical strategy that integrated MD simulations with finite element modeling. This study shows the models developed for FRP composites and the application of MD simulations to understand the structure-property relationship at the fiber-matrix interface. Furthermore, it illustrates how MD simulations can be used to pinpoint the deformation mechanisms at the FRP composite interface. This nano-engineering-based technique is vital for securing the long-term durability and sustainability of FRP applications (Shahsavari et al., 2011).

The molecular modeling of fiber-reinforced polymers (FRPs) employs a hierarchical strategy that reflects the structure and makeup of their constituent parts. Typically, these composites are modeled by combining a polymer matrix, like epoxy, with various fiber types, including carbon (CF), glass (GF), aramid (AF), and basalt (BF), as illustrated in Fig. 2. To simulate carbon fiber, graphite, composed of multiple graphene layers, is frequently used. Graphene's simple and modifiable structure allows for the incorporation of material defects into the model. These atomistic models enable the simulation of the fibers' mechanical properties under manufacturing and service conditions. The resulting composite's mechanical response is influenced by the concentration, aspect ratio, and distribution of the graphene. Furthermore, MD simulations can probe the thermal properties of graphene-epoxy nanocomposites.

For glass fiber, silica serves as the foundational model, being its primary component. However, the chemical complexity of aramid and basalt fibers means that modeling often considers only their fundamental building block, such as Kevlar for aramid fibers. A significant challenge for a comprehensive nano-engineering analysis is the lack of complete atomistic data for these fibers. The modeling of the polymer matrices follows a similar logic, based on their material structures. Given the prevalence of thermosetting polymers, the simulation must accurately represent the crosslinking process. Several computational techniques have been developed to generate realistic crosslinked structures. The crosslink density achieved by these different methods significantly influences the properties of the polymer itself and, consequently, the performance of the final fiber-composite (Heinz et al., 2013).

3.1 FRP Bonded Interface

The performance of structures strengthened with fiber-reinforced polymer (FRP) composites is critically dependent on the interface between the FRP and the substrate. This zone acts as a fundamental junction between two dissimilar materials and often represents the weak point in the system. Flaws at this interface directly compromise the long-term durability of FRP-reinforced concrete. Furthermore, environmental factors like moisture and temperature can cause localized degradation of the FRP bond strength.

Molecular dynamics (MD) simulations, as a nano-engineering tool, make it possible to observe the process of interfacial failure at the atomistic scale. This approach allows researchers to characterize the interface's behavior and pinpoint the molecular mechanisms behind its deformation. Such a bottom-up investigation requires an accurate model of the interfacial structure. A typical model used for this purpose, shown in Fig. 3, represents the bonded system with the epoxy monomer diglycidyl ether of bisphenol A (DGEBA) which contains its essential functional groups and a silica substrate. Silica is chosen to represent the concrete substrate because, as noted previously, it is a major component of concrete aggregates.

Studies have demonstrated that in wet conditions, solvents like water and salt solutions reduce the adhesion energy between the epoxy and silica, thereby weakening the bond. This reduction is a key factor in evaluating the system's long-term endurance. The simulation results not only corroborate the damaging effects of moisture and salt observed in large-scale experiments but also provide a molecular-level explanation for these phenomena (Lothenbach & Nonat, 2015).

The same computational model has been applied to investigate creep behavior at the material interface. This model successfully captures the two distinct modes of molecular movement—chain extension and interfacial sliding—each governed by separate energy barriers. In real-world applications, epoxy forms a crosslinked polymer network, which is accounted for in these simulations. A more detailed model was developed to further analyze moisture intrusion, confirming its consistently detrimental effect on FRP-bonded interfaces. These simulations clarify the molecular interactions at the interface under environmental exposure and enable predictions of adhesive performance.

For wood structures strengthened with FRP, comprehensive mechanical testing is essential. Molecular dynamics (MD) simulations have been used to study how moisture affects the interfacial behavior in FRP-wood systems. A coarse-grained (CG) model was constructed for a multiscale analysis of an epoxy-wood bond,

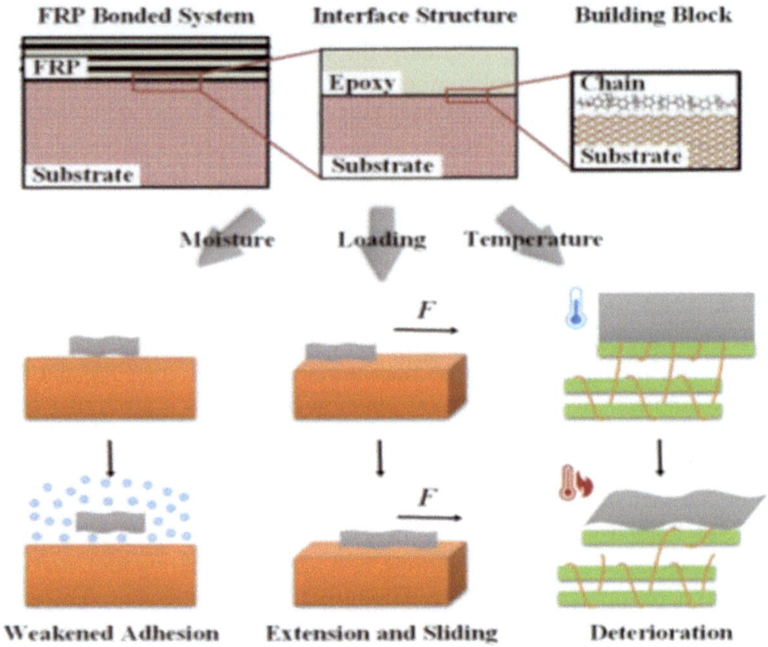

Fig. 3 FRP bonded interfacial structures at different length scale

allowing researchers to study the influence of temperature on the material's structure. The results from modeling the FRP-wood interface provide valuable insights into the mechanisms of debonding and structural degradation, revealing the fundamental deformation processes that are not visible at the macroscopic scale.

The mechanical performance of FRP-bonded systems is also significantly influenced by sustained external loads. Understanding the time-dependent behavior of both the FRP composite and the bonded interface under long-term loading is a critical consideration for design. MD modeling proves particularly valuable here, as it can simulate the creep response at the interface, thereby contributing to a more accurate prediction of the system's long-term durability (Al-Ostaz et al., 2010).

3.2 Bioinspired Design

The sophisticated structures found in natural materials, refined through millennia of adaptation and selection, are responsible for their multifunctional properties. These biological designs provide both functional blueprints and inspiration for engineering advanced material architectures. While human-made materials like concrete and FRP composites rely on artificially synthesized components to enhance their properties, our understanding of the underlying enhancement mechanisms remains limited. In contrast, natural materials have continuously evolved their internal and external structures to effectively withstand environmental stresses.

These biologically optimized structures offer valuable insights for developing novel materials with specialized functions, similar to how radar technology mimics bat echolocation. Biological specimens such as mollusk shells and wood, which demonstrate exceptional mechanical properties, have stimulated contemporary research into bioinspired materials and provided solutions for creating simultaneously strong and tough synthetic materials (Thomas et al., 2010).

By studying these natural systems, researchers have successfully translated biological structures into practical applications. Materials like bone and spider silk have been extensively investigated, with findings subsequently applied to emerging technologies such as 3D printing. Scientists have employed additive manufacturing to create synthetic materials that replicate biomaterial structures, resulting in significantly improved material characteristics. The integration of engineering expertise into construction practices has progressively advanced, recently culminating in the application of 3D printing for building fabrication. This innovative approach utilizes specialized printing equipment and tailored materials. The combination of nano-engineering principles with knowledge of biological materials enables the creation of precisely fabricated, economically viable 3D-printed structures. This technological synergy facilitates construction in challenging or hazardous environments where conventional methods prove impractical.

To fully comprehend the capabilities of natural materials, a thorough investigation of their mechanical performance and structural behavior is essential. These materials typically feature complex, hierarchically organized architectures that have been the subject of numerous research efforts aimed at both understanding and computationally replicating them (Jennings & Bullard, 2011).

The strategic use of hierarchical structures has advanced the development of new composite materials by combining different constituents in innovative arrangements. A crucial step in this process involves identifying the essential structural building blocks present at different size scales. These fundamental units can then inform the design of bioinspired materials or structures tailored to the demands of modern engineering. The nano-engineering perspective is foundational, as mechanical properties originate from molecular-level interactions. This approach allows for the integration of nanoscale characteristics that are dictated by the constituent functional groups. The methodology for this bioinspired design is outlined in Fig. 4. This section offers a detailed overview of nanoscale research on biomaterials and wood. Molecular dynamics (MD) simulations are particularly valuable here, as they provide a way to understand the structural hierarchy of natural materials in ways that experiments alone cannot. The insights gained from the nanoscale can be directly transferred to structural design, creating a powerful link between theoretical design and practical implementation. This is evident in applications inspired by plants and insects—such as hydrophobic surfaces and honeycomb structures—that are now used in coatings, construction, and devices. Recent research has continued to explore the structural and mechanical responses of these biomaterials (Pellenq et al., 2009).

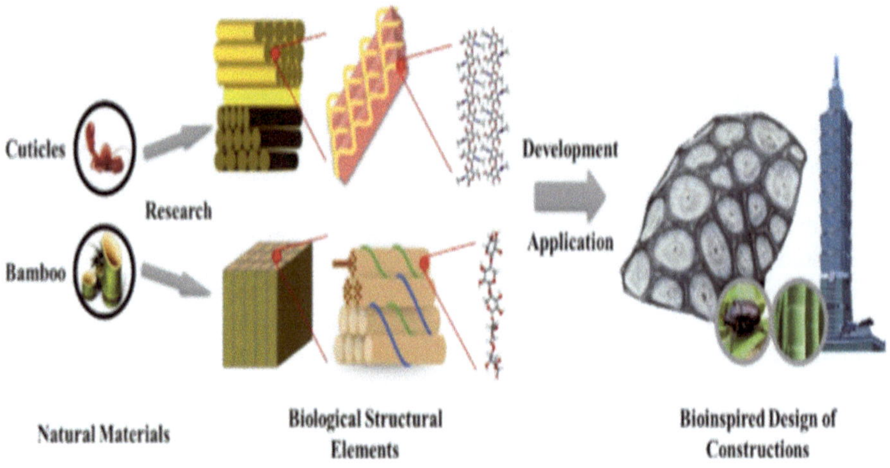

Fig. 4 Schematic diagram showing the process of bioinspired design

4 Conclusion

This chapter has examined advancements in the use of nano-engineering to construction materials, concentrating on three particular material systems. Nano-engineering is an important and relatively new technique that has the potential to be a key technology in creating the building materials of the future. The intricate chemical and physical mechanisms at the nanoscale that control the behavior of macroscopic materials have been made visible in this discipline through the use of molecular dynamics (MD) simulations.

The accuracy of material system modeling and the thoughtful choice of force fields are essential for the success of nano-engineering using MD simulations. The development of models for concrete, FRP composites, as well as bioinspired composites has advanced significantly during the last 10 years, suggesting a promising future. Molecular-level details, deformation processes, and interfacial interactions in these building materials have all been successfully elucidated by MD simulations. However, a significant challenge remains in accurately representing the interactions between different materials. This gap necessitates further research that accounts for intricate microstructures and complex chemical reactions in future work. Currently, the application of MD simulations in engineering is still at an early stage, primarily due to the difficulty in directly applying nanoscale findings to real-world scenarios. Understanding building materials at the nanoscale is made possible by the nano-engineering approach, which allows for a comprehensive comprehension of their mechanical and structural characteristics. The development of building materials in the coming century is expected to depend on multiscale models combined with improved simulation techniques that provide more precise depictions of interactions and structures.

Acknowledgments The author K. Sheela would like to acknowledge the Principal and Management of Sacred Heart College (Autonomous), Tirupattur, Tamil Nadu, India, for providing a financial support of Don Bosco Research Grant [Project ref. No. SHC/DB Grant/2023/04].

Conflict of Interest The authors declare no conflict of interest.

References

Alex, A., Sindu, B. S., & Sasmal, S. (2015). Uniaxial tension and compression studies on cement clinkers using molecular dynamic simulations. *Journal of Structural Engineering, 42*(1), 22–27.

Allen, M. P., & Tildesley, D. J. (1989). *Computer simulation of liquids*. Oxford University Press.

Al-Ostaz, A., Wu, W., Cheng, A.-D., & Song, C. (2010). A molecular dynamics and microporomechanics study on the mechanical properties of major constituents of hydrated cement. *Composites Part B: Engineering, 41*(7), 543–549.

Cygan, R. T., Liang, J. J., & Kalinichev, A. G. (2004). Molecular models of hydroxide, oxyhydroxide, and clay phases and the development of a general force field. *Journal of Physical Chemistry B, 108*(4), 1255–1266.

Heinz, H., Lin, T.-J., Mishra, R. K., & Emami, F. S. (2013). Thermodynamically consistent force fields for the assembly of inorganic, organic, and biological nanostructures: The INTERFACE force field. *Langmuir, 29*(6), 1754–1765.

Huang, J., Wang, B., Yu, Y., Valenzano, L., Bauchy, M., & Sant, G. (2015). Electronic origin of doping-induced enhancements of reactivity: Case study of tricalcium silicate. *Journal of Physical Chemistry C, 119*(46), 25991–25999.

Jennings, H. M., & Bullard, J. W. (2011). From electrons to infrastructure: Engineering concrete from the bottom up. *Cement and Concrete Research, 41*(7), 727–735.

Lothenbach, B., & Nonat, A. (2015). Calcium silicate hydrates: Solid and liquid phase composition. *Cement and Concrete Research, 78*, 57–70.

Manzano, H., Durgun, E., Abdolhosseine Qomi, M. J., Ulm, F.-J., Pellenq, R. J. M., & Grossman, J. C. (2011). Impact of chemical impurities on the crystalline cement clinker phases determined by atomistic simulations. *Crystal Growth & Design, 11*(7), 2964–2972.

Manzano, H., Durgun, E., López-Arbeloa, I., & Grossman, J. C. (2015). Insight on tricalcium silicate hydration and dissolution mechanism from molecular simulations. *ACS Applied Materials & Interfaces, 7*(27), 14726–14733.

Mishra, R. K., Flatt, R. J., & Heinz, H. (2013). Force field for tricalcium silicate and insight into nanoscale properties: Cleavage, initial hydration, and adsorption of organic molecules. *Journal of Physical Chemistry C, 117*(20), 10417–10432.

Padilla Espinosa, I. M., Hodo, W., Rivas Murillo, J. S., Rajendran, A. M., & Mohan, R. V. (2017). Constitutive stiffness characteristics of cement paste as a multiphase composite system — A molecular dynamics-based model. *Journal of Engineering Materials and Technology, 139*(4), 041007–041007.

Pellenq, R. J. M., Kushima, A., Shahsavari, R., Van Vliet, K. J., Buehler, M. J., Yip, S., et al. (2009). A realistic molecular model of cement hydrates. *Proceedings of the National Academy of Sciences, 106*(38), 16102–16107.

Ritschl, F., Fait, M., Fiedler, K., Köhler, J. E., Kubias, B., & Meisel, M. (2002). An extension of the consistent valence force field (CVFF) with the aim to simulate the structures of vanadium phosphorus oxides and the adsorption of n-butane and of 1-butene on their crystal planes. *Zeitschrift für Anorganische und Allgemeine Chemie, 628*(6), 1385–1396.

Shahsavari, R., & Chen, L. (2015). Screw dislocations in complex, low symmetry oxides: Core structures, energetics, and impact on crystal growth. *ACS Applied Materials & Interfaces, 7*(4), 2223–2234.

Shahsavari, R., Pellenq, R. J. M., & Ulm, F.-J. (2011). Empirical force fields for complex hydrated calcio-silicate layered materials. *Physical Chemistry Chemical Physics, 13*(3), 1002–1011.
Shahsavari, R., Tao, L., & Chen, L. (2016a). Structure, energetics, and impact of screw dislocations in tricalcium silicates. *Journal of the American Ceramic Society, 99*(7), 2512–2520.
Shahsavari, R., Chen, L., & Tao, L. (2016b). Edge dislocations in dicalcium silicates: Experimental observations and atomistic analysis. *Cement and Concrete Research, 90*, 80–88.
Sun, H. (1995). Ab initio calculations and force field development for computer simulation of polysilanes. *Macromolecules, 28*(3), 701–712.
Tam, L.-H., Zhou, A., Yu, Z., Qiu, Q., & Lau, D. (2017). Understanding the effect of temperature on the interfacial behavior of CFRP-wood composite via molecular dynamics simulations. *Composites Part B: Engineering, 109*, 227–237.
Thomas, J. J., Jennings, H. M., & Allen, A. J. (2010). Relationships between composition and density of tobermorite, jennite, and nano-scale CaO-SiO2-H2O. *Journal of Physical Chemistry C, 114*(17), 7594–7601.
Van Duin, A. C. T., Dasgupta, S., Lorant, F., & Goddard, W. A. (2001). ReaxFF: A reactive force field for hydrocarbons. *The Journal of Physical Chemistry. A, 105*(41), 9396–9409.
Vandamme, M., & Ulm, F.-J. (2009). Nanogranular origin of concrete creep. *Proceedings of the National Academy of Sciences, 106*(26), 10552–10557.
Wang, Q., Manzano, H., Guo, Y., Lopez-Arbeloa, I., & Shen, X. (2015). Hydration mechanism of reactive and passive dicalcium silicate polymorphs from molecular simulations. *Journal of Physical Chemistry C, 119*(34), 19869–19875.
Wu, W., Al-Ostaz, A., Cheng, A. H.-D., & Song, C. R. (2011). Computation of elastic properties of Portland cement using molecular dynamics. *Journal of Nanomechanics and Micromechanics, 1*(2), 84–90.
Wu, T., Wang, Q., Yao, T., & He, S. (2016). Molecular dynamics simulations of the structural properties of Al2O3-based binary systems. *Journal of Non-Crystalline Solids, 435*, 17–26.
Yu, Z., Xu, Z., & Lau, D. (2014). Effect of acidity on chitin-protein interface: A molecular dynamics study. *BioNanoScience, 4*(3), 207–215.
Zhao, H., Wang, Y., Yang, Y., Shu, X., Yan, H., & Ran, Q. (2017). Effect of hydrophobic groups on the adsorption conformation of modified polycarboxylate superplasticizer investigated by molecular dynamics simulation. *Applied Surface Science, 407*, 8–15.

Ethical and Societal Considerations of Nano Ecosystems

S. Jeyabharathi and M. Sivadharani

Abstract

Nanotechnology offers a treasure trove for improving our future. In this chapter, we glimpse into some of the current innovations of nanotechnology and address both social and ecological challenges simultaneously. In other words, "Nanotech's Double-Edged Sword: A Boon with Potential Risks." These self-sufficient micro-worlds could wreak havoc on natural ecosystems if they're not carefully controlled. Nanotechnology has various applications in different fields, such as construction, agriculture, consumer goods, medicine, energy, and environment. Though it is useful in different fields, the aftermath effect and the lifecycle of the nano products have not been fully assessed. Imagine a scenario where a nano ecosystem designed to clean up pollutants ends up outcompeting native bacteria crucial for nutrient cycling. These particles can potentially enter the soil, water, and air, with unknown long-term ecological consequences. Regulations governing this field's development and use are still evolving. This lack of established standards could lead to irresponsible practices and potential risks. These unintended consequences could be harmful to the environment or even human health. Unlocking the benefits of nano ecosystems will require us to strike a precarious balance. On one hand, we need to maximize their potential to revolutionize fields like medicine and clean energy. On the other hand, we must minimize dangers and not upset the delicate balance of natural environments. This tightrope walk will require careful research, robust regulations, and open communication between scientists, policymakers, and the public. We can only minimize the risks associated with nanoparticles by subjecting them to comprehensive evaluations across their entire lifespan, from design to final elimination.

S. Jeyabharathi (✉) · M. Sivadharani
Department of Microbiology, Cauvery College for Women (Automonous), (Affiliated to Bharathidasan University), Trichy, Tamil Nadu, India
e-mail: jeyabharathi.mb@cauverycollege.ac.in

D. K. Gupta et al. (eds.), *Nano-Ecosystems*,
https://doi.org/10.1007/978-981-95-5033-3_13

Keywords

Nanotechnology · Ethical Consideration · Nano ecosystem · Nanotoxicity · Socioeconomic impact

1 Introduction

Rapidly expanding, nanotechnologies are capturing the attention of many. Nanotechnologies are primarily in the initial stages of exploration and discovery; the variety of specific applications is currently too limited to form a complete picture of the field (Bruce, 2006). Nanotechnology (NT), the control of materials at the nano-level of atoms and molecules, is proving to be a revolutionary influence in a wide range of fields, impacting health, environmental sustainability, social structures, and ethical considerations in various sectors. Ethics is all about figuring out what's right. But sometimes, the path gets twisty. An ethical dilemma pops up when following one moral compass inevitably leads you to bump into another. No matter which direction you choose, you end up doing something that could be considered wrong (Hornyak et al., 2008). Incorporating nanomaterials into consumer products and various industries has prompted inquiries into potential health risks and safety concerns. Scientific studies have demonstrated that certain nanoparticles can infiltrate biological barriers, causing cellular damage and adverse health impacts. Ecosystems can be adversely affected by the accumulation of nanoparticles in the environment during manufacturing, usage, and disposal processes, thereby posing risks to both wildlife and human health. An illustrative example is the disruption caused by silver nanoparticles, commonly found in consumer products like clothing and cosmetics, to aquatic ecosystems.

2 Application of Nanotechnology

The scientific community has allocated resources and time to fully understand and utilize the vast capabilities of NT. It is evident in the creation of nanoassemblies resembling natural elements, such as agricultural products, nanomedicine, and nanotools designed for medical treatment and diagnostic applications in the healthcare field. The use of NT provides a small-scale answer to large-scale issues (Malik et al., 2023) (Fig. 1).

2.1 Nano-enhanced Textiles for Improved Performance

Recent years have witnessed a focus on improving performance characteristics through the use of nano-enhancement. The primary objectives encompassed the attainment of properties such as water and oil resistance, antibacterial efficacy, wrinkle prevention, and self-purification via photocatalytic processes. Single-walled carbon nanotubes are used as an anti-bactericidal in textiles. Direct contact

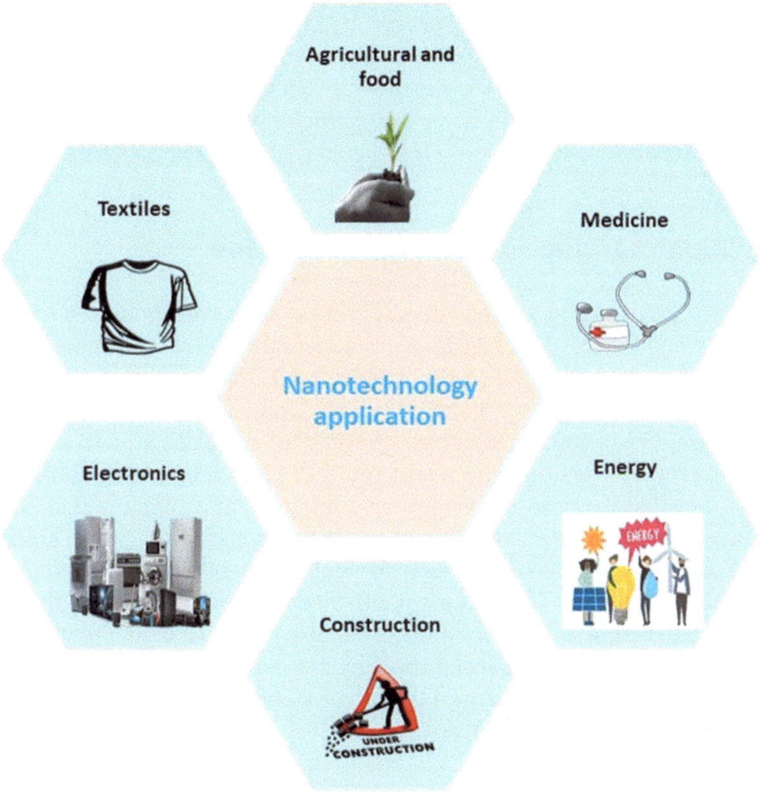

Fig. 1 Application of nanotechnology in different fields

with single-walled carbon nanotubes leads to the destruction of the bacterial cell wall. The fabrication of hydrophobic or super-hydrophobic surfaces commonly involves the use of silica nanoparticles. The increasing research interest in super-hydrophobic surfaces in recent years is mainly driven by their potential applications, such as liquid-repellent protective wear and stain-resistant clothing that prevent the adherence of contaminants. Chitosan nanoparticles were administered onto cotton fabric samples, resulting in enhanced dyeability and thermal stability, as well as providing antibacterial and UV protection properties. By employing dip coating, applied detonation nanodiamonds were successfully incorporated into cotton fabrics, leading to remarkable improvements in tensile strength, abrasion resistance, and thermal stability. A similar approach was utilized to apply nanodiamonds onto wool substrates. As a result, the treated fabric exhibited enhanced durability and significantly enhanced surface wetting properties, especially when the concentration of nanodiamonds in the coating suspension was increased (Krifa & Prichard, 2020). The antibacterial features and UV shielding abilities of ZnO-NP materials for their effective utilization in the textile industry. The antibacterial properties and

UV resistance of ZnO-NP materials make them highly desirable. Nevertheless, the main hurdle is to modify the production methods to ensure they are more eco-friendly and cost competitive. This can be achieved by incorporating safer dyes and chemicals, as well as reducing the expenses associated with effluent treatment and disposal. Fabric embedded with silver nanoparticles effectively eliminates bacteria, resulting in odor-resistant clothing (Huang et al., 2021; Gupta, 2020).

2.2 Supercharging Solar Energy Conversion

NT is like a secret weapon for solar energy! This cutting-edge field of science holds immense potential to revolutionize how we harness the power of the sun. By manipulating materials at the atomic and molecular level, NT can significantly improve various aspects of solar power technology, including capture, conversion, storage, and utilization. The following examples showcase how NT can act as a supercharger for solar energy conversion. The interaction between light and materials can be modified to enable the cost-effective processing of semiconductors into photovoltaic devices. Advancing the productivity of photocatalysts that convert solar radiation into usable energy sources, such as fuels. Achieving greater efficiency in converting chemical fuels into electrical energy and vice versa enables the creation of batteries that are not only more compact but also possess a higher power capacity. Surface efficiency can be enhanced through the implementation of various technologies, including displays, thermos-electric systems, solid-state lighting, and friction reduction methods (Ahmadi et al., 2018).

2.3 Nanotech Revolutionizes Food

NT is making a big impact on farms and food! From super sensors that sniff out disease and pests early on to fertilizers that deliver nutrients directly to plants, nanotechnology is revolutionizing agriculture. This means more food can be grown with less water, fertilizer, and pesticides, which is good for the environment and our wallets. Nanotechnology is also being used to create new food packaging materials that keep food fresher for longer, reducing food waste. Agromicron's spray is a prime example of how scientists can harness the power of biology to create practical tools for everyday use. The spray works by incorporating a clever biodetection method. Here's how it works: the spray contains a special protein that glows in the dark, but only when it comes into contact with harmful germs like *E. coli* or *Salmonella* that might be lurking on your food. This ingenious use of bioluminescence makes it simple to identify contaminated areas, potentially reducing the risk of foodborne illness (Vijayakumar et al., 2022). Super small science is making a big difference in how we eat and stay healthy. Scientists are using nanoparticles to improve food by tinkering with flavor and color, spotting tiny invaders like bacteria, and even zapping them. This tech also helps deliver nutrients throughout the body more effectively, like tiny vitamin submarines. By understanding how food works on a microscopic level, nanotechnology is helping us get the most out of what we eat (Modi et al., 2022).

2.4 Nanotech Boosts Wood

Nanomaterials offer the opportunity to upgrade existing wood-based products. The forest's plentiful supply of lignocellulosic material has sparked a growing fascination with transforming cellulose into nanocellulose. The traits of this material offer immense potential for applications beyond the forest products industry. Nanocellulose exhibits its versatility in the wood-based product industry, where it serves as a reinforcing agent in pulp and paper and wood composite applications, as well as a coating material for wood coatings. Moreover, its multifunctionality allows for its utilization as a component, such as a substrate, stabilizer, or electrode in various non-forest product sectors, including electronics, sensors, batteries, food, pharmaceuticals, and cosmetics. The functionality of wood can be elevated with the addition of nano zinc oxide or nano titanium oxide, resulting in improved durability, fire resistance, UV absorption, and reduced water absorption. Nano zinc oxide coated on wood material for their bioresistance against Gram-positive bacterium *Staphylococcus aureus*, the Gram-negative bacterium *Escherichia coli*, the molds *Aspergillus niger*, and *Penicillium brevicompactum,* as well as the brown-rot fungus *Coniophora puteana*. Other nanoparticles, such as silver, copper, aluminum oxide, titanium oxide, cerium oxide, and silica, are used on wood to enhance its characteristics and wood preservatives (Jasmani et al., 2020).

2.5 Groundwater Remediation

Studies have indicated that iron nanoparticles can be a viable solution for the remediation of organic solvent contamination in groundwater. These nanoparticles exhibit the ability to disperse uniformly within the aqueous environment and catalyze the degradation of the organic compounds present.

2.6 Nano Power Boost for Windmills

The augmentation of electricity production from windmills has been achieved through the implementation of epoxy containing carbon nanotubes. This cutting-edge approach involves utilizing this composite material to fabricate windmill blades, resulting in blades that possess superior strength and reduced weight. As a result, the electricity generated by each windmill is significantly increased (Gupta, 2020).

2.7 Effective Biomedical Treatment

Nanoalloys are used in the biomedical field to get effective results in the treatment. Various biomedical applications benefit from the potential properties of nanoparticles, including gold (Au), silver (Ag), iron cobalt (Fe-Co), and iron platinum (Fe-Pt), titanium dioxide (TiO_2), copper nickel (Cu–Ni), cerium dioxide (CeO_2), silica (SiO_2), and zinc oxide (ZnO). Nanosilver displays unparalleled physical, chemical, and biological properties when compared to its macro-sized silver counterparts. These nanoparticles are used for bioimaging, bioprinting, photoablation, biosensor, anticancer agent, hyperthermia, drug delivery, and regenerative medicine (McNamara & Tofail, 2016). While NPs have been employed for drug delivery in multiple diseases, the main objective has been to enhance cancer treatment.

2.8 Nano Barcodes are Transforming Diagnostics and Food Safety

Nano barcodes are emerging as a rival to nano biosensors in disease detection. These microscopic tags are making waves internationally for their ability to revolutionize both medical diagnostics and food tracking. In the fight against disease, nano barcodes can be used to analyze multiple genes simultaneously, aiding in early and accurate diagnoses. They can also be employed in detailed examinations of tissues at the microscopic level. But their applications extend far beyond the medical field. Nano barcodes are poised to transform the food industry by enabling real-time tracking of food products throughout the entire supply chain, from farm to table. This includes agrifood items and dairy products, ensuring food safety and efficiency. By attaching nano barcodes to food, we can gain valuable insights into their journey, allowing for swift identification and removal of contaminated products. Additionally, nano barcodes can provide data on storage and transportation conditions, helping to optimize the food supply chain and minimize waste.

3 Nanotoxicity and Nanopollutants

Nanotoxicology emerged as a novel subdivision of toxicology with the aim of bridging the knowledge gaps and specifically investigating the potential harmful health consequences associated with nanomaterials (Tiple et al., 2020). The main focus of nanotoxicology lies in evaluating the adverse effects that nanomaterials can impose on both human health and the environment. Nanotoxicology aims to determine the potential dangers of engineered nanomaterials through a collaborative effort involving various disciplines such as toxicology, biology, chemistry, physics, material science, geology, exposure assessment, and pharmacokinetics (Akçan et al., 2020). Nanotoxicity is greatly influenced by the particle size and surface area-to-volume ratio of nanoparticles. This is primarily because surface interactions serve as the predominant mode of interaction between NPs and biological systems.

The reduction in size leads to the exposure of a greater number of active atoms on the surface, consequently enhancing its reactivity. Analysis of endothelial cells in in vitro cytotoxicity studies has demonstrated that amorphous silica nanoparticles with a smaller diameter (14–16 nm) exhibit a higher rate of cell death when compared to their larger counterparts (104–335 nm) (Mohanta & Ahmaruzzaman, 2020).

Nanopollution poses a significant threat to human health. The widespread use of nanoparticles is leading to their increasing presence in the environment, raising concerns about their potential negative impacts. These concerns include the effects of nanopollution on soil health and ecology, crop production and food quality, groundwater purity, and, ultimately, human health. Because of these risks, there is an urgent need to develop safety protocols for different types of nanopollutants to protect our environment and ecological systems (Elsakhawy et al., 2022). Plastic objects are easily found in all aspects of our daily lives, including food containers, water pipes, and the clothing sector. Nanoplastics are formed as a result of the decomposition of plastic waste. The predominant sources of nano plastics are primarily derived from two major sources: primary plastic waste and secondary derivatives. Primary nanoplastics waste is released into the environment in either microscale or nanoscale dimensions. This includes various sources like nanomedicine, nanoimaging, nanosensors, and personal care products. When plastics break down due to physical, chemical, or biological factors, they give rise to secondary derivatives known as microplastics or nanoplastics. As the quantity of plastic waste surges, the existence of micro- and nanoplastics within the food chain gives rise to a perilous situation for human health. The presence of microplastic particles measuring less than 5 mm was identified in 81% of the tap water samples collected from 159 different sources around the world. (Lai et al., 2022). Microplastics are tiny plastic bits showing up everywhere, even in our food. Seafood tops the list, with an average of 1.48 particles per gram. Other foods contain lower amounts: sugar (0.44 particles/g), honey (0.10 particles/g), and salt (0.11 particles/g). Beverages are no exception—alcoholic drinks have 32.27 particles per liter, while bottled water is even higher at 94.37 particles/liter. Tap water is much lower at 4.23 particles/liter. Surprisingly, the air we breathe also contains microplastics, averaging 9.80 particles per cubic meter. Based on this data, people might ingest between 39,000 and 52,000 microplastic particles each year. This can vary depending on age and gender. Inhaling these particles can further increase the annual count to a range of 74,000 to 121,000. So, what can you do to reduce your intake? Switching from bottled water to tap water can make a big difference. Bottled water drinkers might ingest an extra 90,000 particles per year compared to those who stick to tap water (which only adds about 4000 particles) (Cox et al., 2019).

Considering the biological factors, nanotoxicity exerts its effects on aquatic organisms via two distinct mechanisms: (i) The overall capacity of NPs to accumulate within the body through absorption, distribution, metabolism, and excretion (referred to as toxicokinetics or TK) is of utmost importance. (ii) The detrimental consequences arising from the accumulation of nanomaterials at the site of action or target site, termed toxicodynamics (TD), can be substantial. (Chen, 2016). Algae, being a significant group of microorganisms, find extensive application in ecotoxicological studies. The cell wall serves as the primary site for the interaction and uptake of nanoparticles in different species of algae. The bioaccumulation of CuO nanoparticles in green algae *Chlorella pyrenoidosa* was identified when exposed to concentrations of up to 200 mg L^{-1}. This phenomenon occurred due to the adhesion of particles to the cells, followed by their internalization through the process of endocytosis. Approximately 50% of the cellular growth in the green algae *Dunaliella tertiolecta* and the marine diatom *Skeletonema costatum* was inhibited when exposed to silver nanoparticles at concentrations ranging from 0.4 to 6.4 mg L^{-1} in an artificial seawater medium. Diverse nanoparticles have a significant impact on algae cells, resulting in detrimental effects such as cell membrane damage, reduction in chlorophyll content and photosynthetic activity, mitochondrial depolarization, oxidative stress, growth inhibition, alteration in gene expression, and changes in proteomic profile (Abbas et al., 2020). Soil plays a critical role in connecting various parts of our environment. It acts like a bridge between the water (hydrosphere), air (atmosphere), living things (biosphere), and human activities (anthroposphere). Scientists have explored numerous connections between soil and other environmental systems. These include the links between soil, water, and climate change, the impact of soil health on food quality and human well-being, and the two-way relationship between soil microbes and human health. There's even a web connecting clean water, healthy soil, and the plants that nourish us (Elsakhawy et al., 2022).

AgNPs in the environment can harm bacteria that convert ammonia into a usable form for plants (nitrification). This happens because the silver particles release silver ions, which are harmful to these bacteria. CuO NPs can also negatively affect the health of the soil's microbial community. They do this by reducing the activity of enzymes that break down organic matter (hydrolytic activity), lowering the ability of microbes to break down complex molecules (dehydrogenase oxidative potential), and changing the types of microbes present in the soil (community composition) (Biswas & Sarkar, 2019). ZnO NPs' harsher effects were seen when the size ranged 10–25 nm and concentration 500–1000 μg/ml to lettuce (*Lactuca sativa*), corn (*Zea mays*), rapeseed (*Brassica napus*), radish (*Raphanus sativus*), cucumber (*Cucumis sativus*), and perennial ryegrass (*Lolium perenne*). These plants experienced a variety of negative consequences, including stunted root growth, deformed root tips, damage to the root caps and surface layers, enlarged vacuoles, and abnormal cells in the cortex (Ranjan et al., 2021). The consumption of a lone oral dose of zinc oxide (ZnO) nanoparticles may result in adverse effects on the liver, kidneys, and lungs (Vijayakumar et al., 2022) (Table 1).

4 Potential Risks Concerning Both Health and the Environment

Across the diverse stages of manufacturing, packaging, transportation, storage, and waste management, the presence of nanoscale materials and goods can contaminate our air, water, soil, and food sources. This contamination can impact fruits,

Table 1 Some of the ways that nanopollutant can interact with our bodies are tabulated below

Exposure route	Sources of nanopollutants
Respiratory system	Ambient airborne, aerosols, filters
Skin	Wound dressings, textiles, surfaces, cosmetics, devices
Reproductive system	Contraceptive devices, hygiene products
Circulatory system	Nano-drugs, implants, catheters
Gastrointestinal tract	Utensils, coatings, packaging, supplements, filters, hygiene products (Biswas & Sarkar, 2019)

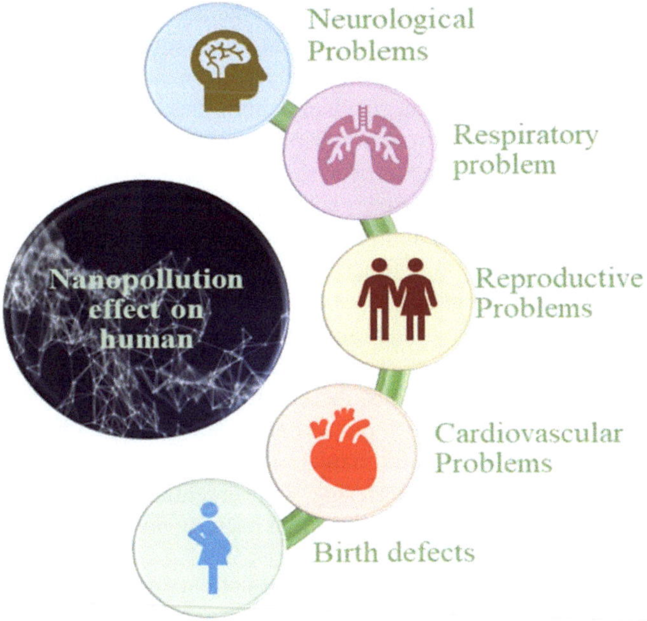

Fig. 2 Nanopollution effect on humans

vegetables, fish, and animal products. Furthermore, nanoparticles have the capability to float in the air for extended durations, traveling considerable distances from their point of origin. When conditions are favorable, these nanoparticles can invade human, animal, and plant cells, resulting in both recognized and unforeseen harmful effects (Asmatulu et al., 2013). Despite the unique qualities NPs possess as a result of their small size, it is crucial to be aware of the potential health risks they may also entail. In 1991, the United States introduced the National Nanotechnology Initiative, acknowledging the importance of nanotoxicity. The initial NNI Strategy for NT-Related Environmental, Health, and Safety (EHS) Research was issued in 2008 and later revised in 2011 (Bharathala, 2019) (Fig. 2).

The nanoparticles, along with atmospheric hydrogen, could reach the stratosphere. This influx of nanoparticles might create conditions for increased water vapor formation. The additional water vapor could act as a coolant in the stratosphere. However, this cooling effect could have unintended consequences. That is, colder temperatures might accelerate certain chemical reactions in the stratosphere, potentially leading to unforeseen environmental changes. Increased water vapor could also contribute to ozone layer depletion through complex chemical interactions (Hussain, 2020). Carbon nanomaterials, such as graphene oxide, single-walled carbon nanotubes, and fullerene (C60), find application in diverse fields. However, it is important to note that these materials possess inherent cytotoxic properties. SWCNTs and C60 are capable of causing the translocation of BAX proteins in mitochondrial membranes, which in turn induces apoptosis and leads to cell toxicity. C60 is known to trigger cytotoxicity and DNA damage in organisms. Additionally, C60 has the ability to cause genetic toxicity in organisms. Unmodified C60 molecules can interact with nucleotides through electron acceptance and exchange, resulting in DNA damage. The mechanism behind these toxic effects on cells primarily involves oxidative stress toxicity (Tiple et al., 2020).

Silver nanoparticle inhalation has been shown to cause adverse effects, including damage to the alveolar-capillary barrier and increased permeability of the epithelial cells, leading to leakage of cell and plasma proteins into the alveolar space. These findings indicate a decline in lung function (Garcés et al., 2021).

In soil, titanium dioxide nanoparticles can have a surprising effect on how organisms break down waste. In small amounts, they appear to act as a booster for the process that breaks down urea, a waste product excreted by many living things. However, at very high concentrations, these particles have the opposite effect, hindering this important waste removal process and even disrupting how bacteria handle nitrogen, an essential element for many biological functions. This disruption can have cascading consequences, interfering with essential functions like breaking down proteins and other organic pollutants that bacteria normally take care of in the environment (Rashid et al., 2021).

Inhaling certain nanoparticles can be hazardous to your lungs. These nanoparticles, like those made from carbon or silica, can irritate lung tissue, triggering inflammation. This inflammation can lead to scarring, a condition known as fibrosis, which can reduce your lungs' ability to function properly. In some cases, the body may even form lumps called granulomas in an attempt to isolate and wall off the inhaled particles. The trouble with nanoparticles doesn't stop at the lungs. Some nanoparticles, such as those made of silver or gold, are small enough to travel throughout the body after inhalation. This raises concerns about their potential to reach and accumulate in sensitive organs like the brain. Even everyday activities can lead to nanoparticle exposure. Nanoparticles used in certain quantum dots, sunscreens, and other products can penetrate the skin. This means these particles have the potential to enter your body and interact with your cells. Some nanoparticles are especially adept at bypassing the body's natural defenses. Manganese dioxide (MnO_2), titanium dioxide (TiO_2), and certain types of carbon nanoparticles can travel directly from your nose to your brain through the olfactory nerve pathway. This pathway is normally responsible for helping you smell, but it also provides a direct route for these tiny particles to reach the brain.

Not all nanoparticles are created equally. Nanoparticles made from certain materials, including titanium dioxide, aluminum oxide ($Al2O_3$), carbon black, cobalt (Co), and nickel (Ni), appear to be more toxic than their larger counterparts. This means that exposure to these nanoparticles may pose a greater health risk (Ray et al., 2009).

Silicon nanoparticles are a cornerstone of modern technology, powering our screens, shrinking electronics, and fueling clean energy solutions. But their usefulness comes with a potential downside. When electronic devices containing these nanoparticles reach the end of their lifespan and decompose in landfills, silicon can leak out. This contamination can spread to surrounding soil and water, impacting delicate ecosystems (Prasad, 2019). Invisible invaders with a hidden threat: Titanium dioxide nanoparticles (TiO_2 NPs) are a common ingredient in sunscreens and protective gear, valued for their shield against harmful UV rays. But these seemingly beneficial particles share a troubling similarity with silver nanoparticles (AgNPs). They escape from products and contaminate our lakes, rivers, and soil. The problem snowballs from there. These microscopic invaders are absorbed by organisms (bioaccumulation) and become more concentrated as they move up the food chain (biomagnification). Their ability to react to sunlight (photocatalytic activity) disrupts the delicate balance of aquatic and soil ecosystems, harming the tiny life forms that form the foundation of the food web. Unfortunately, environmental regulations haven't yet addressed the potential risks posed by these widely used nanoparticles (Prasad, 2019).

5 Nanoparticle Interactions and Toxicity

While NPs offer exciting possibilities for treating and diagnosing diseases, we must carefully consider the unpredictable negative effects they might have on the human body. Unlike larger drug molecules, NPs can travel throughout the body through the bloodstream and lymph systems. This unique characteristic allows them to potentially enter any cell, tissue, or organ, granting them widespread access throughout the body. This accessibility comes at a cost, however, as NPs can interact with and alter important molecules within cells. These alterations can disrupt the normal functions of cells and potentially lead to harm in entire organs. Therefore, it is crucial to thoroughly examine not only the positive results of using NPs in medicine, but also the possible unpredictable negative consequences of their action on the human body. Research using mice has shown an effect of CdTe/ZnS quantum dots. When the cores of cdte/zns quantum dots (qds) degrade, cadmium and tellurium ions are released and tend to accumulate in different organs and tissues. Cadmium primarily builds up in the liver, kidneys, and spleen, while tellurium accumulates almost exclusively in the kidneys.

Nanoparticles (NPs) can harm cells in several ways:

- *They can create oxidative stress:* NPs trigger the formation of harmful molecules like free radicals, which damage cells.
- *They can disrupt cell membranes:* NPs can poke holes in cell walls, leading to malfunctioning.
- *They can mess with cell structure:* NPs can damage the internal framework of cells, hindering the movement of materials and cell division.
- *They can damage DNA:* NPs can interact with DNA, increasing mutations and potentially leading to cancer.
- *They can impair energy production:* NPs can harm mitochondria, the power plants of cells, disrupting energy balance.
- *They can block waste disposal:* NPs can interfere with the formation of cleaning crews inside cells, leading to a buildup of harmful materials and cell death.
- *They can disrupt communication:* NPs can alter proteins in cell membranes, hindering the flow of essential materials in and out of cells.
- *They can trigger inflammation:* NPs can confuse cells, leading to inflammation and damage to tissues and organs (Sukhanova et al., 2018).

6 Mitigate the Risks Associated with Nanotechnology

No matter how they're incorporated into a product, all materials eventually become waste and need recycling or disposal (Kelechukwu, 2017). The potential to decrease nanotoxicity is substantial through the surface coating of nanoparticles using a range of surfactants, polymeric compounds, natural organic matters, and similar substances. Assessing the harmful effects of CdO nanoparticles on zebrafish with and without Cd-citrate coordination polymer coating. The implementation of a surface coating resulted in reduced toxicity, as indicated by a lower level of oxidative stress. This reduction can be attributed to the controlled dissolution of Cd_2^+ ions, facilitated by the carbon surface coverage, in contrast to the original CdO nanoparticles (Mohanta & Ahmaruzzaman, 2020).

When organic material is heated with limited oxygen in a special container, it breaks down into three products: biochar (a charcoal-like substance), bio-oil (a liquid fuel), and syngas (a gas mixture). Biochar itself is a mix of unburnt organic materials and mineral ash, left behind from the breakdown process. The exact makeup of biochar depends on the type of organic matter used and the heating conditions. Focusing on cost and sustainability: There's a strong need to investigate further how biochar can be cheaply and sustainably modified to become a more effective tool for tackling silver nanoparticle pollution. Highlighting practical application: We should delve deeper into how biochar can be modified in an affordable and environmentally friendly way to be used for real-world applications in reducing silver nanoparticle contamination. Researchers have been diligently investigating methods for developing biochar that integrates graphene. This innovative material has exhibited significant potential in capturing a wide range of nano pollutants from the environment. By embracing biosynthetic pathways for the design of nanomaterials and ensuring proper chemical treatment prior to their release, we can effectively mitigate any potential harm to the environment (Islam et al., 2021).

Green synthesis stands out as a safer, cleaner, and eco-friendly method for nanoparticle production. It utilizes natural resources and avoids harsh chemicals, making it a more responsible choice for both human health and the environment. Copper oxide nanoparticles (Cu-NPs) pollutants can be used as a powerful weapon in the fight against fungal diseases that plague our favorite crops. From bananas and tea to cocoa, coffee, citrus fruits, and beyond, these nanoparticles offer protection against a range of fungal foes, including powdery mildew and rust blight. Studies show promising results, particularly against the notorious *Phytophthora infestans* that attacks tomato plants. Cu-NPs also show potential in preventing bacterial blight in pomegranates by inhibiting the growth of *Xanthomonas axonopodis pv. Punicae* (Krishnani et al., 2022).

The research conducted by Gui et al. (2021) represents a significant leap forward in the development of next-generation filtration technologies. Their work paves the way for the design of filters that can effectively remove nanoparticles from wastewater. These advanced filters may also have the potential to be applied in industrial settings for separating gas from solids or liquids in environments characterized by harsh and corrosive conditions. This three-dimensional reticular architecture design provides exceptional stability, allowing the membrane to be reused for at least 200 wastewater treatment cycles. With a pore size of around 10 micrometers, it acts as a robust mechanical filter. Notably, the membrane achieves an impressive filtration efficiency, removing over 99.989% of nanoparticles from wastewater. To build public trust and foster internal understanding, nanotechnologists need a common ethical framework. While a rigid professional structure might not be ideal, a loose collaboration among practitioners could be beneficial for this developing field. This would help establish shared identity and ethical principles for nanotechnology (Shew, 2009). Policymakers across the West are going all in on "responsible research and innovation" (RRI). This means they're telling scientists and engineers to huddle up more with social scientists and humanities scholars. The big idea here is to make sure new technologies lift up everyone in society, don't cause any unwanted problems, and keep the public happy. By bringing in these different perspectives, they hope to guarantee that shiny new inventions benefit everyone, avoid any unforeseen snags, and keep the public in the loop (Viseu, 2015).

7 Ethical Considerations of Nanotechnology

Ethics, fundamentally, is about understanding right and wrong in order to live a virtuous life. Technology significantly shapes our experiences and the way we navigate the world. This powerful influence means technology has a deep impact on the ethical considerations we face. In simpler terms, the tech we use and how we interact with it shape our moral compass (Sandler, 2009). The emerging materials are predominantly fabricated through new techniques, resulting in the absence of specific rules and regulations established to oversee manufacturing processes for the safety of workers, scientists, and engineers (Asmatulu et al., 2013). The global surge in NT applications across various industries has sparked a wave of innovation in processes and products. This breakneck progress, however, comes with a significant hurdle: precisely gauging both the promising opportunities and the potential risks that remain shrouded in uncertainty when implementing this technology.

In light of these discussions, a key question remains: what role should law and regulation play in shaping the future of nanotechnology? The answer is that no country has all-encompassing laws for managing nanoscale materials. However, international efforts are underway to establish legal frameworks for nanotechnology.

The biggest ethical concern right now is the lack of safety standards for nanoparticles. This puts both people and the environment at risk. Governments should take a cautious approach ("precautionary principle") and establish regulations before widespread production and sale of nanotech products. This protects workers, consumers, and the environment. Ignoring this could lead to health problems, environmental damage, and even public backlash against all nanotechnology, good or bad. That could waste research funding and hinder the potential benefits of this technology. Because this is such a pressing issue, individual governments need to start creating regulations within their own countries, while also working together internationally to establish global standards (Schummer, 2007). The passage argues for establishing core ethical boundaries for human enhancement technologies, regardless of cultural differences in self-harm tolerance. These boundaries should be enforced through laws in all countries. Here are the key points:

1. *Informed Consent:* No one should be pressured or incentivized to participate in enhancement experiments. People must be fully aware of the risks involved before making a free choice.
2. *Vulnerable Populations:* People with mental disabilities or limited education should be protected from exploitation and cannot be used for enhancement experiments.
3. *Minimizing Risk to Others:* Enhancement procedures should not put anyone other than the participant at risk. This eliminates experiments like brain surgery enhancements, which could potentially harm others due to mental changes.
4. *Healthcare Resource Allocation:* If some enhancements are permitted, regulations should ensure they don't overload healthcare systems and limit access to essential medical care (Schummer, 2007).

There's an urgent need for regulatory bodies to develop reliable methods and procedures to assess the risks of using nanoparticles. This is crucial because of four immediate challenges:

1. *Rapidly Evolving Technology:* Nanoparticle technology is advancing quickly, making it difficult to keep up with potential risks.

2. *Uncertain Future Applications:* We don't yet know all the ways nanoparticles might be used, which makes it hard to predict future risks.
3. *Understanding and Verifying Safety Measures:* The methods used to ensure the safety of nanoparticles need to be clear, well-understood, and demonstrably effective.
4. *Developing Clear Public Policy:* Governments need to establish clear policies on nanoparticle use to protect public health and the environment.

Nanoparticles can enter our bodies in various ways, like breathing them in, being delivered as medicine, or even through our digestive system. These particles have unique properties that influence how they interact with our body. For instance, depending on their makeup and how they interact with fats, they might bypass important barriers and reach sensitive organs like the brain. While nanoparticles offer exciting possibilities, some of the materials used in their construction, like nickel or certain plastics, can be harmful at high concentrations. This interaction with our bodies also depends on how easily the particles break down. If they don't degrade naturally (nonbiodegradable), they can build up and potentially cause negative health effects (Souto, 2020).

The ethical concerns surrounding NT research and development go beyond the simple fear of its misuse or use for harmful purposes. The development and use of nanotechnologies should prioritize economic and environmental responsibility. This means minimizing energy consumption and resource use throughout the production lifecycle. Additionally, wherever possible, products should be designed for reuse or recycling to reduce waste and promote a circular economy. Military applications of nanotechnologies should be strictly limited to defensive purposes, with ensuring public safety as the top priority. While some details may be classified for national security reasons, there should be a strong commitment to transparency. This includes informing the public about the potential risks and benefits of military nanotechnologies, as well as establishing clear democratic oversight mechanisms to ensure the responsible development and use of these powerful technologies (Souto, 2020). When evaluating the ethical implications of NT, groups or commissions need to agree on a common set of core values. These values should guide their decisions on specific issues related to:

1. Protecting human health
2. The value of human life
3. Environmental sustainability
4. Economic and social impacts
5. The advancement of scientific research
6. Individual freedom of choice
7. Personal privacy
8. Relationships within society (domestic relations)
9. International relations (global cooperation)
10. Our understanding of humanity (identity, nature, and the human condition)

By considering these core values, ethical evaluations can address the complexities surrounding NT and its impact on various aspects of human life (Patenaude et al., 2015).

8 Societal Considerations

Nanotechnology's impact goes far beyond medicine. This revolutionary technology has the potential to significantly reshape how we interact with each other, how our political systems function, and even the global economic landscape. It's clear that the benefits and challenges won't be equally distributed. In the twenty-first century, developed nations with strong industrial bases see nanotechnology as essential for global competition. This focus on nanotech by wealthier countries raises concerns about a widening gap between developed and developing nations. Before discussing the impact of nanotechnology, it's important to clarify the distinction between developed and developing countries. Development isn't a simple on/off switch. Countries are classified based on a set of criteria, often including economy, measured by gross national product (GNP); politics, stable governments and democratic processes; and social factors, higher standard of living and less inequality.

These are just some of the key factors used to categorize countries along this development spectrum. Developing countries often struggle to meet basic needs like clean water, education, and healthcare. This lack of basic necessities often translates to limited resources for research and development (R&D). However, there are exceptions. Some developing countries, like China, have significantly increased their R&D spending in recent years. Despite potential financial resources, a major obstacle remains: a lack of experience in managing technological advancements. This inexperience can hinder a developing country's ability to effectively utilize and benefit from nanotechnology. To truly understand the impact of nanotechnology on developing countries, we need to analyze it through a socioeconomic lens. This means looking beyond the technology itself and considering its wider social and economic implications. The way to do this is by breaking down the "life cycle" of a technology:

1. *R&D and Innovation*: The initial phase involves generating ideas and developing technology through research and development.
2. *Production and Consumption*: This phase focuses on manufacturing technology, marketing it, and integrating it into society.
3. *End-of-Life Management*: The final phase considers how to dispose of waste products generated by technology in a responsible way.

By examining nanotechnology through these stages, we can better understand the potential challenges and opportunities it presents for developing countries. Nanotechnology research is a hotbed of activity, with funding steadily increasing. In the United States alone, government spending on nanotech research reached $2.1 billion in 2012. This brings the total US investment since 2001 to a staggering $16.5

billion. Globally, the focus extends beyond just the technology itself. Since 2005, an estimated $575 billion has been directed towards research on the environmental, health, and defense implications of nanotechnology. Additionally, a significant $390 billion has been dedicated to exploring the ethical, legal, and social considerations surrounding this powerful technology. The trend isn't limited to the United States. Recognizing the potential of nanotechnology, other countries are also ramping up their investments. In 2008, the European Union and Japan committed $1.7 billion and $950 million, respectively, to nanotech research. That same year, China, Korea, and Taiwan also joined the game, allocating $430 million, $310 million, and $110 million, respectively. This global surge in funding highlights the growing importance and potential of nanotechnology research (Gökçay & Arda, 2015).

There's a global effort to ensure the safe use of nanotechnology in food and agriculture. International organizations like the EU's Scientific Committees (EUSCA), the FAO/WHO Joint Meeting on Pesticide Residues (JMPR), and the US Food and Drug Administration (USFDA) are all involved. These groups work together to create guidelines, policies, and regulations that minimize any risks associated with nanotech products. Other organizations like the International Organization for Standardization (ISO) and the Organization for Economic Cooperation and Development (OECD) also play a role in setting international standards (Saritha et al., 2022).

8.1 UNESCO: Leading the Way in Ethical Nanotechnology Governance

Recognizing the ethical considerations surrounding nanotechnology, UNESCO established the International Bioethics Committee (IBC). This committee brings together 36 experts from diverse backgrounds and regions of the world. The IBC serves several key functions:

1. *Knowledge Sharing*: The committee fosters the exchange of ideas and experiences related to the ethical implications of life sciences research.
2. *Policy Advice*: The IBC provides recommendations to policymakers to guide ethical decision-making in science and technology.
3. *Bridging the Gap*: The committee acts as a bridge between policymakers and the public, ensuring open communication.
4. *Public Awareness*: UNESCO recognizes the importance of raising public awareness about nanotechnology, particularly among people with varying educational backgrounds.

8.2 Proactive Approach to Ethical Issues

Given the rapid pace of nanotechnology development, UNESCO emphasizes the need for proactive consideration of ethical issues. By conducting ongoing research, they aim to identify both the potential benefits and drawbacks of nanotechnology before widespread social concerns or ethical dilemmas arise.

8.3 A Global Framework for Governance

Looking ahead, UNESCO is planning to establish a framework for the governance of science and technology, particularly in developed countries that rely on resources from developing nations. This framework aims to promote responsible and equitable practices within the global scientific community (Gökçay & Arda, 2015).

In an effort to promote global consistency and improve public health practices around the world, the World Health Organization (WHO) has established a Guideline Development Group. This group of experts will collaborate to develop a set of internationally applicable guidelines on exposure assessment standards and regulations. These guidelines will provide a comprehensive framework for countries to design and implement effective exposure assessment programs, ensuring a more standardized approach to measuring exposure to environmental hazards and risks (Babatunde et al., 2020).

9 Conclusion

The potential of nano ecosystems to revolutionize various fields, from medicine and environmental remediation to manufacturing and energy production, is undeniable. However, the ethical and societal considerations surrounding their development and use demand careful attention. The research presented here highlights the potential health risks associated with nanoparticles, including lung damage, widespread distribution within the body, and even brain penetration. These findings necessitate a multi-pronged approach. Environmental quality is diminishing as a consequence of human actions and natural phenomena. Natural ecosystem erosion is negatively impacting the quality of the environment. Further research is crucial to fully understand the long-term effects of nano ecosystems on human health and the environment. While initial studies provide valuable insights, extended investigations are needed to assess the potential for chronic health problems or unforeseen ecological consequences. Rigorous risk assessments that consider the entire life cycle of nano ecosystems, from creation and use to disposal and potential environmental interactions, should be conducted before widespread application.

The precautionary principle should guide development. This means erring on the side of caution if potential risks are unclear. Strict regulations and safety protocols should be established to minimize potential harm. These regulations should be adaptable to incorporate new scientific findings and evolve alongside the technology. International collaboration is necessary to ensure consistent safety standards and prevent a race to the bottom, where environmental or health risks are prioritized for economic gain. Transparent communication with the public is essential. Scientists and policymakers must engage in open dialogue to educate the public about the potential benefits and risks of nano ecosystems. Public trust and participation are vital for responsible development. This means creating accessible information channels, fostering public discourse, and incorporating public concerns into decision-making processes. By fostering a culture of open communication, scientists, policymakers, and the public can work together to ensure the responsible development and use of nano ecosystems. By prioritizing safety, ethical considerations, and long-term planning, we can unlock the potential of nano ecosystems while minimizing the risks. This journey requires collaboration between scientists, policymakers, industry leaders, and the public to ensure a future where these microscopic marvels serve humanity for the better. In this collaborative effort, the public plays a critical role in ensuring that the development of nano ecosystems aligns with societal values and prioritizes the well-being of future generations.

Hence, our advice is for upcoming research to center on grasping the possible hazards and risks of chronic exposure to a variety of nanomaterials at relevant concentrations.

References

Abbas, Q., Yousaf, B., Ullah, H., Ali, M. U., Ok, Y. S., & Rinklebe, J. (2020). Environmental transformation and nano-toxicity of engineered nano-particles (ENPs) in aquatic and terrestrial organisms. *Critical Reviews in Environmental Science and Technology, 50*(23), 2523–2581. https://doi.org/10.1080/10643389.2019.1705721

Ahmadi, M. H., Ghazvini, M., Alhuyi Nazari, M., et al. (2018). Renewable energy harvesting with the application of nanotechnology: A review. *International Journal of Energy Research*, 1–24. https://doi.org/10.1002/er.4282

Akçan, R., Aydogan, H. C., Yildirim, M. Ş., Taştekin, B., & Sağlam, N. (2020). Nanotoxicity: A challenge for future medicine. *Turkish Journal of Medical Sciences, 50*(4), 1180–1196. https://doi.org/10.3906/sag-1912-209

Asmatulu, R., Zhang, B., & Asmatulu, E. (2013). Safety and ethics of nanotechnology. In *Nanotechnology safety* (pp. 31–41). https://doi.org/10.1016/B978-0-444-59438-9.00003-5

Babatunde, D. E., Denwigwe, I. H., Babatunde, O. M., Gbadamosi, S. L., Babalola, I. P., & Agboola, O. (2020). Environmental and societal impact of nanotechnology. *IEEE Access, 8*, 4640–4667. https://doi.org/10.1109/ACCESS.2019.2961513

Bharathala, S. (2019). Biomedical applications of nanoparticles. In *Nanotechnology in modern animal biotechnology* (pp. 113–132). https://doi.org/10.1016/B978-0-12-818823-1.00008-9

Biswas, J. K., & Sarkar, D. (2019). Nanopollution in the aquatic environment and ecotoxicity: No Nano issue! *Current Pollution Reports, 5*, 4–7. https://doi.org/10.1007/s40726-019-0104-5

Bruce, D. (2006). Ethical and social issues in nanobiotechnologies: Nano2Life provides a European ethical 'think tank' for research in biology at the nanoscale. *EMBO Reports, 7*(8), 754–758. https://doi.org/10.1038/sj.embor.7400762

Chen, W.-Y. (2016). Toxicokinetic modeling challenges for aquatic nanotoxicology. *Frontiers in Marine Science, 2*, 114. https://doi.org/10.3389/fmars.2015.00114

Cox, K. D., Covernton, G. A., Davies, H. L., Dower, J. F., Juanes, F., & Dudas, S. E. (2019). Human consumption of microplastics. *Environmental Science & Technology, 53*(12), 7068–7074. https://doi.org/10.1021/acs.est.9b01517

Elsakhawy, T., Omara, A. E., Amer, M., El-Ramady, H., Prokisch, J., & Brevik, E. (2022). A diagrammatic mini-review on soil-human health-nexus: With focus on soil nano-pollution. *Environment, Biodiversity and Soil Security, 6*(2022), 327–338. https://doi.org/10.21608/jenvbs.2022.173272.1199

Garcés, M., Magnani, N. D., Pecorelli, A., Calabró, V., Marchini, T., Cáceres, L., Pambianchi, E., Galdoporpora, J., Vico, T., Salgueiro, J., Zubillaga, M., Moretton, M. A., Desimone, M. F., Alvarez, S., Valacchi, G., & Evelson, P. (2021). Alterations in oxygen metabolism are associated to lung toxicity triggered by silver nanoparticles exposure. *Free Radical Biology & Medicine, 166*, 324–336. https://doi.org/10.1016/j.freeradbiomed.2021.02.008

Gökçay, B., & Arda, B. (2015). Nanotechnology, nanomedicine; ethical aspects. *Revistaromana de bioetica, 13*(3) http://www.bioetica.ro/index.php/arhiva-bioetica/article/view/829/pdf

Gui, W., Liu, J., Song, X., Zhang, H., Lin, J., & Luan, B. (2021). A new microfiltration membrane with three-dimensional reticular architecture for nano-pollutants removal from wastewater. *Progress in Natural Science: Materials International, 31*(3), 414–419.

Gupta, S. K. (2020). Study of nanotechnology and its application. *Journal of Physics & Optics Sciences*. SRC/JPSOS/112. https://doi.org/10.47363/JPSOS/2020(2)107

Hornyak, G. L., Tibbals, H. F., Dutta, J., & Moore, J. J. (2008). Introduction to nanoscience and nanotechnology. CRC press.

Huang, R., et al. (2021). Progress ofzinc oxide-based nanocomposites in the textile industry. *IET Collaborative Intelligent Manufacturing, 3*(3), 281–289. https://doi.org/10.1049/cim2.12029

Hussain, C. M. (2020). *The ESLI handbook of nanotechnology: Risk safety, ESLI and commercialization* (1st ed.). Wiley.

Islam, M. A., Jacob, M. V., & Antunes, E. (2021). A critical review on silver nanoparticles: From synthesis and applications to its mitigation through low-cost adsorption by biochar. *Journal of Environmental Management, 281*, 111918. https://doi.org/10.1016/j.jenvman.2020.111918

Jasmani, L., Rusli, R., Khadiran, T., et al. (2020). Application of nanotechnology in wood-based products industry: A review. *Nanoscale Research Letters, 15*, 207. https://doi.org/10.1186/s11671-020-03438-2

Kelechukwu, E. (2017). Social, legal, ethical, health, safety and environmental aspects of nanotechnology.

Krifa, M., & Prichard, C. (2020). Nanotechnology in textile and apparel research—An overview of technologies and processes. *The Journal of the Textile Institute*. https://doi.org/10.1080/00405000.2020.1721696

Krishnani, K. K., Boddu, V. M., Chadha, N. K., Chakraborty, P., Kumar, J., Krishna, G., & Pathak, H. (2022). Metallic and non-metallic nanoparticles from plant, animal, and fisheries wastes: Potential and valorization for application in agriculture. *Environmental Science and Pollution Research International, 29*(54), 81130–81165. https://doi.org/10.1007/s11356-022-23301-4

Lai, H., Liu, X., & Qu, M. (2022). Nanoplastics and human health: Hazard identification and biointerface. *Nanomaterials (Basel), 12*(8), 1298. https://doi.org/10.3390/nano12081298

Malik, S., Muhammad, K., & Waheed, Y. (2023). Nanotechnology: A revolution in modern industry. *Molecules (Basel, Switzerland), 28*(2), 661. https://doi.org/10.3390/molecules28020661

McNamara, K., & Tofail, S. A. M. (2016). Nanoparticles in biomedical applications. *Advances in Physics: X, 2*(1), 54–88. https://doi.org/10.1080/23746149.2016.1254570

Modi, S., Prajapati, R., Inwati, G. K., Deepa, N., Tirth, V., Yadav, V. K., Yadav, K. K., Islam, S., Gupta, P., Kim, D.-H., et al. (2022). Recent trends in fascinating applications of nanotechnology in allied health sciences. *Crystals, 12*, 39. https://doi.org/10.3390/cryst12010039

Mohanta, D., & Ahmaruzzaman, M. (2020). *The ELSI handbook of nanotechnology (Risk, safety, ELSI and commercialization) Addressing Nanotoxicity* (pp. 103–112). https://doi.org/10.1002/9781119592990.ch6

Patenaude, J., Legault, G.-A., Beauvais, J., Bernier, L., Beland, J.-P., Boissy, P., Chenel, V., Daniel, C.-E., Genest, J., Poirier, M.-S., et al. (2015). Framework for the analysis of nanotechnologies' impacts and ethical acceptability: Basis of an interdisciplinary approach to assessing novel technologies. *Science and Engineering Ethics, 21*, 293–315. https://doi.org/10.1007/s11948-014-9543-y

Prasad, R. (2019). [Nanotechnology in the life sciences] microbial nanobionics (Volume 2, basic research and applications). *Nanotoxicity: Sources and Effects on Environment*, 169–179. https://doi.org/10.1007/978-3-030-16534-5_9

Ranjan, A., Rajput, V. D., Minkina, T., Bauer, T., Chauhan, A., & Jindal, T. (2021). Nanoparticles induced stress and toxicity in plants. *Environmental Nanotechnology, Monitoring and Management, 15*, 100457. https://doi.org/10.1016/j.enmm.2021.100457

Rashid, M. M., Forte Tavčer, P., & Tomšič, B. (2021). Influence of titanium dioxide nanoparticles on human health and the environment. *Nanomaterials (Basel), 11*, 2354. https://doi.org/10.3390/nano11092354

Ray, P. C., Yu, H., & Fu, P. P. (2009). Toxicity and environmental risks of nanomaterials: Challenges and future needs. *Journal of Environmental Science and Health, Part C: Environmental Carcinogenesis and Ecotoxicology Reviews, 27*(1), 1–35. https://doi.org/10.1080/10590500802708267

Sandler, R. (2009). Nanotechnology: The social and ethical issues.

Saritha, G. N. G., Anju, T., & Kumar, A. (2022). Nanotechnology – big impact: How nanotechnology is changing the future of agriculture? *Journal of Agriculture and Food Research, 10*, 100457.

Schummer, J. (2007). Identifying ethical issues of nanotechnologies. In *Nanotechnologies, ethics and politics* (pp. 79–98).

Shew, A. (2009). Nanotechnology's future: Considerations for the professional. In F. Allhoff & P. Lin (Eds.), *Nanotechnology & society*. Springer. https://doi.org/10.1007/978-1-4020-6209-4_7

Souto, E. B. (2020). Ethical issues in research and development of nanoparticles. *Drug Delivery Aspects, 4*, 157–168. https://doi.org/10.1016/B978-0-12-821222-6.00007-5

Sukhanova, A., Bozrova, S., Sokolov, P., et al. (2018). Dependence of nanoparticle toxicity on their physical and chemical properties. *Nanoscale Research Letters, 13*, 44. https://doi.org/10.1186/s11671-018-2457-x

Tiple, A. D., Badwaik, V. J., Padawd, S. V., Chaudhary, R. G., & Singh, N. B. (2020). A review on nanotoxicology: Aquatic environment and biological system. *Materials Today: Proceedings*. https://doi.org/10.1016/j.matpr.2020.05.755

Vijayakumar, M. D., Surendhar, G. J., Natrayan, L., Patil, P. P., Ram, P. M. B., & Paramasivam, P. (2022). Evolution and recent scenario of nanotechnology in agriculture and food industries. *Journal of Nanomaterials, 2022*, 1280411.

Viseu, A. (2015). Caring for nanotechnology? Being an integrated social scientist. *Social Studies of Science, 45*(5), 642–664. https://doi.org/10.1177/0306312715598666

Challenges and Future Prospective of Nano-ecosystem Engineering

Archana Sundaram, Selvendran Thamaraiselvan,
Siva Vijayakumar Tharumasivam, and Janani Chandran

Abstract

The concept of 'nano' refers to the manipulation or self-assembly of individual atoms or molecular structures to create novel solutions with vastly different properties. An ecosystem is a self-contained unit of living things that includes producers (plants), consumers (animals) and decomposers (microbes). The conceptualization of 'nano-ecosystem' necessitates the study of microorganisms, such as bacteria, fungi and algae, at the nanoscale level within an ecosystem. It also involves the study of nanoscale interactions between microorganisms and the environment, which influences crucial environmental processes. The notion of nano-ecosystem engineering implies the deliberate manipulation and design of nano matters at an extremely smaller scale using nanotechnology. It also entails controlling and optimizing the functions of an ecosystem from nutrient cycles to the interactions between organisms. Nano-ecosystem engineering holds numerous significance in creating resilient environment which benefits both humans and the environment in the aspects of environmental sustainability by offering innovative solutions for addressing environmental challenges, such as pollution remediation, water purification and habitat restoration, optimizing resource efficiency, improving agricultural productivity, conservation of biodiversity by controlling ecosystem dynamics and other biotechnological importance that enables the development of novel techniques ranging from drug delivery to biological production. Likewise, this nano-ecosystem engineering possesses future prospectives, including medicine, environmental remediation

A. Sundaram · S. Thamaraiselvan · J. Chandran (✉)
Department of Biotechnology, Srimad Andavan Arts and Science College (Autonomous), Trichy, Tamil Nadu, India

S. V. Tharumasivam
Dhanalakshmi Srinivasan University, Trichy, Tamil Nadu, India

D. K. Gupta et al. (eds.), *Nano-Ecosystems*,
https://doi.org/10.1007/978-981-95-5033-3_14

and energy production. On the other hand, nano-ecosystem engineering grips to several challenges regarding the potential long-term impacts on ecosystems and society. This chapter will provide a comprehensive knowledge of how nano-ecosystem engineering contributes to shaping the future and the challenges evolved in it.

Keywords

Challenges · Engineering · Interaction · Microorganisms · Nano-ecosystem

1 Introduction

The term 'nano' has been used in various contexts, including science, technology and everyday life. It is derived from the Greek word 'nanos', which means 'dwarf'. In the application of science and technology, nano refers to the scale of nanometres, which are extremely smaller units of measurement (Mallakpour et al., 2021). Eventually, one nanometre is equal to one billionth of a metre or 0.000000001 metres. Therefore, nanotechnology is a field that deals with the manipulation and control of matters at the nanoscale (Gupta, 2023). It also involves working with materials and structures that are less than 100 nanometres in size in at least one dimension. Thus, nanotechnology has gained significant attention and has huge applications in several fields, including medicine, electronics, energy and material science (Ariga, 2021; Qiu et al., 2020). Nanostructures, which are the materials at this scale, exhibit unique properties due to their smaller size. Nanotechnology also involves the development of nanoscale materials and structures to create new products and methods contributing to the field of science and technology. Nanotechnology can involve top-down approaches, where larger materials are manipulated and scaled down to the nanoscale, or bottom-up approaches, where the materials are built from individual atoms or molecules (Abid et al., 2022; Andreo et al., 2022).

The ecosystem is a complex system that consists of living organisms, their physical environment and the interactions between them within a particular unit of space. It is a dynamic, interconnected web of relationships where the organisms and their environment, influence and depend on each other (Nunes et al., 2020). Ecosystems may vary in size, ranging from smaller ones including a nano-ecosystem. It is a network of interconnected nanodevices and nanomaterials that works synergistically to perform complex functions at the nanoscale level (Rafique et al., 2023). This signifies the concept of nano-ecosystem engineering, which manipulates the unique properties of nanotechnology in many ways, such as enhanced electrical, chemical and mechanical characteristics to achieve advanced capabilities in various fields like medicine, electronics and environmental monitoring (Malik et al., 2023). The integrating components, such as nanoparticles, nanotubes, nanowires, nanosensors and quantum dots to enable innovative solutions that can enhance drug delivery, improve diagnostic tools and create more efficient electronic devices and develop new materials with superior performances (Su & Song, 2021; Welch et al.,

2021). This chapter aims to explore the future prospects of nano-ecosystem engineering, delving into the innovative applications, challenges and ethical considerations that accompany this nascent field.

2 Nano-ecosystem Engineering and Synthetic Ecologies

Microbial ecosystems play a pivotal role in nature and their engineering for industrial, medical and biotechnological applications with a significant focus for synthetic biologists and biological engineers has always been forefront (Rastegari et al., 2020). Thus, the recent advancements in both natural and synthetic microbial ecosystem engineering. Underscores key design principles for forward engineering and explores theoretical and quantitative models to introduce novel experimental and manipulation tools (Liang et al., 2022). It also ensures the potential applications of microbial ecosystem engineering to reach positive future prospects. The strategy of engineering individual microbes overcomes the results due in fragile, homogenous populations struggling to persist, especially in diverse and unpredictable environments (Sudheer et al., 2020). Conversely, the engineered microbial ecosystems exhibit greater robustness and the capacity to perform complex tasks with the spatial and temporal precision essential for truly programmable biology (Fig. 1). This shift from focussing on an individual microbe to creating resilient microbial ecosystems cam enable more sophisticated and reliable biological applications.

The design and manipulation of microbial communities help in understanding the dynamic interactions within the ecosystem and in the forecasting of their behaviours, which vary under conditions (van de Water et al., 2022). The advanced experimental tools and techniques, such as CRISPR-based gene editing and high-throughput sequencing, facilitate precise modifications and in-depth analysis of microbial ecosystems. The potential applications of microbial ecosystem engineering are vast and varied. In industry, engineered microbial consortia can enhance biofuel production, waste treatment and the synthesis of valuable chemicals. In medicine, they can be used to develop probiotics that maintain or restore healthy microbiota, combat pathogens or deliver therapeutic agents (Durán-Vinet et al., 2021; Saradadevi et al., 2021). In agriculture, engineered microbial ecosystems can improve soil health, promote plant growth and increase crop yields. By this approach, the designs of robust ecosystems are capable of thriving in complex environments and performing intricate biological functions. This is a novel approach to unlock new frontiers in synthetic biology and biotechnology, which are paving the way for an innovative solution to global challenges (Eskandar, 2023).

3 Nano Engineering in Agro-ecosystem

The exceptional reactivity and absorption capacity of nanomaterials surpass that of bulk materials and owing to their expansive surface area and high number of particles per unit mass (Ding et al., 2023). Although metal nanoparticles are increasingly

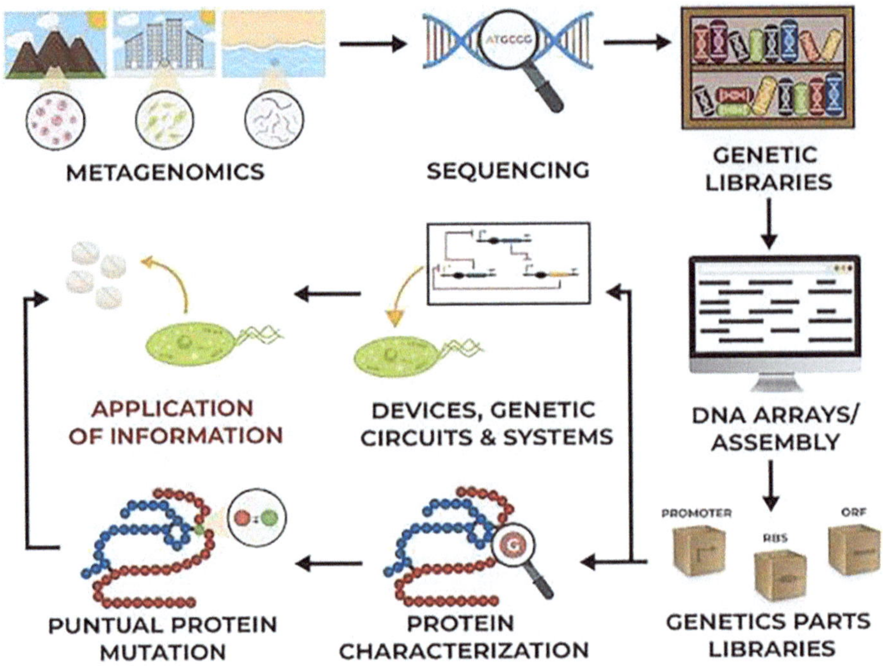

Fig. 1 Nano-ecosystem engineering and synthetic ecology (Govindasamy et al., 2022)

prevalent across various industries, there is a burgeoning interest in identifying the most biologically and environmentally safe methods for their production (Fig. 2). While traditional methods for producing nanoparticles rely on chemical and physical approaches, recent years have seen a fervent exploration of plant extracts and diverse microorganisms as alternative means of production (Devi et al., 2024). The integration of nanotechnology into agricultural systems is poised to revolutionize the industry. This cutting-edge technology encompasses an array of tools for disease treatment and detection, serving as sensors, fertilizers and a sophisticated delivery system for gene transfer (Malik et al., 2023). However, the potential adverse effects of large-scale application of this technology on biological and other systems cannot be overlooked. Therefore, an in-depth exploration of various approaches to nanoparticle production and their positive and negative implications, particularly within the agricultural ecosystem is encouraged. Furthermore, there is a compelling need to anticipate the potential effects of nanotechnology on broader environmental issues in the near future (Burketová et al., 2022).

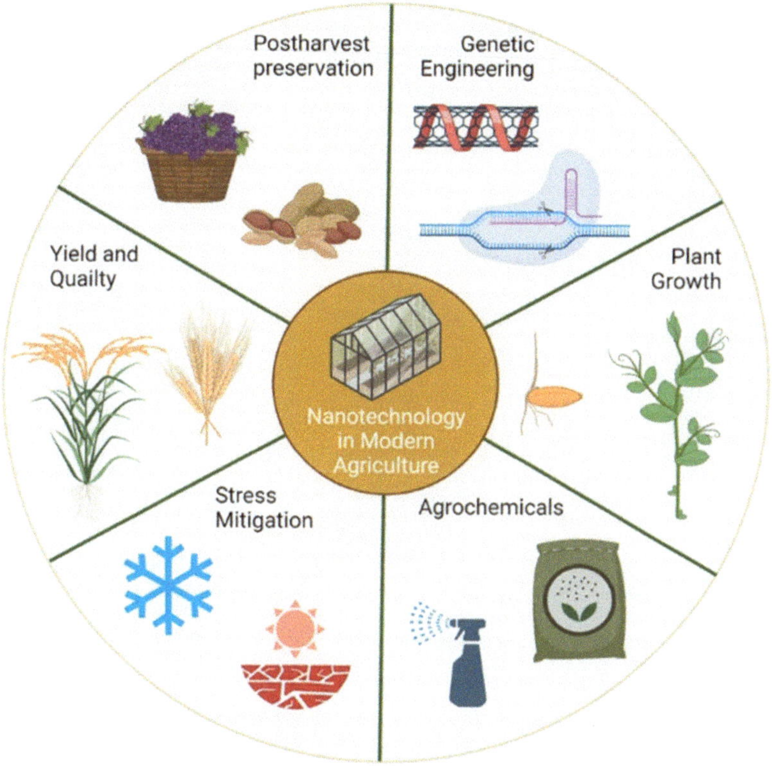

Fig. 2 Nano engineering in agrosystem (Jiang et al., 2021)

4 Nano-ecosystem Engineering in Environmental Transformation

The rapid advancement of nanotechnology has led to the widespread integration of Engineered Nanoparticles (ENPs) into a diverse array of consumer products, which results in the substantial release of ENPs into various environmental compartments (Abbas et al., 2020; Ahmed et al., 2021). This release of ENPs throughout the life cycle of nano-enabled products has raised significant concerns regarding environmental health and safety (Fig. 3). Several studies have revealed that the primary mechanism of nano-toxicity involves Reactive Oxygen Species (ROS)-mediated oxidative stress, which can occur through physical damage via direct contact or the release of toxic ions following the dissolution process of ENPs (Sánchez et al., 2021). It also delves into the uptake, transformation and toxicity of ENPs at physio-morphological, biochemical and molecular levels in primary producers within terrestrial environments, especially plants. In light of the potential environmental risks associated with ENPs, the various sustainable approaches for the effective management and treatment of nano-wastes were needed. These approaches include

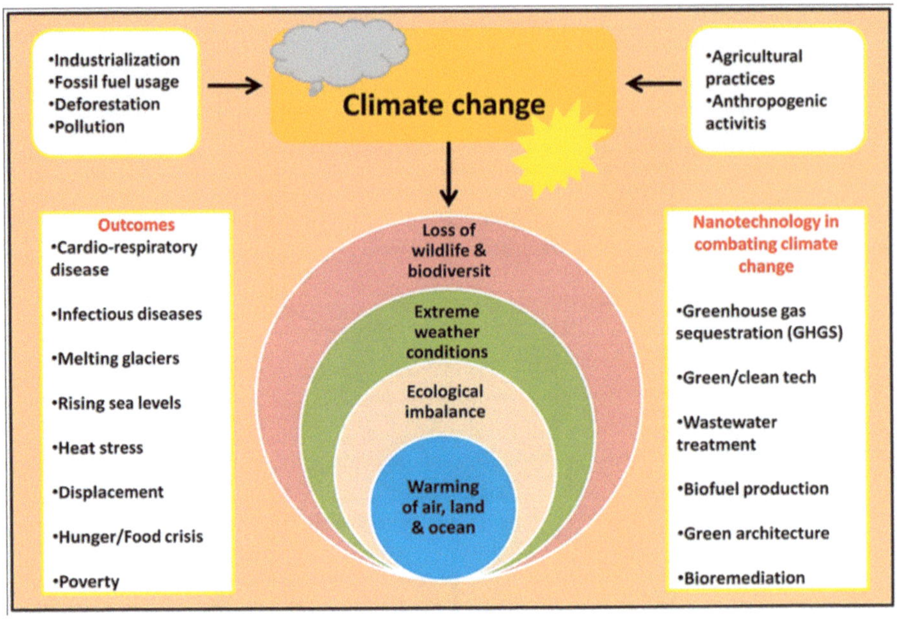

Fig. 3 Nano-ecosystem engineering in environmental transformation (Chausali et al., 2023)

biogenic synthesis, the role of clay minerals, biochar application, bioremediation and legislative measures. The overarching goal of these proposed measures is to maximize the benefits of nanotechnology while minimizing negative outcomes (Goh, 2022). The increasing production and use of ENPs in various consumer products have raised concerns about their potential environmental impact. Thus, the a need for a cohesive understanding of the fate and behaviour of ENPs in different environmental compartments, as well as their potential effects on natural systems and human health (Martínez et al., 2020). It also underscores the importance of developing risk assessment guidelines and addressing the long-term accumulation characteristics of ENPs in ecosystems. Thus, a comprehensive examination of the current state of knowledge regarding the ecotoxicity of ENPs offers valuable insights into the potential risks and benefits associated with the widespread application of nanotechnology. It emphasizes the importance of adopting sustainable approaches to mitigate the environmental impact of ENPs and maximize the positive outcomes of nanotechnology (Tran et al., 2023).

5 Emerging Trends in Nano-bioengineering

Microorganisms established their presence on Earth before the evolution of multicellular organisms. The advent of microscopy revealed their existence and shed light on the myriad processes they govern, many of which are directly relevant to human interests. One particularly intriguing process is their ability to thrive in the presence of toxic metals (Baluška et al., 2023). It appeared that metal ions were simply sequestered into inclusion bodies or cell surfaces, facilitating their conversion into non-toxic nanostructures. However, the emergence of genome sequencing techniques highlighted the genetic composition of these microbes as a pivotal aspect of these phenomena. The discovery of Metal Resistance Genes (MRGs) in these microbes unveiled the complex regulation of these processes (Gillieatt & Coleman, 2024). Since most MRGs are encoded in plasmids, they can be horizontally transferred. The discovery of nanoparticles and their diverse applications, which range from polymer chemistry to drug delivery, sparked a surge in demand for innovative nanoparticle synthesis techniques. It is now well established that microbial synthesis of nanoparticles offers numerous advantages over existing chemical methods (Vishwanath & Negi, 2021). However, it is the deliberate utilization of biotechnology, molecular biology, metabolic engineering, synthetic biology and genetic engineering tools that has revolutionized the realm of microbial nanotechnology. The in-depth exploration of the micro and even nanoscale organization of microbial life also captivated biologists and engineers, leading to the development of molecular motors that mimic bacterial flagellar motors (Harun-Ur-Rashid et al., 2023). Thus, the paramount significance and vast untapped potential of bioengineering tools in harnessing the field of microbial nanoparticle synthesis. Also, to underscore the specific modifications oriented towards the application that can be implemented at various stages of nanoparticle synthesis. Thus, the role of these nanoparticles in the natural ecosystem (Mysore et al., 2024).

6 Nano-ecosystem Engineering in Shaping the Future

Nanotechnology has permeated various facets of human life, revolutionizing industries, healthcare and environmental conservation (Fig. 4). The intersection of nanotechnology with ecosystem engineering presents a frontier of opportunity with the potential to transform our approach to environmental sustainability (Palit & Hussain, 2022). The integration of nanotechnology into ecosystem engineering holds promise for a myriad of innovative applications. Nano-enabled solutions offer precise control over environmental processes and pave the way for enhanced water purification, soil remediation and air quality management. Nanomaterials can be tailored to encapsulate pollutants, catalyse the degradation of contaminants and facilitate the efficient delivery of nutrients to support plant growth. The development of nanobiosensors enables real-time monitoring of environmental parameters, empowering proactive intervention and management of ecosystem health (Muneer, 2023; Pratama & Rahayu, 2024).

Despite its potential, the consideration of nano-ecosystem engineering is nothing without challenges. The unintended consequences of nanomaterial release into ecosystems pose ecological and human health risks. The potential for bioaccumulation and biomagnification of nanoparticles necessitates a thorough understanding of

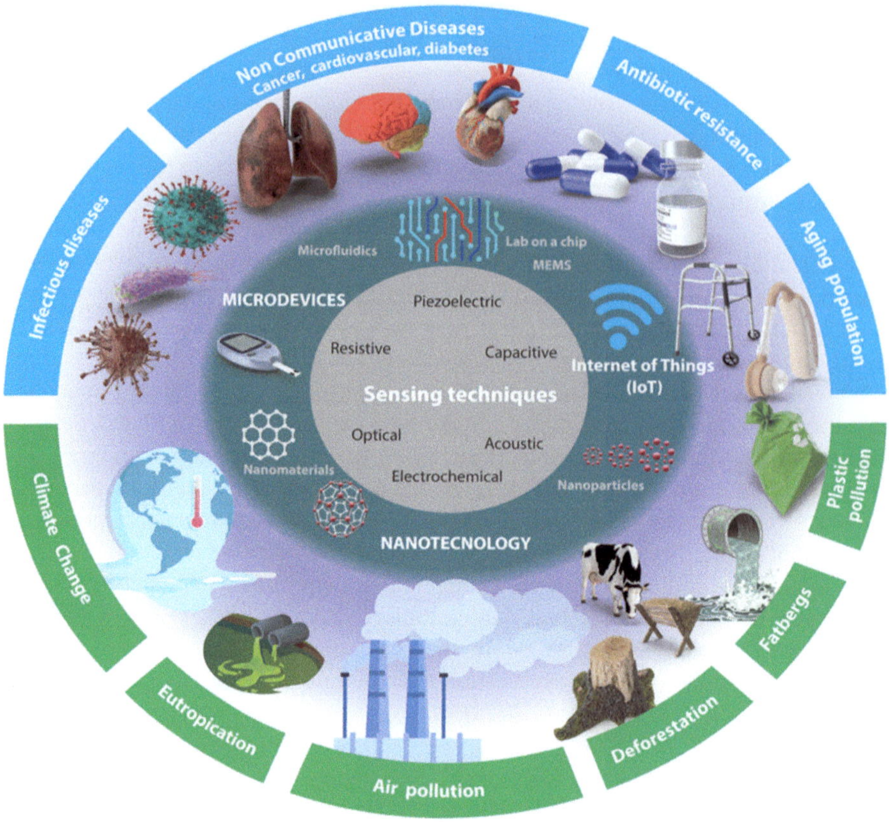

Fig. 4 Future in nano-ecosystem engineering. (Tovar-Lopez, 2023)

their fate and behaviour in natural systems (Pérez-Hernández et al., 2021). Moreover, ethical considerations surrounding the deliberate manipulation of ecosystems at the nano-scale demand careful reflection. Balancing innovation with prudence, transparency and public engagement is crucial to the responsible advancement of nano-ecosystem engineering. The convergence of nanotechnology with bioremediation, phytoremediation and ecological restoration has yielded pioneering approaches to environmental stewardship (Malik et al., 2022). Nano-enabled bioremediation harnesses the catalytic properties of nanomaterials to accelerate the degradation of organic contaminants, while nano-phytoremediation manipulates the targeted delivery of nanoparticles to enhance plant-based remediation processes. The development of nanobots capable of autonomously restoring damaged ecosystems represents a paradigm shift in ecological restoration efforts, holding the potential to mitigate the impact of environmental degradation (Arora et al., 2024; Rajput et al., 2024).

7 Regulatory and Ethical Considerations in Nano-ecosystem Engineering

Nano-ecosystems, which involve the use of nanomaterials and nanoscale technologies, present distinct regulatory and ethical challenges that necessitate careful consideration and management. It is crucial to establish comprehensive safety guidelines and standards for the development, manufacturing and utilization of nanomaterials and nano-devices (Inshakova & Inshakova, 2020). Thorough risk assessments should be conducted to evaluate the potential environmental, health and safety impacts of nano-ecosystems throughout their lifecycle. The roles and responsibilities of regulatory agencies must be clearly defined to oversee nano-ecosystem applications across various sectors, such as healthcare, environment and consumer products (Whittemore, 2023). Measures should be implemented to control exposure to nanomaterials during production, use and disposal to minimize potential risks to workers, consumers and the environment. The protocols for monitoring and reporting adverse effects associated with the deployment of nano-ecosystems should be developed to ensure timely responses and mitigation (Sousa et al., 2020).

8 Ethical Framework of Nano-ecosystem Engineering

Nano-ecosystem engineering offers a pathway to bolster ecosystem resilience in the face of environmental stressors (Fig. 5). By using nanomaterials to modulate nutrient cycling, enhance soil fertility and mitigate the impacts of climate change, ecosystems can be empowered to adapt and thrive in dynamically changing environments (Okem et al., 2024). The precision and versatility of nanomaterials enable the fine-tuning of ecological processes and foster adaptive responses that promote ecosystem stability and biodiversity conservation. Thus, the future of nano-ecosystem engineering holds immense potential to redefine our approach to environmental conservation and ecosystem management (Iqbal et al., 2023). Embracing innovation while upholding ethical principles and environmental stewardship, we can chart a course towards a harmonious coexistence with nature, using the transformative power of nanotechnology to engineer resilient, adaptive and sustainable ecosystems for future generations. As nano-ecosystem engineering progresses, the ethical framework of nanosystem engineering is imperative to establish a robust ethical and regulatory framework that safeguards environmental integrity and human well-being (Sable et al., 2024). The proactive risk assessments are the essential components of responsible innovation. The development of guidelines for the safe deployment and disposal of nano-enabled ecosystem interventions is paramount in mitigating potential risks and ensuring long-term environmental sustainability (Corsi et al., 2023). Addressing concerns about the collection, storage and use of personal data in nano-ecosystem technologies, especially within healthcare and biometric applications, is crucial. Involving stakeholders, including communities, researchers, policymakers and industry, in ethical discussions and decision-making processes concerning the development and deployment of nano-ecosystems is

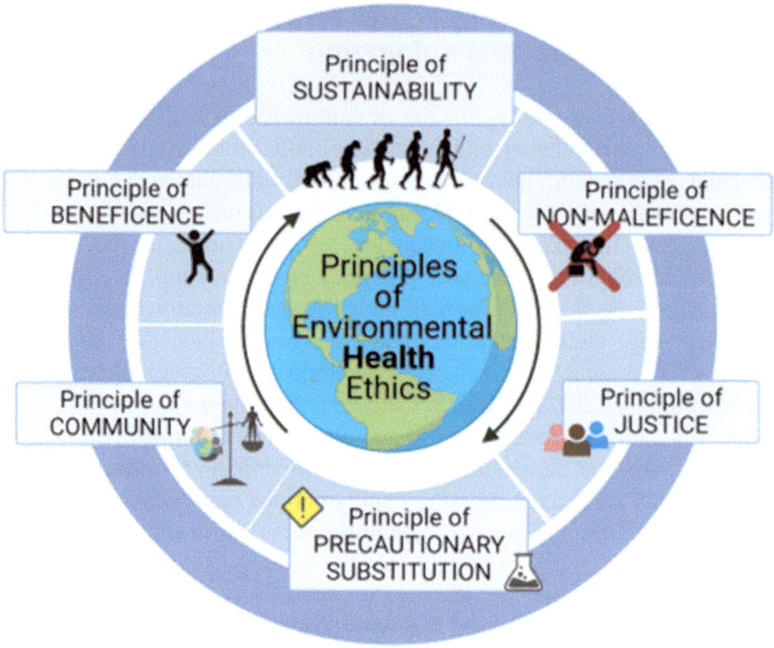

Fig. 5 Ethical framework of nano-ecosystem engineering

essential. Ensuring fair access to the benefits of nano-ecosystems while reducing potential disparities in technology adoption and impact is important (Sable et al., 2024).

Promoting transparency in nano-ecosystem engineering practices, such as material composition, manufacturing processes and potential risks, is necessary. Conducting ethical reviews and assessments of nano-ecosystem research and applications helps maintain ethical standards and societal values. Sharing best practices, research findings and regulatory experiences supports global understanding and effective governance of nanotechnologies. Collaborating with international partners to align regulatory frameworks and standards for nano-ecosystem engineering across borders is important (Whittemore, 2023). Enhancing public awareness and understanding of nano-ecosystems, their potential benefits and associated risks through education and outreach initiatives is crucial. Providing ethical training and guidance for researchers, engineers and practitioners involved in nano-ecosystem engineering fosters responsible innovation and decision-making (Alqahtani et al., 2022). Addressing regulatory and ethical considerations is essential to ensure the safe, sustainable and ethical development and deployment of nano-ecosystems.

Establishing regulatory frameworks and guidelines for the safe handling, transportation, use and disposal of nanomaterials and nano-devices is necessary. Implementing monitoring programs to track nanoparticle concentrations in environmental matrices and ensure compliance with regulatory limits and safety thresholds is important (Nielsen et al., 2021). Investigating the long-term effects of chronic nanoparticle exposure on ecosystems and human health is also essential. Developing sustainable nanotechnology solutions that minimize environmental impacts while maximizing benefits across applications is critical. Additionally, enhancing public awareness and engagement regarding nano-ecosystems, their environmental and health impacts and responsible use practices is vital (Pokrajac et al., 2021).

9 Challenges of Nano-ecosystem Engineering

Nano-ecosystem engineering represents a burgeoning field at the intersection of nanotechnology and environmental science, which offers innovative solutions for sustainable ecosystem management. This nascent discipline is not without its challenges (Palit & Hussain, 2022). This chapter aims to elucidate the multifaceted challenges encountered in nano-ecosystem engineering, encompassing technical, ethical and regulatory considerations that underpin the responsible development and application of nanotechnology in ecosystem engineering. The challenges in nano-ecosystem engineering are multifaceted, encompassing technical, ethical, regulatory and societal dimensions. Addressing these challenges requires a holistic and interdisciplinary approach that integrates scientific rigor, ethical considerations and group engagement (Vega-Baudrit et al., 2023). By navigating these challenges with prudence and foresight, nano-ecosystem engineering holds the potential to usher in a new era of sustainable ecosystem management, using the transformative power of nanotechnology for the benefit of both ecosystems and human societies (Shukla et al., 2023).

10 Environmental Impact and Ecotoxicity

Nano-ecosystems, leveraging nanomaterials and nanoscale technologies, offer significant advancements across various industries (Fig. 6). However, they also pose potential environmental and health impacts that necessitate careful consideration and proactive management. Nanoparticles released into the environment can interact with ecosystems, potentially affecting organisms such as aquatic life, soil microbes and plants (Ebi & Hess, 2020). Some nanoparticles may accumulate in organisms through the food chain, leading to ecological disruptions and biomagnification. Nanoparticles' small size and unique properties can lead to their persistence in the environment, influencing long-term environmental health (Grillo et al., 2021). Nano-ecosystems used in agriculture, industry and consumer products may enter water systems through runoff or improper disposal, impacting water quality. Nanoparticles in soil can affect soil microbial communities, nutrient cycling processes and overall soil health.

Nanoparticles released during manufacturing, use or disposal processes can contribute to air pollution, affecting respiratory health and atmospheric dynamics (Kalwani et al., 2022). Nanoparticles may undergo transformations in the

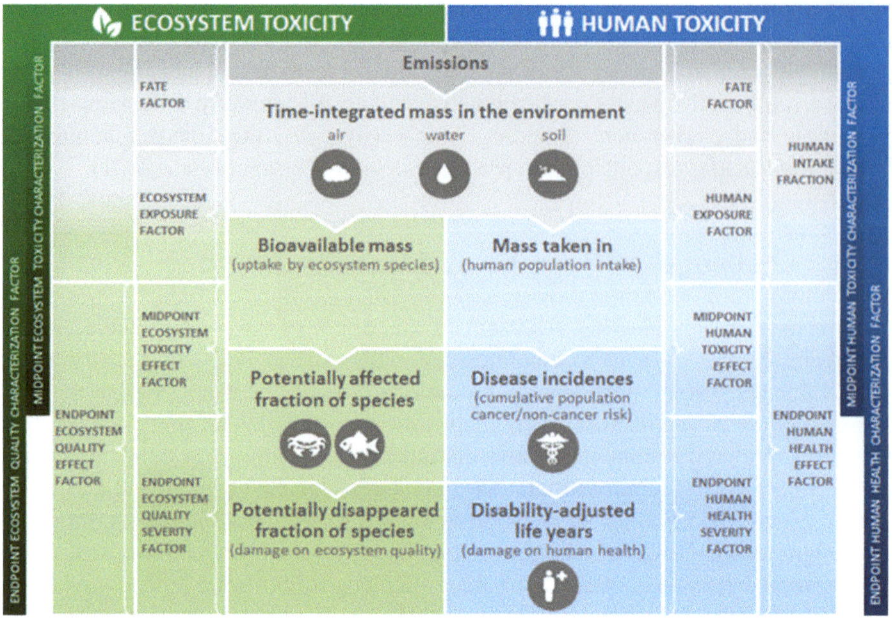

Fig. 6 Comparison between ecosystem toxicity and human toxicity. (Image courtesy: DTU hosting new USEtox International Centre-DTU Management)

atmosphere, potentially influencing climate and regional air quality. Inhalation, ingestion and dermal contact are primary routes through which nanoparticles can enter the human body. Nanoparticles may induce adverse health effects depending on their chemical composition, size, surface characteristics and exposure duration. Potential impacts include respiratory issues, cardiovascular effects and neurological disorders (Zlokovic et al., 2020). Comprehensive characterization of nanoparticles is essential for understanding their behaviour, fate and potential impacts in environmental and biological systems. Developing models to predict nanoparticle exposure scenarios and assessing associated risks to human health and the environment (El-Kady et al., 2023), implementing measures to minimize nanoparticle release, exposure pathways and environmental impacts through engineering controls, regulations and best practices (Svendsen et al., 2020).

One of the primary challenges in nano-ecosystem engineering pertains to the environmental impact and ecotoxicity of nanomaterials. The deliberate release of engineered nanoparticles into natural ecosystems raises concerns regarding their potential adverse effects on ecological balance and biodiversity (Pérez-Hernández et al., 2021). The intricate interactions between nanoparticles and living organisms, including bioaccumulation and biomagnification, necessitate a comprehensive understanding of the fate, behaviour and long-term implications of nanomaterials in diverse ecosystems. Conducting comprehensive risk assessments of nano-ecosystem interventions is a critical challenge that necessitates a multidisciplinary approach (Singh et al., 2023). The dynamic and complex nature of nanomaterial interactions with ecosystems demands robust risk assessment methodologies to identify and mitigate potential environmental and human health risks. Also, the development of effective mitigation strategies to minimize the unintended consequences of nano-ecosystem interventions is paramount for ensuring the responsible and sustainable application of nanotechnology in ecosystem engineering (Corsi et al., 2023).

Public perception and engagement play a pivotal role in shaping the trajectory of nano-ecosystem engineering. Building public awareness, fostering dialogue and addressing concerns related to the potential risks and benefits of nano-ecosystem interventions are essential for garnering societal acceptance and support. Engaging diverse groups, including communities, policymakers and environmental advocacy groups, in transparent and inclusive discussions is crucial for navigating the societal dimensions of nano-ecosystem engineering (Pateman et al., 2021). The technological limitations and uncertainties inherent in nano-ecosystem engineering pose significant challenges to the development and application of nanotechnology in ecosystem management. Addressing the scalability, reproducibility and long-term efficacy of nano-ecosystem interventions while navigating the inherent uncertainties associated with complex ecological systems requires a concerted effort to advance the technological frontiers of nano-ecosystem engineering (Ganie et al., 2021).

11 Conclusion

The future prospects and challenges of nano-ecosystem engineering present a complex and dynamic landscape that requires a balanced approach to harness the transformative potential of nanotechnology for sustainable ecosystem management. The innovative applications of nanotechnology in ecosystem engineering hold promise for addressing pressing environmental challenges, yet the multifaceted challenges underscore the need for a comprehensive and responsible approach to the advancement of this nascent field. The future prospects of nano-ecosystem engineering are characterized by pioneering innovation and transformative potential. The integration of nanotechnology with ecosystem engineering offers a pathway to revolutionize environmental conservation, resource management and ecological restoration. The precision and versatility of nanomaterials enable targeted interventions to address environmental stressors, enhance ecosystem resilience and mitigate the impact of pollution on natural systems. The development of nano-enabled bioremediation, phytoremediation and ecological restoration technologies holds promise for catalyzing paradigm shifts in environmental stewardship. The future prospects also include the creation of resilient, adaptive and sustainable ecosystems that thrive in the face of dynamic environmental challenges. The promising future prospects and the challenges of nano-ecosystem engineering are multifaceted, encompassing environmental, ethical, regulatory and technological dimensions. The environmental impact and ecotoxicity of nanomaterials pose complex challenges that demand a thorough understanding of their interactions with living organisms and ecosystems. Navigating the regulatory landscape and governance of nanotechnology in ecosystem engineering requires the establishment of robust guidelines and standards to ensure the responsible deployment and management of nano-ecosystem interventions. Ethical considerations surrounding informed consent, public engagement and transparency present challenges that necessitate careful reflection and inclusive decision-making processes. Moreover, addressing the technological limitations and uncertainties inherent in nano-ecosystem engineering requires a concerted effort to advance the scientific frontiers and develop scalable, reproducible and effective nano-ecosystem interventions. The convergence of future prospects and challenges in nano-ecosystem engineering underscores the imperative of balancing innovation with responsibility. Embracing the transformative potential of nanotechnology while upholding ethical principles, environmental integrity and human safety is paramount for the responsible advancement of this nascent field. By fostering interdisciplinary collaborations, integrating engagement and navigating the societal dimensions of nano-ecosystem engineering, it is possible to cultivate a culture of responsible innovation that prioritizes environmental sustainability and societal well-being.

The future prospects and challenges are the opportunities for collaboration and dialogue to address the multifaceted dimensions of nano-ecosystem engineering. Engaging diverse groups, including scientists, policymakers, industry, communities and environmental advocacy groups in transparent and inclusive discussions is essential for navigating the ethical, regulatory and societal dimensions of nano-ecosystem interventions. The fostering of open dialogue and public engagement is crucial for building trust, addressing concerns and garnering societal acceptance of nano-ecosystem engineering endeavours. By embracing a collaborative and inclusive approach, it is possible to navigate the challenges and manipulate the future prospects of nano-ecosystem engineering for the benefit of both ecosystems and human societies. As nano-ecosystem engineering advances, there is a pressing need to evolve the ethical and regulatory framework to address the unique challenges and opportunities presented by nanotechnology in ecosystem management. Developing standardized regulations, guidelines and risk assessment methodologies tailored to the characteristics of nanomaterials is essential for ensuring environmental integrity and human safety. The fostering of a culture of responsible innovation that prioritizes ethical considerations, transparency and public engagement is crucial for navigating the ethical dimensions of nano-ecosystem engineering. By proactively addressing the ethical and regulatory challenges, it is possible to pave the way for the responsible and sustainable application of nanotechnology in ecosystem engineering. The future prospects and challenges of nano-ecosystem engineering underscore the transformative potential and complexity of this nascent field. By embracing innovation while upholding responsibility, fostering collaboration and dialogue and evolving the ethical and regulatory framework, it is possible to navigate the challenges and manipulate the future prospects of nano-ecosystem engineering for the benefit of both ecosystems and human societies. This requires a holistic and interdisciplinary approach that integrates scientific rigor, ethical considerations and group engagement to chart a course towards a harmonious coexistence with nature, using the transformative power of nanotechnology for the sustainable management of ecosystems.

References

Abbas, Q., Yousaf, B., Ali, M. U., Munir, M. A. M., El-Naggar, A., Rinklebe, J., & Naushad, M. (2020). Transformation pathways and fate of engineered nanoparticles (ENPs) in distinct interactive environmental compartments: A review. *Environment International, 138*, 105646.

Abid, N., Khan, A. M., Shujait, S., Chaudhary, K., Ikram, M., Imran, M., & Maqbool, M. (2022). Synthesis of nanomaterials using various top-down and bottom-up approaches, influencing factors, advantages, and disadvantages: A review. *Advances in Colloid and Interface Science, 300*, 102597.

Ahmed, T., Noman, M., Manzoor, N., Ali, S., Rizwan, M., Ijaz, M., & Li, B. (2021). Recent advances in nanoparticles associated ecological harms and their biodegradation: Global environmental safety from nano-invaders. *Journal of Environmental Chemical Engineering, 9*(5), 106093.

Alqahtani, M. S., Abbas, M., Abdulmuqeet, M., Alqahtani, A. S., Alshahrani, M. Y., Alsabaani, A., & Ramalingam, M. (2022). Forecasting the post-pandemic effects of the SARS-CoV-2 virus using the bullwhip phenomenon alongside use of nanosensors for disease containment and cure. *Materials, 15*(14), 5078.

Andreo, J., Ettlinger, R., Zaremba, O., Pena, Q., Lächelt, U., de Luis, R. F., & Wuttke, S. (2022). Reticular nanoscience: Bottom-up assembly nanotechnology. *Journal of the American Chemical Society, 144*(17), 7531–7550.

Ariga, K. (2021). Nanoarchitectonics: what's coming next after nanotechnology? *Nanoscale Horizons, 6*(5), 364–378.

Arora, D., Arora, A., Panghal, V., Singh, A., Bala, R., Kumari, S., & Kumar, S. (2024). Unleashing the feasibility of nanotechnology in phytoremediation of heavy metal–contaminated soil: A critical review towards sustainable approach. *Water, Air, & Soil Pollution, 235*(1), 57.

Baluška, F., Miller, W. B., & Reber, A. S. (2023). Cellular and evolutionary perspectives on organismal cognition: From unicellular to multicellular organisms. *Biological Journal of the Linnean Society, 139*(4), 503–513.

Burketová, L., Martinec, J., Siegel, J., Macůrková, A., Maryška, L., & Valentová, O. (2022). Noble metal nanoparticles in agriculture: Impacts on plants, associated microorganisms, and biotechnological practices. *Biotechnology Advances, 58*, 107929.

Chausali, N., Saxena, J., & Prasad, R. (2023). Nanotechnology as a sustainable approach for combating the environmental effects of climate change. *Journal of Agriculture and Food Research, 12*, 100541.

Corsi, I., Venditti, I., Trotta, F., & Punta, C. (2023). Environmental safety of nanotechnologies: The eco-design of manufactured nanomaterials for environmental remediation. *Science of the Total Environment, 864*, 161181.

Devi, L., Kushwaha, P., Ansari, T. M., Kumar, A., & Rao, A. (2024). Recent trends in biologically synthesized metal nanoparticles and their biomedical applications: A review. *Biological Trace Element Research, 202*(7), 3383–3399.

Ding, Y., Wang, C., Zeng, M., & Fu, L. (2023). Atomic manufacturing of advanced nanomaterials. *Advanced Materials, 37*, 2306689.

Durán-Vinet, B., Araya-Castro, K., Chao, T. C., Wood, S. A., Gallardo, V., Godoy, K., & Abanto, M. (2021). Potential applications of CRISPR/Cas for next-generation biomonitoring of harmful algae blooms: A review. *Harmful Algae, 103*, 102027.

Ebi, K. L., & Hess, J. J. (2020). Health risks due to climate change: Inequity in causes and consequences: Study examines health risks due to climate change. *Health Affairs, 39*(12), 2056–2062.

El-Kady, M. M., Ansari, I., Arora, C., Rai, N., Soni, S., Verma, D. K., & Mahmoud, A. E. D. (2023). Nanomaterials: A comprehensive review of applications, toxicity, impact, and fate to environment. *Journal of Molecular Liquids, 370*, 121046.

Eskandar, K. (2023). Revolutionizing biotechnology and bioengineering: Unleashing the power of innovation. *Journal of Applied Biotechnology and Bioengineering, 10*(3), 81–88.

Ganie, A. S., Bano, S., Khan, N., Sultana, S., Rehman, Z., Rahman, M. M., & Khan, M. Z. (2021). Nanoremediation technologies for sustainable remediation of contaminated environments: Recent advances and challenges. *Chemosphere, 275*, 130065.

Gillieatt, B. F., & Coleman, N. V. (2024). Unravelling the mechanisms of antibiotic and heavy metal resistance co-selection in environmental bacteria. *FEMS Microbiology Reviews, 48*, fuae017.

Goh, N. S. (2022). *Advancing nanomaterial usage in agriculture: Engineering nanomaterials towards nanomaterial-mediated plant genetic engineering*. University of California.

Govindasamy, R., Gayathiri, E., Sankar, S., Venkidasamy, B., Prakash, P., Rekha, K., & Thiruvengadam, M. (2022). Emerging trends of nanotechnology and genetic engineering in cyanobacteria to optimize production for future applications. *Life, 12*(12), 2013.

Grillo, R., Fraceto, L. F., Amorim, M. J., Scott-Fordsmand, J. J., Schoonjans, R., & Chaudhry, Q. (2021). Ecotoxicological and regulatory aspects of environmental sustainability of nanopesticides. *Journal of Hazardous Materials, 404*, 124148.

Gupta, P. K. (2023). *Nanotoxicology in nanobiomedicine*. Springer.

Harun-Ur-Rashid, M., Jahan, I., Foyez, T., & Imran, A. B. (2023). Bio-inspired nanomaterials for micro/nanodevices: A new era in biomedical applications. *Micromachines, 14*(9), 1786.

Inshakova, E., &Inshakova, A. (2020). Nanomaterials and nanotechnology: Prospects for technological re-equipment in the power engineering industry. In IOP conference series: Materials science and engineering 709, 3, 033020.

Iqbal, B., Li, G., Alabbosh, K. F., Hussain, H., Khan, I., Tariq, M., et al. (2023). Advancing environmental sustainability through microbial reprogramming in growth improvement, stress alleviation, and phytoremediation. *Plant Stress, 100283*.

Jiang, M., Song, Y., Kanwar, M. K., Ahammed, G. J., Shao, S., & Zhou, J. (2021). Phytonanotechnology applications in modern agriculture. *Journal of Nanobiotechnology, 19*, 1–20.

Kalwani, M., Chakdar, H., Srivastava, A., Pabbi, S., & Shukla, P. (2022). Effects of nanofertilizers on soil and plant-associated microbial communities: Emerging trends and perspectives. *Chemosphere, 287*, 132107.

Liang, Y., Ma, A., & Zhuang, G. (2022). Construction of environmental synthetic microbial consortia: Based on engineering and ecological principles. *Frontiers in Microbiology, 13*, 829717.

Malik, J. A., Goyal, M. R., & Wani, K. A. (Eds.). (2022). Bioremediation and phytoremediation technologies in sustainable soil management: volume 1: Fundamental aspects and contaminated sites. CRC Press.

Malik, S., Muhammad, K., & Waheed, Y. (2023). Nanotechnology: A revolution in modern industry. *Molecules, 28*(2), 661.

Mallakpour, S., Hussain, C. M., & Kaushik, S. (2021). Nanoproducts: Biomedical, environmental, and energy applications. In *Handbook of Consumer Nanoproducts* (pp. 1–26).

Martínez, G., Merinero, M., Pérez-Aranda, M., Pérez-Soriano, E. M., Ortiz, T., Villamor, E., & Alcudia, A. (2020). Environmental impact of nanoparticles' application as an emerging technology: A review. *Materials, 14*(1), 166.

Muneer, N. (2023). Emerging materials and technologies for the next generation of engineering solutions. *Liberal Journal of Language and Literature Review, 1*(01), 39–46.

Mysore, T. H. M., Patil, A. Y., Hegde, C., Sudeept, M. A., Kumar, R., Soudagar, M. E. M., & Fattah, I. M. R. (2024). Apatite insights: From synthesis to biomedical applications. *European Polymer Journal, 209*, 112842.

Nielsen, M. B., Baun, A., Mackevica, A., Thit, A., Wallinder, I. O., Gallego, J. A., & Hansen, S. F. (2021). Nanomaterials in the European chemicals legislation–methodological challenges for registration and environmental safety assessment. *Environmental Science: Nano, 8*(3), 731–747.

Nunes, F. C., de Jesus Alves, L., de Carvalho, C. C. N., Gross, E., de MarchiSoares, T., & Prasad, M. N. V. (2020). Soil as a complex ecological system for meeting food and nutritional security. In *Climate change and soil interactions* (pp. 229–269). Elsevier.

Okem, E. S., Iluyomade, T. D., & Akande, D. O. (2024). Nanotechnology-enhanced roadway infrastructure in the US: An interdisciplinary review of resilience, sustainability, and policy implications. *World Journal of Advanced Engineering Technology and Sciences, 11*(2), 397–410.

Palit, S., & Hussain, C. M. (2022). Green Nanoproducts: A far-reaching review. In *Handbook of consumer Nanoproducts* (pp. 17–34). Springer.

Pateman, R. M., Dyke, A., & West, S. E. (2021). The diversity of participants in environmental citizen science. *Citizen Science: Theory and Practice., 6.*

Pérez-Hernández, H., Pérez-Moreno, A., Sarabia-Castillo, C. R., García-Mayagoitia, S., Medina-Pérez, G., López-Valdez, F., et al. (2021). Ecological drawbacks of nanomaterials produced on an industrial scale: Collateral effect on human and environmental health. *Water, Air, and Soil Pollution, 232*, 1–33.

Pokrajac, L., Abbas, A., Chrzanowski, W., Dias, G. M., Eggleton, B. J., Maguire, S., & Mitra, S. (2021). Nanotechnology for a sustainable future: Addressing global challenges with the international network4sustainable nanotechnology. *ACS Nano, 15*, 18608.

Pratama, A., & Rahayu, S. (2024). Biosensors enabled by nanotechnology: Advancing precision agriculture through IoT monitoring. *Journal of Intelligent Connectivity and Emerging Technologies, 9*(1), 25–39.

Qiu, L., Zhu, N., Feng, Y., Michaelides, E. E., Żyła, G., Jing, D., & Mahian, O. (2020). A review of recent advances in thermophysical properties at the nanoscale: From solid state to colloids. *Physics Reports, 843*, 1–81.

Rafique, S., Hanif, M., Hussain, A., Rafeeq, H., & Ahmad, S. (2023). Environmental remediation and protection. In *Bionanocatalysis: From design to applications* (pp. 451–476). Elsevier.

Rajput, V. D., Singh, A., Ghazaryan, K., Alexiou, A., Al-Tawaha, S., & Mohammad, A. R. (Eds.). (2024). Harnessing nanoOmics and nanozymes for sustainable agriculture. IGI Global.

Rastegari, A. A., Yadav, A. N., & Yadav, N. (2020). *New and future developments in microbial biotechnology and bioengineering: Trends of microbial biotechnology for sustainable agriculture and biomedicine systems: Diversity and functional perspectives*. Elsevier.

Sable, H., Kumar, V., Singh, V., Rustagi, S., Chaudhary, V., & Pandit, S. (2024). Nanosystems-enhanced electrochemical biosensors for precision in one health management. *Journal of The Electrochemical Society, 171*(3), 037527.

Sánchez, M. B., Fito-López, C., & Cajaraville, M. P. (2021). A life cycle perspective of the exposure to airborne nanoparticles released from nanotechnology enabled products and applications. In *Health and environmental safety of nanomaterials* (pp. 173–194). Woodhead Publishing.

Saradadevi, G. P., Das, D., Mangrauthia, S. K., Mohapatra, S., Chikkaputtaiah, C., Roorkiwal, M., & Mohannath, G. (2021). Genetic, epigenetic, genomic and microbial approaches to enhance salt tolerance of plants: A comprehensive review. *Biology, 10*(12), 1255.

Shukla, S., Khan, R., Saxena, A., Patel, A., Selvam, S., & Hussain, C. M. (2023). Future of modern society: Sustainability in green nanotechnology. In *Nanoremediation* (pp. 393–410). Elsevier.

Singh, S., Prasad, S. M., & Bashri, G. (2023). Fate and toxicity of nanoparticles in aquatic systems. *ActaGeochimica, 42*(1), 63–76.

Sousa, S. P., Peixoto, T., Santos, R. M., Lopes, A., Paiva, M. D. C., & Marques, A. T. (2020). Health and safety concerns related to CNT and graphene products, and related composites. *Journal Of Composites Science, 4*(3), 106.

Su, M., & Song, Y. (2021). Printable smart materials and devices: Strategies and applications. *Chemical Reviews, 122*(5), 5144–5164.

Sudheer, S., Bai, R. G., Usmani, Z., & Sharma, M. (2020). Insights on engineered microbes in sustainable agriculture: Biotechnological developments and future prospects. *Current Genomics, 21*(5), 321–333.

Svendsen, C., Walker, L. A., Matzke, M., Lahive, E., Harrison, S., Crossley, A., & Spurgeon, D. J. (2020). Key principles and operational practices for improved nanotechnology environmental exposure assessment. *Nature Nanotechnology, 15*(9), 731–742.

Tovar-Lopez, F. J. (2023). Recent progress in micro-and nanotechnology-enabled sensors for biomedical and environmental challenges. *Sensors, 23*(12), 5406.

Tran, T. K., Nguyen, M. K., Lin, C., Hoang, T. D., Nguyen, T. C., Lone, A. M., & Nguyen, D. D. (2023). Review on fate, transport, toxicity and health risk of nanoparticles in natural ecosystems: Emerging challenges in the modern age and solutions toward a sustainable environment. *Science of the Total Environment*, 169331.

van de Water, J. A., Tignat-Perrier, R., Allemand, D., & Ferrier-Pages, C. (2022). Coral holobionts and biotechnology: From blue economy to coral reef conservation. *Current Opinion in Biotechnology, 74*, 110–121.

Vega-Baudrit, J. R., Camacho, M., & Araya, A. (2023). Regulating nanotechnology: Ensuring responsible and safe innovation in the advancement of science and technology, Ciencia. *Tecnología y Salud, 10*(2), 177–191.

Vishwanath, R., & Negi, B. (2021). Conventional and green methods of synthesis of silver nanoparticles and their antimicrobial properties. *Current Research in Green and Sustainable Chemistry, 4*, 100205.

Welch, E. C., Powell, J. M., Clevinger, T. B., Fairman, A. E., & Shukla, A. (2021). Advances in biosensors and diagnostic technologies using nanostructures and nanomaterials. *Advanced Functional Materials, 31*(44), 2104126.

Whittemore, A. A. (2023). Perceived business skills needs of MFTs: Implications for a doctoral program.

Zlokovic, B. V., Gottesman, R. F., Bernstein, K. E., Seshadri, S., McKee, A., Snyder, H., & Chen, J. (2020). Vascular contributions to cognitive impairment and dementia (VCID): A report from the 2018 National Heart, Lung, and Blood Institute and National Institute of Neurological Disorders and Stroke Workshop. *Alzheimer's & Dementia, 16*(12), 1714–1733.

MIX
Papier aus verantwortungsvollen Quellen
Paper from responsible sources
FSC® C105338

If you have any concerns about our products,
you can contact us on
ProductSafety@springernature.com

In case Publisher is established outside the EU,
the EU authorized representative is:
Springer Nature Customer Service Center GmbH
Europaplatz 3, 69115 Heidelberg, Germany

Printed by Libri Plureos GmbH
in Hamburg, Germany